Warnings and Product Liability

Warnings and Product Liability

Lessons Learned from Cognitive Psychology and Ergonomics

Sanne Pape

eleven
international publishing

Published, sold and distributed by Eleven International Publishing
P.O. Box 85576
2508 CG The Hague
The Netherlands
Tel.: +31 70 33 070 33
Fax: +31 70 33 070 30
e-mail: sales@budh.nl
www.elevenpub.com

Sold and distributed in USA and Canada
International Specialized Book Services
920 NE 58th Avenue, Suite 300
Portland, OR 97213-3786, USA
Tel.: 1-800-944-6190 (toll-free)
Fax: +1 503 280-8832
orders@isbs.com
www.isbs.com

Eleven International Publishing is an imprint of Boom uitgevers Den Haag.

ISBN: 978-94-90947-42-2

Printed in The Netherlands

Table of Contents

Abbreviations

All ER	All England Law Reports
ANSI	American National Standards Institute
BGH	Bundesgerichthof (German Supreme Court)
DCC	Dutch Civil Code
DCCP	Dutch Code of Civil Procedure
ECJ	European Court of Justice
ECR	European Court Reports
EPLD	European Product Liability Directive 85/374/EEC
EPLR	European Product Liability Review of Hogan Lovells
EU	European Union
EWHC	High Court of England and Wales
GFL	General Food Law 178/2002/EC
GPSD	General Product Safety Directive 2001/95/EC
HR	Hoge Raad (Dutch Supreme Court)
ISO	International Standard Organisation
JA	Jurisprudentie Aansprakelijkheid
LJN	'Landelijk JurisprudentieNummer'
NJ	Nederlands Juristenblad
NJF	Nederlandse Jurisprudentie Feitenrechtspraak
NJW	Neue Juristische Wochenschrift
NJW-RR	Neue Juristische Wochenschrift Rechtsprechungsreport Zivilrecht
OGH	Oberster Gerichtshof (Austrian Supreme Court)
OJ	Official Journal of the European Union
OLG	Oberlandesgericht (German Court of Appeal)
PIQR	Personal Injury and Quantum Reports
QB	Queen's Bench
RAPEX	Community's rapid alert system fof non-food consumer products
Rb.	Rechtbank
REACH	Registration Evaluation and Authorisation of Chemicals
TSD	Toy Safety Directive 2009/48/EC
VersR	Zeitschrift für Versicherungsrecht, Haftungs- and Schadenrecht

1 Introduction

1.1 Point of Departure

European Union (EU) citizens are injured each year whilst using products.[1] One way to prevent the harm arising from accidents involving consumer products is by giving product warnings. Product liability law can contribute to this by the way in which the liability requirements in the context of warnings are framed and applied. A case in point is the European Product Liability Directive for defective products (hereinafter EPLD or the Directive) that holds producers liable for damage caused by a defect in their products.[2] Under this Directive, several product warning issues can arise with which civil courts need to deal when confronted with a claim of a consumer who suffered injuries during the use of a product. For example, courts may have to address the adequacy of a given product warning to determine whether a product is defective or they may have to form an opinion about whether a product warning should have been disclosed in instances when it was absent.

Underlying these legal warning issues is a number of presumptions about how humans behave and interact with products and with their warnings. Primarily, it presumes that warnings can be effective in modifying user behaviour. But is this a valid presumption? Relative to this is the manner in which courts or litigants evaluate product warnings in European product liability law. Of course, it is common sense that the size of a warning can be considered as relevant for the assessment of a warning's adequacy, but is this a well-considered basis for legally judging a warning? Judges are no experts on how a warning should be designed to compel consumers to use products safely and it seems that they have difficulty adjudicating warning issues effectively.

In order for courts to rule more consistently and effectively in warning issues, a solution resides in the use of guidelines in product liability law that are based on empirical evidence on how humans interact with products and with warnings. Available insights from cognitive psychology and ergonomics (human factors) on people's abilities and limitations with regard to processing warning information, as well as insights on how the design of products can contribute to preventing accidents are crucial for understanding the positive and

1 See the EU Injury Database, a systematic injury surveillance system that collects accident and injury data from selected emergency departments of Member State hospitals, including data related to home and leisure accidents where products are involved. See the recent report EuroSafe, KfV & DG Sanco November 2009.

2 Directive 85/374/EEC of the European Parliament and of the Council of 25 July 1985 on the approximation of the laws, regulations and administrative provisions of the Member States concerning liability for defective products (*OJ* 1985, L 210/29).

negative effects that warnings can have on consumer behaviour and for improving the EU level of product safety. There is a considerable body of warning research that has specifically studied the effective processing of warnings and there is accompanying academic warning research literature that can both provide useful insights on whether warnings work, under what conditions warnings are likely to be successfully processed and when warnings are unlikely to be effective. Such insights can thus serve as a source of inspiration for framing the requirements on warnings set by product liability law in such a way that they reflect the way consumers deal with warnings, in order to avoid unnecessary accidents.

1.2 AIM

The objective of this monograph is to use available insights from cognitive psychology and ergonomics as a stepping stone for proposing a toolkit of recommendations that should guide European civil courts and litigants on how to deal best with important warning issues within the liability framework of the EPLD. In this research project, I will analyse per warning issue the value of the insights from the warning research literature for European product liability law focusing on the liability test of the Directive. The value of the insights lies in that they can provide detailed explanations supporting why European product liability law holds its explicit or implicit view on these warning issues. Furthermore, the insights can contribute to a better understanding of the warning issues in European product liability law and moreover they may provide support for suggested improvements on how European product liability law – in particular the liability system of the Directive – should deal with warnings.

In order to reach this objective, I have divided my project into three parts.[3] Chapter 2 is an introductory chapter on warnings from a legal perspective. Chapter 3 discusses warnings from a behavioural perspective. Chapter 4 embodies the central part of the book where these perspectives are combined and where my recommendations are given.

Based on the discussion of warnings from a legal perspective and a behavioural perspective in chapters 2 and 3, respectively, I have identified the following warning issues – which will be dealt with in more detail in chapter 4 – as relevant for European product liability law:

- What is a product warning?;
- Why warn?;
- What risks need a warning?;
- When should consumers be warned in relation to other design solutions?;
- How should consumers be warned?

3 See paragraph 1.5 for a more detailed presentation of the structure of this book.

1.3 Methodology

1.3.1 *A Behavioural Approach*

As can be gathered from the previous section, this book utilises what I would like to call a 'behavioural approach' to European product liability law.[4] The behavioural approach underlying this monograph stems from undertaking a multidimensional approach towards doctrines pertaining to civil law, as advocated by Dutch scholars Van Boom, Giesen and Verheij.[5] Although the approach of using insights from one social science or more disciplines to analyse the law is not new, their book, containing contributions from several authors, was one of the first in the Dutch domain of civil law that expressly took this approach. The basic aim of the approach entails that lawyers lean on knowledge from social sciences, such as psychology and sociology, in order to analyse from a legal perspective the presumptions about human behaviour made in private law by courts and/or the legislator and the behavioural effects of private law based on the behavioural presumptions.[6] A lawyer who uses this approach is not required to have a degree in psychology or another behavioural science to be able to examine and use the empirical and theoretical literature that can provide insights that could be interesting to evaluate civil law topics.[7]

1.3.2 *Disciplines of the Warning Research*

Within this behavioural approach, I seek guidance from the sciences of cognitive psychology and ergonomics, as these disciplines are especially important for understanding human behaviour in relation to the use of products and product warning messages. These are considered briefly below.[8] More specifically, I mainly use the insights derived from the warning studies and academic research literature that has evolved over the last 30 years. It comprises theoretical models of warnings as well as research into warnings. The research

4 This thesis is part of the research programme "Behavioural approaches to tort and contract law" of the Erasmus School of Law. The programme analyses specific features of tort and contract law by making use of insights of behavioural sciences. For more information on this research programme see <www.esl.eur.nl/home/research/research_programmes/behavioural_approaches_to_contract_and_tort_relevance_for_policymaking/>.

5 Van Boom, Giesen & Verheij 2008. See also Giesen 2005.

6 Behavioural presumptions are defined by these authors as the explicit or implicit conceptions about how individuals or corporations think, decide, deal with uncertainty and how they will respond to legislative intervention. Behavioural effects are described as the consequences, intended or unintended, as a result of legislative intervention of legislators and courts, and the interaction between the intervention and the addressee: Van Boom, Giesen & Verheij 2008, p. 21.

7 Van Boom, Giesen & Verheij 2008, p. 34.

8 For a more detailed description of these disciplines see paragraph 3.2.

activity has been substantial and various articles, book sections and books have been published providing qualitative and quantitative reviews of the warning research covering a certain period.[9]

Cognitive psychology is a science that is essential for a thorough legal analysis of warning issues as it can explain how people interact with product warnings. *Cognitive psychology* studies cognitive processes of the human mind, such as perception, understanding, thought, memory and decision making. It is concerned with how people attend to and gain information; how that information is stored and processed by the brain; how people solve problems; think and formulate language and how these processes may be manifested as behaviours.[10] A well-accepted theoretical model within cognitive psychology is the information processing model that assumes that cognition can be explained by a model that is based on a series of sequential stages through which information is processed.[11] This model has also been popular for use as a basis for describing and understanding the processing of warnings.[12] As will become clear, this model therefore also plays a central role in this monograph.

Ergonomics is the second discipline that is used as an information source. Basically, ergonomics is a discipline that deals with the interactions between people and machines.[13] The International Ergonomics Association[14] defines *ergonomics* as: 'The scientific discipline concerned with the understanding of interactions among humans and other elements of a system, and the profession that applies theory, principles, data and methods to design in order to optimise human well-being and overall system performance'.[15]

It follows that ergonomics or human factors, terms that can be used interchangeably, is an applied science that uses knowledge of various disciplines, such as engineering and psychology, to focus on the interactions between people and technological artefacts, including products.[16] Ergonomists adapt products to people, based upon the physiological and psychological capacities and limitations of humans, whereas engineers improve products from the viewpoint of mechanical and electrical design and psychologists only study the mind and behaviour of people.[17] This discipline is of relevance for the legal analysis of this thesis, as it can shed light specifically on matters involving the interaction

9 See *e.g.* Ayres e.a. 1989; DeJoy 1989; Lehto & Miller 1986; Lehto & Papastavrou 1993; Stewart & Martin 1994; Silver & Braun 1999; Cox III e.a. 1997; Rogers, Lamson & Rousseau 2000; Stewart, Folkes & Martin 2001; Argo & Main 2004; Kalsher & Williams 2006; Lesch 2006 and Laughery 2006; Wogalter, DeJoy & Laughery 1999b; Wogalter 2006.

10 Solso, Maclin & Maclin 2008, p. 10.

11 Solso, Maclin & Maclin 2008, p. 14.

12 For more information on theoretical models of the warning process see chapter 3.

13 Payne & Wenger 1998, p. G-6.

14 For more information, see <www.iea.cc/>.

15 Karwowski 2006, p. 4.

16 Payne & Wenger 1998, p. 16; Helander 2006, p. 5.

17 Stanton 1998a, p. 1.

between humans and products and their warnings, such as how consumers perceive the hazards that are attached to products and how the design of products can contribute to reducing injuries on the basis of such knowledge.

1.3.3 Legal Domain

As regards the legal domain that is the subject of my investigations, I analyse the topic of warnings within the context of European product liability law. The focus of this doctoral thesis is on the application of the Directive's liability test, namely, the defectiveness test, in relation to product warnings.

The legal analysis is divided into two parts. Chapter 2 introduces the legal domain with a discussion on product warnings within the context of European product liability law. This chapter serves as a basis for chapter 4, where a detailed analysis of warnings and product liability is made on using the insights from cognitive psychology and ergonomics.

Chapter 2 gives a concise analysis of the state-of-the-art with regard to the liability requirements pertaining to warnings under European product liability law. This chapter explores how European product liability law deals with warnings by studying European academic literature and European case law from a number of Member States including the Netherlands, England and Germany. After describing the product liability theories based on fault and strict liability that are common in the Netherlands, the emphasis is put on the provisions of the strict liability regime of the Directive, especially the application of defectiveness in the context of product warnings.

The overview does not purport to reflect an extensive analysis of the product liability systems and case law of all Member States. Instead, examples are drawn from various jurisdictions, two in particular: The Netherlands on the one hand and England and Wales on the other. The choice for England and Wales hardly needs explaining. Within the European legal systems, the English common law system of torts serves as a prototypical specimen of a standards-oriented legal system. The choice for The Netherlands is justified by the fact that the Dutch law of torts is in many ways prototypical of continental code-based tort law. Although a small jurisdiction, the Dutch legal system is based on one of the most recent Civil Codes in Europe, which combines elements of both French and German tort law. As such, it may serve as a rough proxy for other legal systems in Europe. Hence, notwithstanding substantial differences in detail between the various tort law systems, the development of product liability law in the Netherlands, especially with regard to product warnings, is used as a springboard for testing my arguments on the role of cognitive psychology and ergonomics in European product liability law.

1.3.4 *Assumption of Preventive Effect of Product Liability*

In this book, I proceed from the assumption that product liability rules can contribute to the prevention and reduction of accidents involving the use of products and thus to a higher level of product safety in Europe. This assumption, borrowed from the law and economics approach on accident law,[18] is based on the notion that, because producers know that they are held liable for the damage done if they have violated a requirement defined by product liability law, they will be encouraged to exercise appropriate care to prevent accidents from happening in the first place. In European tort law, it is generally agreed that compensation of victims after the event is the primary aim of tort law. Nevertheless, injury prevention, *i.e.* providing precautionary incentives before the event, has also been suggested as a function of tort law.[19] Unfortunately, there is a lack of empirical evidence that confirms this preventive function of European tort law, including European product liability law.[20]

As regards product liability, the preamble of the Directive does not refer explicitly to accident prevention as an objective. It does refer to consumer protection, but this is more related to ensuring a solid mechanism for European consumers to claim compensation after an accident has occurred, as the preamble states that approximation of the laws of the Member States concerning product liability is necessary because the existing divergences entail a differing degree of protection of the consumer against injury or property damage.[21] The preamble also stipulates that approximation is necessary because it may distort competition and affect the movement of goods within the common market. The European Commission's green paper provides a more substantiated view. It points out that the Directive seeks to protect victims and to promote improvements in product safety within the internal market through a regulatory framework that is as consistent as possible and based on a fair apportionment of the risks inherent in modern production.[22] It can thus be argued that the provisions of the Directive are also geared towards increasing the level of European product safety, albeit implicitly.[23]

This goal of accident prevention is not equal to achieving maximum safety; that is unattainable. The goal is to reach an optimal level of safety, whereby the interests of producers have been taken into account and whereby consumers carry a responsibility with regard to safe product use. It follows from the Directive's text that the Directive supports

18 Calabresi 1970.

19 *E.g.* Cane 2006, p. 479, 424 ff; Van Boom 2006a; Van Boom 2006b, p. 18. In the Netherlands see *e.g.* Spier e.a. 2009, p. 7; Visscher 2005, p. 16.

20 Empirical studies have been done in the United States indicating that there is a preventive effect of American tort law, including product liability law. See *e.g.* Dewees, Duff & Trebilcock 1996; Cane & Kritzer 2010.

21 Recital 1.

22 European Commission 1999.

23 Cf. Faure 2000, p. 469; Reich 1986; Grubb & Howells 2007, p. 15.

this point of view. Its requirements are formulated in such a way that, according to the European Commission, a fair balance between the interests of consumers and of producers is reached.

1.3.5 *Relevance*

There are two main reasons for taking a behavioural approach to the assessment of warning issues within the liability standard of the Directive. These relate to (1) the empirical origin of the insights and (2) the need for more guidance in European product liability law.

First and foremost, seeking guidance from these insights benefits product safety. Based on the aforementioned assumption that the (warning) obligations imposed under product liability function as an incentive for producers, it follows that understanding human behaviour in relation to product warnings will most likely lead to fewer occurrences of death and personal injuries associated with product use and to fewer liability claims. The use of empirical evidence on how individuals use products, how they perceive the risks attached to their use and how they process warning information provide insights into the ways in which producers' obligations that pertain to product warnings need to be framed so that they reflect the way consumers deal with warnings and unnecessary accidents can as a result thereof be avoided. For example, by gaining a better understanding of how people process information, predictions can be made of how the format and content of warnings should be designed in order to be effectively processed by warning recipients and end in safe behaviour. And as a consequence, these predictions can provide lessons to European product liability law on framing the warning requirements that producers must comply with.

It must be remarked here that the empirical findings of the warning studies and the accompanying warning research literature are related to safety science: improving safety is viewed as the ultimate goal of consumer product warnings and therefore, the goal of many empirical warning studies is to identify and determine the impact of factors that influence the stages of the warning process and to improve the design of warnings in order to enhance the likelihood that the warning will be complied with.[24] Hence, it might be argued that in general warning researchers have a blinkered view; they strive to find design solutions to maximize the behavioural effectiveness of warnings. As noted above, the Directive does, however, not purport to have a one-sided approach that favours consumers only. This view can be deduced from several provisions. For example, a product having caused damage to a person is not a sufficient requirement to impose liability under the Directive's regime; the damage must be caused by defectiveness. Other evidence can be found in the presence of a number of defences that a defendant producer can invoke to

24 See Wogalter, Conzola & Smith-Jackson 2002, p. 226.

free himself from liability and in the presence of a provision related to the expiration date of the liability of producers. The underling principle for this approach under the Directive is the principle of a fair apportionment of risk between the injured person and the producer set forth in the seventh recital in the preamble to the Directive.

The second reason for taking a behavioural approach is that insights from the empirical warning studies and the research literature can result into tools that can be valuable for the legal assessment of key warning issues. Guidance in this field is welcome. The Directive implicitly refers to the relevance of product warnings by stipulating that the presentation of the product can be a decisive factor for establishing a defect. However, an explicit textual reference to warnings is absent and there is also not much guidance from the text of the Directive to rely upon when assessing warning issues. In addition, neither the European Court of Justice (ECJ) nor the European Commission has given an opinion on the manner in which warning issues ought to be assessed in light of the wording of the Directive. Moreover, the number of liability cases related to warnings in Europe is far from abundant. The number of published decisions specifically dealing with product warnings after the implementation of the Directive is scarce in the Netherlands. On the other hand, there is ample literature on European product liability and on the Directive's implementation into the laws of the Member States. Many have written about this topic. It must be borne in mind though, that this literature can date from the time of the Directive's enactment; that this literature is often less accessible because of a language barrier, and that even if it is written in English, it usually concerns country reports that only slightly touch upon the topic of product warnings. All together, there is, to my knowledge, no comprehensive book available that is completely devoted to the topic of warnings in European product liability law. Let this thesis be the first. Having said that, there are a couple of well-written, comprehensive English books on comparative product liability in Europe. Also, the European Product Liability Reviews of the European law firm Hogan Lovells offer a peek inside product liability and are used as an information source for this monograph. Furthermore, the European literature often makes reference to case law and the provisions of the US Restatement of the Law (Third), Torts when discussing the European situation. In view of the paucity of European case law and literature relative to the United States, the American experience can be instructive.[25] Since this thesis takes the approach of turning to the disciplines of cognitive psychology and ergonomics for direction with respect to the assessment of liability in product warning cases, reference to American product liability is only made insofar as it complements the discussion.

25 Miller & Goldberg 2004, p. V; Grubb & Howells 2007, p. xi.

1.3.6 A Note on the Use of the Warning Research in a Legal Setting

Before drawing any inferences for product liability from the warning research, a few comments that relate to measurement issues and future research directions seem appropriate at the outset of this thesis.[26]

First of all, it needs mentioning that the warning research has come a long way. Despite the complexity of how humans process warnings, there is now a substantial body of research on the design and evaluation of warnings. This research has identified a number of influential variables (*i.e.* factors) that pertain to the effective processing of warnings, and design implications have been derived from the empirical evidence that form valuable tools for improving warning effectiveness.

Nonetheless, a full understanding of warning effectiveness is far from complete. There are inconsistencies and gaps in the findings that need to be addressed in future studies. Potentially relevant variables have either not yet or only rarely been studied. Furthermore, future research needs to be conducted to better understand the relative benefits of the variables as individual variables as well as their combined effects.[27] In addition, an overview of the warning studies shows that researchers have focused more on examining the effects of warning design changes on the earlier stages of information processing, such as attention and comprehension, and less on their impact on attitudes and beliefs, motivation and actual behavioural compliance. More research is needed in these areas.

Researchers have also stipulated that, in order to move forward, future research should address a number of methodological issues.[28] Rogers, Lamson and Rousseau have developed a set of recommendations to tackle methodological problems when doing empirical research. By taking account of the methodological considerations when designing measures for warning effectiveness, it is expected that the research will yield more valid and reliable data.[29] One of these concerns is, for instance, the inappropriate selection of the dependent measure by warning researchers which, as a result, can limit the conclusions of a study. Another limitation concerns the validity and reliability of data. For example, subjective measures like self-reports or questionnaires can be influenced by factors related to the subjects, such as decreased memory.[30] In addition, researchers have criticized the absence of a control condition, the condition in which no warning is given, in a number of experiments. This makes it impossible to answer the question whether including a warning would be superior to having no warning or to which degree compliant behaviour was going to

26 The research methods used in warning studies and methodological considerations of the warning research are further discussed in paragraph 3.5. Future research directions are discussed in more detail in paragraph 3.6.12.

27 Rogers, Lamson & Rousseau 2000, p. 133.

28 Rogers, Lamson & Rousseau 2000, p. 132.

29 Smith-Jackson & Wogalter 2006, p. 30.

30 Rousseau & Wogalter 2006, p. 149; Kalsher & Williams 2006, p. 326; Young & Lovvoll 1999, p. 31.

occur anyway.[31] Nevertheless, such studies remain valuable, because they can show the impact of manipulated variables on warning compliance. Furthermore, because a control condition is lacking, it makes it difficult to compare the effects of variables on compliance across studies and hence to generalise the findings regarding a certain variable.[32]

All in all, it is recognised here that the warning research literature, with its theoretical perspectives of the warning process and the research findings, does not present a complete picture of the interaction between humans and warnings. It is clear that in spite of the challenges that are present, a body of research findings and warning design implications has emerged that can be used as a foundation to evaluate law.

1.4 Scope

Because defining the scope of the research is essential for writing a legal monograph – especially one that takes a behavioural approach to law – the following topics are beyond the scope of this doctoral thesis.

This monograph covers the legal domain of civil law, *i.e.* product liability. I neither address the question regarding what type of law would provide incentives to producers to exercise appropriorate care to prevent accidents involving products nor whether a regulatory approach is superior to a tort law approach when analysing the way in which the product liability requirements pertaining to product warnings should be framed in view of the empirical insights. It would be an interesting topic for future research to study how the public law policy of the EU with regard to products can contribute to improving the level of safety of products in the EU.

In addition, this book only deals with the interpretation of the liability test of the Directive. This means that this book does not examine other requirements set by the Directive, such as causation, burden of proof, statutory defences, definitions of product, producer and damage or contributory negligence, unless this is of interest for the interpretation of warnings under the defectiveness test. It also follows from this that the issue of after-sale warnings, which relates to the subsequent duty to warn of dangers after the product is put on the market, is not addressed in this book, as the Directive merely measures liability for defective products at the time the product was put into circulation and not at a later date. Lastly, the relevance of the precautionary principle for determining product liability is also not discussed here.

31 Edworthy & Adams 1996, p. 10; DeJoy 1989; Cox III e.a. 1997, p. 201.

32 Adams & Edworthy 1995, p. 8.

1.5 Reader's Guide

Having identified both the objective and domain of the research – against the background of an assumption on my part of the preventive effect of product liability – I can now give a more detailed outline of how this research is built up.

The paper is organised around the following three chapters: a chapter on warnings from a legal perspective; a chapter on warnings from a behavioural perspective; and the central chapter which combines these perspectives and where the findings from the warning research literature are used to analyse the law. More specifically, the insights are used to explain and to understand the presumptions that European product liability law has about the interaction between warnings and human behaviour and to construct arguments for how the warning issues should be dealt with under the Directive's liability system.

⚠ WARNING

To reduce the risk of experiencing information processing problems whilst reading this book, please follow these reading tips:

1. **If you have a civil law background, continue with chapter 3 before reading chapter 4.**
2. **If you have a background in psychology, continue with chapter 2 before reading chapter 4.**
3. **If you prefer a quick read, continue with chapter 4 and use chapters 2 and 3 as reference work.**

I start my exploration in chapter 2 by introducing the topic of product warnings from a product liability perspective. Chapter 2 contains a concise discussion of the current state of the two product liability theories in tort (fault-based liability and strict liability) that are popular in the Netherlands (paragraphs 2.2 and 2.3). The central part of the chapter (paragraph 2.4) explores the consensus on product warning requirements for determining defectiveness under the Directive's liability system on the basis of studying case law of a number of Member States and legal academic literature. Its focus is on interpreting the defectiveness standard, especially in relation to warnings, and not on other requirements of the Directive. It has not been my intention to provide an extensive chapter on product liability in Europe. Therefore, readers of this thesis with a legal background, especially in

European civil liability law, may take the advice upon themselves to skip this part and move on to read chapter 3.

In chapter 3, I analyse the topic of warnings from the perspectives of cognitive psychology and ergonomics. I provide an in-depth overview of the available findings of the warning research and the associated academic warning research literature on a number of relevant warning topics that have received attention in the warning research and academic literature. Several warning research studies are described for illustrative purposes. Besides the paragraphs in chapter 3 that discuss relevant warning issues, there are paragraphs that fulfil a supportive role to better understand the warning research that has been done. For example, the chapter contains a paragraph that discusses the disciplines that are involved in the warning research (paragraph 3.2) and a paragraph on the research methods that warning researchers have employed to achieve empirical findings (paragraph 3.5). Chapter 3 is – in contrast to chapter 2 – not concise and for a good reason. The target audience of this monograph are those with a degree in law, who will not have knowledge of the topics described in chapter 3. Nevertheless, this book can also be of interest to people who are involved in the type of research described in chapter 3. For those interested who do have a background in psychology, reading this chapter may be superfluous.

Chapter 4 is the central chapter of this monograph. I apply the insights thus gained to product liability and on the basis thereof make recommendations with respect to how European civil courts and litigants should deal best with the relevant warning issues in the context of the defectiveness test of the Directive 85/374/EEC. The warning issues include: 1) What is a product warning?; 2) Why warn?; 3) What risks need a warning?; 4) When should consumers be warned in relation to other design solutions?; and 5) How should consumers be warned? The chapter is written in such a way that I believe it is accessible to those with a legal background who have not read chapters 2 and 3. To prevent unnecessary repetition as much as possible, I have made references to chapters 2 and 3 in chapter 4 for readers who are in need of or are interested in a more detailed discussion.

2 Product Warnings and European Product Liability

2.1 Introduction

Accident statistics show that people hurt themselves whilst using products, for example because there was no warning that told users about the risk or because the product did not function as it should. A person who has suffered personal injuries when using a (consumer) product has several legal remedies at his disposal to claim compensation. In general, consumers in Europe can base their action on contractual liability, fault-based liability in tort and strict liability in tort. The latter two theories are most important as consumers and producers are usually not bound by contract. For this reason, this thesis only explores fault-based liability in tort and strict liability in tort as regards the liability of producers for their products. The strict liability regime embodies the provisions of the EPLD that are implemented into the laws of Member States.[1] In many European countries, the Directive is the primary cause of action in product liability litigation.[2] Nevertheless, fault-based liability still maintains its effect, especially for claims that fall outside the scope of the Directive or as an alternative claim next to a primary action based on strict liability.[3] In the Netherlands, product liability claims are mainly based on the Directive's liability regime implemented in Articles 185 to 193 of Book 6 of the 'Burgerlijk Wetboek', *i.e.* the Dutch Civil Code (DCC), or on the fault liability regime of Article 6:162 DCC.[4] The basis of these theories laid down in liability law and relevant case law are dealt with in short in the following paragraphs.[5]

As I have explained in the introductory chapter of the thesis, the objective of this chapter is to look at product warnings from a liability law perspective. This chapter first gives a sketch of the Dutch product liability regime based on fault in paragraph 2.2 and

---

1 Directive 85/374/EEC of the European Parliament and of the Council of 25 July 1985 on the approximation of the laws, regulations and administrative provisions of the Member States concerning liability for defective products (*OJ* 1985, L 210/29).

2 See *e.g.* Campbell 2007; Fairgrieve 2005; Hodges 1993a.

3 See Article 13 of the Directive. This can be the case in the event that the consumer suffered economic loss (Article 9 of the Directive) or because the periods stipulated in Article 10 expired. Another reason can be that the product was put into circulation before the date on which the Directive's provisions entered into force (Article 17).

4 See *e.g.* Giesen & Loos 2001; Van Dam 2005.

5 For a more complete discussion on product liability in Europe, see Grubb & Howells 2007; Miller & Goldberg 2004; Whittaker 2005; Fairgrieve 2005. For a discussion in English of product liability in the Netherlands, see Van Dam 2005; Giesen & Loos 2001.

secondly of the Dutch strict product liability regime in paragraph 2.3. Since the strict product liability regime embodies the implementation of the EPLD, the discussion of the Dutch strict liability regime can be regarded as part of European product liability law. Paragraph 2.4 forms the central part of the chapter, as it specifically addresses the interplay between product warnings and product liability. It explores the consensus on warning requirements under European product liability law. Attention is given to the two claims with respect to product warnings that are normally distinguished in product liability. These are (1) a failure to warn/absence of a product warning and (2) the inadequacy of a given product warning. Because the Directive does not mention the word 'product warning' once, the warning requirements are derived from the wording of the law, case law and academic literature. The discussion is supplemented with relevant case law from other European countries that have been mentioned in European product liability literature.[6]

2.2 PRODUCT LIABILITY BASED ON FAULT

2.2.1 *Introduction*

Before getting to the heart of this chapter, namely paragraph 2.4, the theories of fault-based and strict liability need some further explaining. This paragraph discusses product liability based on fault. In the following paragraphs, I first provide a general introduction into the legal requirements of fault-based liability under Dutch law (paragraph 2.2.2). After that, specific attention is paid to the important requirement of the yardstick for assessing wrongful conduct (paragraph 2.2.3). Next, paragraph 2.2.4 addresses the wrongful conduct of producers in particular and the applicable unwritten duties for producers that have come forth under Dutch case law. The final paragraph (2.2.5) provides a summary.

2.2.2 *Fault-Based Liability under Dutch Law in a Nutshell*

The cornerstone of fault-based liability in European tort law is the standard of reasonable conduct of the tortfeasor.[7] The general rules of fault-based liability in the Netherlands are laid down in Article 6:162 DCC.

6 The main sources that I have used in this respect are Howells & Borghetti 2010; Grubb & Howells 2007; Fairgrieve 2005; Miller & Goldberg 2004 and issues of the *Hogan Lovells European Product Liability Review*.

7 See *e.g.* Van Dam 2006 for a discussion of European tort law in English.

Article 6:162 DCC reads as follows:

> 1) A person who commits an unlawful act toward another which can be imputed to him must repair the damage which the other person suffers as a consequence thereof.
> 2) Except where there is a ground for justification, the following acts are deemed to be unlawful: the violation of a right, an act or omission violating a statutory duty or a rule of unwritten law pertaining to proper social conduct.
> 3) An unlawful act can be imputed to its author if it results from his fault or to a cause for which he is answerable according to law or common opinion.[8]

Hence, fault-based liability consists of two main elements: the wrongfulness of the act and the attributability/imputability of the act to the wrongdoer. Wrongfulness refers to the act or omission and the requirement of imputability concerns the wrongdoer. The most important ground for attributing the act to the wrongdoer is that the wrongdoer can be blamed for his wrongful behaviour.[9] The actor is at fault or blameworthiness can be established if the wrongdoer could and should have acted differently. Case law shows that blameworthiness is measured objectively: the standard is that of a normal, reasonable person. Even though the two elements are theoretically distinguished from each other, this dividing line is vague in practice. In most cases, whenever wrongfulness has been determined, the element of blameworthiness usually presents no difficulties.[10] In sum, to establish product liability based on fault, a producer must have acted wrongfully and he/she must also be at fault or the act must be attributable to the producer.

Under fault-based liability, it is the injured party who is required to prove the requirement of imputability on the part of the injuring party as well as proof of the damage and the causal link between the damage and the wrongful act. The general rule is laid down in Article 150 of the 'Wetboek van Burgerlijke Rechtsvordering', *i.e.* the Dutch Code of Civil Procedure (DCCP). The article stipulates that it is in principle the claimant who has to prove the facts substantiating the legal grounds of his claim, if and insofar as the defendant states a reasoned denial of these facts. Nevertheless, there are general procedural rules on evidence and burden of proof that may alleviate or even shift this burden.[11]

Article 6:162 DCC constitutes the basis for an action under the general tort rules, including a negligence claim against producers. The starting principle underlying the fault-

8 The translation of this article is taken from Haanappel & MacKaay 1990.

9 Article 6:162(3) DCC. Under this article, there are three grounds for imputation: the person can be blamed for his act ('schuld', *i.e.* fault), or (the cause of) his act must be imputed to him, either on a statutory basis, or plainly because the 'verkeersopvattingen' thus demand (that is, an unwritten source of legal and moral opinion, as it is expressed in case law).

10 Spier e.a. 2009, p. 80.

11 See *e.g.* Spier 1998 and Van Boom 2005 for a discussion of fault-based liability under Dutch law in English.

based liability theory is liability for own acts or omissions. According to paragraph 2 of Article 6:162 DCC three sorts of a *wrongful act* , *i.e.* onrechtmatige daad, can be distinguished: (a) conduct which infringes a person's right; (b) conduct contrary to a statutory duty; and (c) any act or omission which violates a rule of unwritten law regarding a proper social conduct.

The first unlawfulness/wrongfulness category covers conduct which is wrongful because it violates a subjective right. Important rights are property rights (*e.g.* a person's right of ownership, the rights of products of the mind) and personality rights (*e.g.* the right to the protection of privacy or of physical integrity). The second category, violating a statutory duty, has a broad scope of application; it includes acts or omissions contrary to different sorts of statutory provisions, laid down in Acts of Parliament and subordinate legislation as well as in EC legislation. The third category of wrongful acts, the so-called 'standard of due care', refers to unwritten standards of due care generally accepted in society. Different types of improper social conduct fall within this category. These types have been sorted into general subcategories, such as creating a nuisance for neighbouring property, the duty of care of a professional vis-à-vis its client, or creating dangerous situations. In practice, the latter category has shown to be most important. Unfortunately, deciding whether someone has acted in conflict with what is generally accepted according to unwritten law is generally considered burdensome for courts, seeing these standards are not laid down in law. Case law can provide some guidance. It must be borne in mind though that each case is judged on its own merits.[12]

As far as the relationship between the three grounds of wrongfulness is concerned, the starting point of the national legislator is that they have independent significance as they are mentioned separately in the article. Generally, wrongful conduct can often be placed under more than one ground. Behaviour which infringes a person's right can often simultaneously be regarded as a violation of a statutory duty, especially when such a right has a legal basis. Similarly, when behaviour is wrongful because it violates a statutory duty or a person's right, this conduct will usually be contrary to unwritten standards as well. The latter category is considered as most useful and most significant, because it fulfils an additional role; whenever conduct cannot be considered as a violation of a person's right or of a statutory duty, it is possible that the act or omission is nonetheless contrary to unwritten standards in society.[13]

A final note worthy to mention is the principle of relativity of Article 6:163 DCC. It stipulates that an action on the basis of a wrongful act must be rejected if the standard breached does not serve to protect against damage such as that suffered by the person suffering the loss. Article 6:163 DCC is of particular importance for the violation of a

12 Spier e.a. 2009, p. 44.
13 Spier e.a. 2009, p. 26.

statutory duty. It is generally agreed that conduct contrary to a statutory duty is in principle wrongful. This however does not hold in the event that the statutory provision does not have the purpose to protect both the interests of the victim in question and the type of damage suffered.[14]

2.2.3 The Standard of Wrongful Endangerment

2.2.3.1 The *Kelderluik* Factors

Traditionally, the most important subcategory coming under the third type of wrongful acts – the unwritten standards of due care generally accepted in society – concerns the unwritten safety standards that are applicable in situations of 'gevaarzetting', meaning endangerment or increased danger. Endangerment means that a person creates a dangerous situation which may harm people or goods without taking the necessary safety measures.[15]

Since the landmark *Kelderluik* decision of the Dutch Supreme Court of 1965, it has been established case law that the wrongfulness of breaching safety standards depends on a balancing test of the circumstances of the case.[16] The Dutch Supreme Court held that only in light of the circumstances of the case at hand, it can be determined if and to what extent it can be required of someone, who creates a situation that is dangerous to others who do not exercise the required amount of attention and caution, that he/she takes into account the possibility that this attention and caution will not be taken and that he/she consequently adopts certain precautionary measures. The Dutch Supreme Court proceeded with providing four important circumstances that should be considered when assessing whether there is wrongful endangerment. These are termed the *Kelderluik* factors and they have been cited by the Dutch Supreme Court and lower courts on a regular basis ever since. They consist of:

- the probability that potential victims are careless and inattentive;
- the likelihood that this leads to accidents;
- the gravity of the consequences of the accident;

14 Spier e.a. 2009, p. 25. See on this topic: Lindenbergh 2007.

15 Spier e.a. 2009, p. 48. A person who perceives a dangerous situation can act wrongfully as well, when this person has been aware of the seriousness of the danger, but omits to warn or to eliminate the dangerous situation, see *e.g.* HR 22 november 1974, *NJ* 1975, 149 (*Struikelende broodbezorger*).

16 HR 5 november 1965, *NJ* 1966, 136 (*Kelderluik*). In the *Kelderluik* case, a chauffeur of the Coca Cola Company named Sjouwerman was carrying crates of bottles out of the cellar of a café and left the kelderluik, *i.e.* the cellar trapdoor open. The victim Duchateau had not noticed the open trapdoor and fell down into the hole when he hurried to the toilet. Duchateau holds the Coca Cola Company liable for the damages suffered as his employee Sjouwerman acted wrongfully by leaving the trapdoor open without taking any preventive measures. The Supreme Court concludes that the Court of Appeal was correct in its assessment that Sjouwerman acted wrongfully by not taking into account the possibility that guests would be inadvertent and by failing to take (sufficient) precautionary measures. Furthermore, there is a 50% contributory negligence on the part of the victim.

- the costs of taking precautionary measures.

The first three factors concern the extent of the risk (*i.e.* probability and gravity of the hazard). The latter refers to the amount of precaution that needs to be taken in order to provide adequate protection against the risk. This factor of the costs of taking precaution relates to whether, in the absence of the precautionary measure, the measure was possible take and customary as well as to the inconvenience in terms of time, effort and financial costs of adopting the measure.

Providing warnings is an example of such a precautionary measure. Courts assess whether the level of precaution taken by the wrongdoer is sufficient in view of the level of risk. More specifically, to arrive at a fair judgement courts need to examine what precautionary measure(s) was (were) taken and what precautionary measure could and should have been taken.[17] Hence, it can be said that Dutch courts use a risk-utility analysis to determine unlawfulness in situations of endangerment. In this respect, it has been argued in the literature that the *Kelderluik* standard is the Dutch equivalent of the American formula of Judge Learned Hand to determine negligence.[18]

The list of factors is non-limitative. In addition to these factors, other factors evidenced by case law and literature may be pertinent as well in order to arrive at the opinion that the dangerous conduct is unlawful. These include for instance the nature of the damage, the nature and utility of the conduct of the injuring party, the foreseeability of the danger to the injuring party, the capacity of the potential victims, the degree in which the victims' behaviour contributed to the manifestation of the risk.[19] These factors are not always relevant in every case and their weight differs. The factors that are indicated to be relevant in a specific case need to be balanced against each other to assess whether the dangerous behaviour is wrongful. The value which needs to be attached to the one or other depends on the context of the case in which an unwritten safety standard was breached.[20]

Given the casuistic nature of the cases brought before the courts, it is difficult to provide general ideas as regards how to best arrive at the opinion that the conduct in the concrete situation was unlawful. Nevertheless, the legal literature provides several general rules of thumbs derived from case law that can be used as starting points when assessing a case.[21]

In general, the degree of probability that accidents occur is a fundamental circumstance. It is settled case law that an act or omission cannot be considered wrongful by the mere fact that there is a possibility that damage to people or goods can occur as a result of the

17 Spier e.a. 2009, p. 48; Giesen 2005, p. 27.
18 Van Dam 2000, p. 174; Van Dam 2006, p. 195; De Mot, Canta & Gangapersadsing 2004; Van Maanen 2008.
19 Van Dam 2000, p. 173 ff; Jansen (Onrechtmatige daad I), Art. 162, aant. 88; Bolt, Spier & Haazen 1996, p. 163.
20 Hartlief 2004; Spier e.a. 2009, p. 50.
21 Spier e.a. 2009, p. 47, 48; Van Maanen 2008.

act or omission and that the risk is materialised. Risks are to a certain degree acceptable in society. According to the Dutch Supreme Court, wrongful endangerment by virtue of Article 6:162 DCC is established if the degree of probability of damage resulting from the act or omission is so high that the wrongdoer ought to have refrained from such conduct.[22] Note that this does not preclude accepting liability in cases where the probability of an accident is low. Given that the factors of a specific case act as communicating vessels, the unlawfulness may then be rooted in the significance of other relevant factors, for example the seriousness of the consequences, *e.g.* severe personal injuries, and/or the circumstance that adopting a certain measure is customary and simple to do.[23] Another rule of thumb is that the larger the seriousness and extent of the damage, particularly personal injury, the higher the duty of due care. The same is true when the conduct becomes more dangerous. Finally, it is said that if the adoption of possible precautions is less inconvenient in terms of costs, time, effort and so forth, the obligation to take precautionary measures increases. Consequently, the failure to adopt the relatively simple and inexpensive measure in these circumstances will in principle be considered wrongful.[24]

2.2.3.2 Expectations of Behaviour

It can be said that the degree of safety in a dangerous situation which must be taken into consideration depends on the expectations both parties, the injuring party and the injured party, are entitled to have about behaviour in the context of that specific situation.[25] As with the basic principle that it is acceptable for the injuring party to create a certain degree of risk, a certain degree of careless and inattentive behaviour on the part of potential victims can generally also be regarded as permitted.[26]

The significance of anticipating the behaviour of others has long since been recognised in Dutch liability law. As follows from the *Kelderluik* judgement of 1965, it depends on the circumstances of the case whether and to what extent someone who creates a situation that is dangerous for people who do not exercise the required caution, is obliged to take into account the possibility that this needed caution and attentiveness will not be exercised and to take certain precautionary measures in view of that. Factors that may be of relevance here are the (objective) knowledge of the injuring party and the capacity of the potential victims. This factor of the probability of carelessness of others is closely connected to the probability of accidents as careless and inattentive behaviour influences the likelihood that accidents occur and that damage is suffered. Because of this relationship, these two

22 See *e.g.* HR 9 december 1994, *NJ* 1996, 403 (*Zwiepende tak*); HR 7 april 2006, *NJ* 2006, 244 (*Uienrot*).
23 Van Dam 2000, p. 184; Van der Wiel 2007, p. 178.
24 Spier e.a. 2009, p. 47; Jansen (Onrechtmatige daad I), Art. 162, aant. 88.3.
25 Hartlief 2004, p. 870; cf. Spier e.a. 2009, p. 54. See *e.g.* the *Jetblast* case discussed below.
26 Jansen (Onrechtmatige daad I), Art. 162, aant. 88.8.

Kelderluik factors are often considered together in Dutch case law in terms of the probability of accidents or the probability of damage.[27]

The importance of anticipating the behaviour of others has been followed by the Dutch Supreme Court in specific fields of liability related to accidents. Within the course of Article 6:162 DCC, it is established case law that the liability of a road supervisor, such as a municipality, rests among other things on the question whether the municipality has taken into account the fact that not all road users exercise the necessary caution.[28] A city's local government has the responsibility to ensure that a public road is kept in good repair, which also entails the duty to take care that the conditions of a road do not endanger the safety of people and goods. This implies, according to the Dutch Supreme Court in the *Bussluis* ruling, that in the event that for the benefit of traffic measures the municipality lays out a road in such a way that without safety measures it is dangerous to persons and goods, the municipality must ensure that by adopting adequate safety measures, such as warning signs, the safety of persons and goods is sufficiently guaranteed. The municipality must include in the considerations that not all road users will constantly exercise the needed due care and attention. The Dutch Supreme Court closes with the statement that if the safety cannot be sufficiently guaranteed, the municipality should waive this layout of the road. Equally, in the field of work safety and fault liability, the Dutch Supreme Court has held that given the circumstance of the use of a machine or tool on a daily basis, an employer has the responsibility to consider that an employee will not always exercise the due care needed to prevent occupational accidents.[29] In the area of fault-based product liability, the Dutch Supreme Court provided a comparable ruling. This case is discussed in paragraph 2.2.4.

> Dutch *Bussluis* case
> In this case, a taxi driver caused an accident by driving into a 'bussluis', *i.e.* a hole built in the road as a way of closing a bus lane to traffic other than public transport. The accident happened in November 1984 and resulted in car damage and personal injury of a passenger. Two traffic signs were placed at the road side, both containing the written message that traffic was not allowed, only public transport. Another sign contained a pictogram of a car turning over into a hole and underneath the caption 'Bussluis'. Furthermore, the word 'BUS' was painted onto the road surface. This traffic measure was built in December 1983 to ensure that that specific part of the road was closed to traffic except public transport which was expressed by the traffic signs. This traffic measure was

27 Jansen (Onrechtmatige daad I), art. 162, aant. 88.1; Van Dam 2000, p. 184.

28 HR 20 maart 1992, *NJ* 1993, 547 (*Bussluis*).

29 See *e.g.* HR 14 april 1978, *NJ* 1979, 245 (*Messaoudi/Hoescht*); HR 22 maart 1991, *NJ* 1991, 420 (*Roeffen/Thijssen*).

not generally known to road users at the time of the accident. The taxi firm brought an action for damages against the municipality stating that the latter had acted wrongfully by constructing the hole in the road surface.
The Court of Appeal held that the municipality acted without due care owed to the taxi firm, because it had created a dangerous traffic situation as a result of building and maintaining the bussluis. The presence of the specific warning sign 'Bussluis' was insufficient because the sign was not immediately recognisable as a warning sign. Its colours and shape corresponded with traffic signs that are being placed to provide general information. Furthermore, the injured party stated that the phenomenon of a 'bussluis' was not generally known at that time, which was not contradicted by the injuring party. The Supreme Court concluded that the Court of Appeal's judgement was correct. Neverthless, the Court of Appeal's rejection of the municipality's defence that there were circumstances that could be imputed on the victim was not considered correct. The Court of Appeal proceeded on the assumption that the taxi driver disobeyed the traffic signs that prohibited driving there and hence, that he violated the traffic regulations. In view of that circumstance, the Court's decision that there was no contributory negligence on the part of the injuring party is not understandable according to the Supreme Court.

Although court decisions have stated that consideration should be given to the circumstance that potential victims may behave without the required caution and attention, the amount of care that is needed is not unlimited and depends on the specific circumstances of the case.

In the Dutch *Gekantelde vrachtwagen* case, the Supreme Court gave an indication of the type of behaviour of potential victims that falls outside the scope of fault-based liability.[30] This case concerned the liability of a road supervisor, *i.e.* the water authority, on the basis of Article 6:162 DCC.

Dutch *Gekantelde vrachtwagen* case
This particular case concerned a driver who drove a heavily loaded lorry onto a road which was partially surfaced. The width of the road varied from 2 to 2.5 meters. When arriving at a smaller section of the road, the lorry overturned and ended up in a ditch alongside the road. The road was open to all traffic and contained no traffic sign that warned against the fact that the road was unfit for heavy transport. The victim instituted proceedings against the road supervisor, *i.e.* the water authority, on the basis of Article 6:162 CC and pleaded

30 HR 26 september 2003, *NJ* 2003, 660 (*Gekantelde vrachtwagen*); see Giesen 2004.

> that the supervisor had committed a wrongful act by failing to warn against the dangerous road situation or by failing to take other precautionary measures to prevent accidents.
> Both the court of first instance and the Court of Appeal established liability of the road supervisor. The Court of Appeal held that there was a dangerous traffic situation that had to be dealt with, especially since this was possible in a rather simple way. The Supreme Court quashed the appellate court's decision and ruled that in the context of the circumstances of this specific situation, the supervisor had no duty to warn with regard to the danger of using the road for heavy transport. The judgement of the Court of Appeal was an incorrect interpretation of the law or the Court of Appeal insufficiently motivated it.

Starting with an explicit reference to the *Kelderluik* and the *Bussluis* judgements, the Dutch Supreme Court holds that the liability of the road supervisor is established only if the supervisor should have taken into account that such a driver of a heavy lorry would use the road and that this driver would (and could) not suspect that the road was unfit for that lorry and thus dangerous. According to the Dutch Supreme Court, it follows from this that there is no liability if, based on the circumstances of the specific case, it ought to have been immediately obvious to a driver who is not continuously cautious and attentive that this road was unfit to be used by a heavily loaded lorry. One could submit that this ruling implies that a person with a duty to care is entitled to expect from people that they properly respond to dangers which are or should have been immediately evident to them. Someone is not obliged to adjust his/her duty of due care to an abnormal or improbable form of careless and inattentive behaviour of others. Hence, even warnings are not needed in such circumstances. The potential victims' own responsibility precludes accepting liability.[31]

2.2.3.3 Warnings: The Extent and Content of the Duty to Warn Adequately

There are several ways to exercise care in order to eliminate or reduce the potential negative consequences of a dangerous situation. Providing a warning is one means of precaution. Accordingly, the duty to warn adequately is a duty of due care. As a result, whether there is a duty to warn should be assessed on weighing the concrete circumstances of the case.

Naturally, to determine whether there is a failure to warn involves that courts should consider the *Kelderluik* factors to balance the size of the risk against the burden of providing a warning. The above-mentioned factors pertaining to the burden of taking precaution, *i.e.* whether a warning is actually possible to take, whether using a warning is customary and the costs in time, money and effort of providing a warning, are of importance for assessing a failure to warn against the risk. Warnings are a popular precautionary measure

31 Cf. Giesen 2004, p. 38; Hartlief 2004, p. 871.

to adopt. It seems that courts generally consider providing warnings a simple and easy way of exercising due care and therefore often not too high of a cost to adopt. As illustrated above, improbable behaviour of potential victims can bar a successful claim that is based on a failure to warn. Furthermore, case law shows that the obviousness of a risk or whether it is generally known often also play a role in determining whether a warning ought to have been provided or not.[32]

In cases where a warning was given and yet damage occurred, liability commonly revolves around answering the question whether the adopted measure of warning can be considered sufficient. The content of the warning and the extent of this duty depend on the circumstances of the case. An important ruling of the Dutch Supreme Court governing the adequacy of warnings under Article 6:162 DCC is the recent *Jetblast* case.[33] The *Jetblast* case involves the liability of an airport for providing an inadequate warning against the danger of low-flying aircrafts. The judgement of the Dutch Supreme Court states that to answer the question whether a warning can be considered an adequate measure to protect against a certain danger, it is of decisive importance whether it is to be expected that the warning will lead to an act or omission as a result of which the danger is avoided.

Dutch *Jetblast* case

As a result of the jet blast from a Boeing 747 during taking-off, a woman was blown over and hit her head on a rock. The woman was standing behind a fence that formed the dividing line between the site of the airport and Maho Beach on Sint Maarten and was looking at the departing aircraft. A sign attached to the fence warned against the danger of jet blasts by means of a pictorial of an airplane taking off with the English wording 'WARNING!', underneath which was the sentence 'low flying and departing aircraft can cause physical injury'. The woman instituted her claim against the airport on the basis of Article 6:162 DCC and alleged that the airport acted wrongfully by failing to provide a safe environment for the visitors of Maho Beach. The Court of Appeal rejected the victim's claim.

The Court of Appeal argued that in this specific case, the question is whether the warning signs are sufficiently adequate. The Court of Appeal decided that the warning signs are adequate, given that as a result of the sign people could have been informed of the danger. The injured party brought an appeal in cassation.

Having repeated the *Kelderluik* ruling, the Supreme Court formulated the above-mentioned yardstick for assessing whether a warning is an adequate

32 See *e.g.* HR 26 september 2003, *NJ* 2003, 660 (*Gekantelde vrachtwagen*).

33 HR 28 mei 2005, *NJ* 2004, 105 (*Jetblast*). See Giesen 2005; Hartlief 2004; Haak 2006.

safety measure. The Supreme Court held that the Court of Appeal mistakenly decided that the warning signs were adequate safety measures. The Supreme Court considered that the Court of Appeal made an incorrect interpretation of the law if it judged that a warning sign can be regarded an adequate safety measure to prevent the public against the danger if the public can know of this danger as a result of the sign. To answer the question whether a warning can be regarded as an adequate measure to protect against a certain danger, it is of decisive importance whether it is to be expected that the warning will lead to an act or omission as a result of which the danger is avoided. Furthermore, the Supreme Court stated that in the event that the Court of Appeal did not fail to recognise this standard, the court insufficiently motivated that this standard was met under the established circumstances of the case. The Court of Appeal established that the airport could have expected tourists to stand close to the fence to watch the airplanes taking off and that in spite of the warning signs, the majority of the tourists gather at that place to watch the aircrafts. According to the Supreme Court, the Court of Appeal also failed to take into account the victim's assertion that tourists could not conclude sufficiently from the wording of the sign precisely which danger was present.[34]

34 HR 28 mei 2005, *NJ* 2004, 105 (*Jetblast*). In the judgement of the Court of Appeal that followed the decision of the Supreme Court, the court rules as follows. When applying the Supreme Court's standard regarding the adequacy of warnings, the Court of Appeal argues that the probability that people do not pay attention to the signs (and the probability that these signs do not lead to conduct that prevents the danger) is high. This is shown by the circumstance that tourists constantly stand close to the fence to watch the airplanes taking off. Furthermore, the warning signs do not present the danger of 'physical injury' in a very penetrating way and the signs do not adequately make clear which type of physical injury is involved. The signs also do not convey that there can be a powerful moving of air that can blow people who stand nearby over with force. In addition, the Court of Appeal presumes that the probability of accidents is low. Nevertheless, one can get severely injured or die and that circumstance counts heavily. The Court of Appeal consequently rules that the airport was obliged to take safety measures and that the warning signs were not adequate in this respect. The following question that needs to be answered according to the Court of Appeal is whether the cost of taking further precautionary action must be considered too high so that this cannot be required from the airport. If the answer is in the affirmative, the next question is whether this circumstance discharges the airport from liability. The court rules that the answer is not in the affirmative and as a result, the second part of the question does not need to be addressed. The Court of Appeal states that the warning signs could have been made more clear, *e.g.* by means of a pictogram, and more salient. The Appellate Court also takes account of the possibility, mentioned by the victim, of having a supervisor at the place, as it can be assumed that most tourists will follow the direct warning of a supervisor. For these reasons, the airport is liable under Article 6:162 DCC. A final question addressed by the Court of Appeal is whether there is contributory negligence on the part of the victim. This is not the case according to the Court of Appeal, as the warning signs were not very clear about the type of injuries that people could suffer from a jet blast. In addition, these consequences were not generally known. The Court of Appeal considers that the airport is fully liable to pay the damages caused in the concrete case, see Gemeenschappelijk Hof Nederlandse Antillen en Aruba 18 maart 2005, *NJ* 2005, 302.

A special note can be made with respect to the relationship between warnings and other precautionary measures to prevent the risk from occurring. In certain instances, for example if the injury is severe or if certain careless behaviour is to be expected of people, it is questionable whether the given warning alone is sufficient to provide the required level of precaution or whether physical precautionary measures to prevent the dangerous situation should have been taken by the wrongdoer.[35]

A guiding decision in this regard is the recent Dutch Supreme Court's ruling pertaining to an employer's fault liability concerning dangerous industrial machinery.[36] In this *Multivac-machine* decision, the Supreme Court was assigned to judge the scope of the employer's duty of due care under fault liability.[37]

> Dutch *Multivac-machine* case
> An employee sustained injuries to his hand while working with a wrapping machine for meat products. The worker had put his left hand into a narrow opening in the Multivac machine after it failed to wrap a meat product in foil. When he was trying to grab the foil inside the machine, his hand got under the machine's press. This resulted in three fingertips needing to be amputed. The machine was furnished with protective hoods, a stop button, and a warning sticker reading 'DANGER AMPUTATION DANGER. Do not remove protective hoods. Do not reach under the protective hoods unless the main switch is off', and displayed a pictogram of a crushed hand underneath a bar. The employer informed the machine manufacturer of the accident, whereupon the manufacturer affixed an extra safety strip to completely close off the opening in the machine.
> The Supreme Court held that to answer the question whether the employer is liable on the basis of Article 7:658 DCC for the employee's accident in spite of the taken measures, the underlying principle is that the purpose of this article is not to provide an absolute guarantee for the protection against danger. Under this article, the employer must take those measures that are reasonably needed to prevent that the employee suffers damage in the course of his/her profession. What is reasonable to expect from the employer depends on the circumstances of the case. The Supreme Court subsequently held that especially in the event of employment involving a hazardous machine, warning of risks that may arise from the use of a machine, oral or written warnings, symbols, or instructions for use, will not always be sufficient. This corresponds with the circumstance

35 See *e.g.* HR 20 maart 1992, *NJ* 1993, 547 (*Bussluis*).
36 HR 11 november 2005, *NJ* 2008, 460 (*Multivac-machine*).
37 Article 7:658 DCC.

> that an employer is obliged to consider the empirical fact that the daily use of a machine or tool may lead to careless behaviour on the part of an employee, even though warnings are present. For this reason, it may be reasonably expected from an employer that he examines whether adequate preventive measures are possible or whether the machine can operate more safely. If this is not the case, an employer must look closely at whether it is possible to warn against the danger in an adequate and effective manner. In this regard, it is deemed important to consider the probability that potential victims do not exercise the required care and attention, the likelihood that this leads to accidents, the gravity of the consequences of the accident and the costs of taking precautionary measures. In addition, it must be taken into consideration that if, in the absence of measures that can prevent the danger, it is only possible to warn against the danger, it is generally not sufficient that the user of the machine can be aware of the danger. To answer the question whether a warning can be regarded as an adequate measure to protect against a certain danger, it is of decisive importance whether it is to be expected that the warning will lead to an act or omission as a result of which the danger is avoided. When it becomes manifest that more effective measures to prevent an accident were possible, it must be examined why nevertheless taking this measure at that time could not be required from the employer. To assess this, the aforementioned factors are of relevance and also to what extent the adoption of this precautionary measure was obvious to the employer and his experts prior to the occurrence of an accident.[38]

The aforementioned decision of the Dutch Supreme Court implies that the inclusion of warnings onto a machine will not necessarily lead to the outcome that the machine is reasonably safe for use. Instead, an employer should first focus his attention on the safety of the design of the machine.

2.2.4 *Wrongful Conduct of Producers: Unwritten Obligations under Dutch Case Law*

A specific category of fault-based liability related to accidents is the liability of producers. Naturally, a duty of due care rests upon producers and the alleged violation of this duty can be assessed under the general liability rules of Article 6:162 DCC. Wrongful behaviour of producers that results in damage can be committed by violating written or unwritten safety standards of due care that are generally accepted in society. Written safety standards

38 HR 11 november 2005, *NJ* 2008, 460 (*Multivac-machine*).

that are applicable to products are mainly laid down in European public law provisions and subsequently transposed into domestic law.[39] Breaching these product safety provisions under public law can lead to a wrongful act, provided that Article 6:163 DCC is met. In general, the wrongfulness of violating a product safety provision depends on the detail with which the provision is described. In the event that the standard clearly defines the product safety requirement that has to be met, the breach is self-evident. Moreover, it can be said that compliance with regulations issued by the authorities is not a defence in itself under fault-based liability.[40] Even if a producer has complied with relevant safety provisions under public law, wrongfulness by virtue of Article 6:162(1) DCC can still be established in the event of a breach of an unwritten safety standard of due care. Nevertheless, the circumstance of compliance or non-compliance with the relevant product-related public provision is a factor to be taken into account when assessing whether an unwritten standard of due care has been violated.[41]

In view of what is said above, the mere fact that a product caused damage to a consumer is not sufficient for imposing wrongfulness. The wrongfulness of the conduct of producers should be assessed by weighing the relevant (*Kelderluik*) factors of a concrete case.

In 1989, the Dutch Supreme Court provided guidance to courts by anticipating the defectiveness test set forth in the EPLD to determine whether the producer of the medicine Halcion was liable under Article 6:162 DCC.[42] The general rule that can be deduced from this decision is that it is wrongful to put a defective product into circulation, proceeding on the basis that a product is defective if it does not offer the safety a user is entitled to

39 For more on this subject, see paragraph 4.2.

40 See HR 30 juni 1989, *NJ* 1990, 652 (*Halcion*). The mere registration of a medicine did not free the producer of liability under Article 6:162 DCC.

41 Dommering-Van Rongen 2000, p. 64.

42 The court anticipated the test of defectiveness, as the European Product Liability Directive was not implemented into Dutch law until September 1990 and it entered into force in November 1990. In this case, several consumers sustained severe personal injuries as a result of the use of a sleeping drug called Halcion.

expect, taking all circumstances into account.[43] Hence, the criterion of defectiveness has been declared applicable to assess the conduct of producers under Article 6:162 DCC.[44]

In another case, the Dutch Supreme Court ruled that the wrongfulness of the conduct of a producer should be answered on the basis of the knowledge and insights at the time of the dangerous conduct.[45]

> Dutch *Eternit* case
>
> Eternit, a producer/supplier of asbestos boards, is held liable for the damage suffered by a woman. In 2002, the female victim was diagnosed with the disease mesothelioma, a form of cancer that is caused by having been exposed to asbestos. The disease is probably caused by having shaken out the clothing of her brothers who had sawed asbestos boards to build a shed behind her parents' home. This happened in the 1970s.
>
> The victim alleged that Eternit is liable on the basis of Article 6:162 DCC, because Eternit failed to warn users of the health dangers of the asbestos boards, *e.g.* on the packages of the asbestos boards, in spite of the circumstance that

43 HR 30 juni 1989, *NJ* 1990, 652 (*Halcion*). In a later case (HR 6 december 1996, *NJ* 1997, 219 (*Du Pont/Hermans*)) and repeated in HR 22 oktober 1999, *NJ* 2000, 159 (*Koolhaas/Rockwool*), the Supreme Court used a dissimilar standard to assess the wrongfulness of the conduct of a producer. The damage in this case consisted of economic loss of a non-consumer. In the opinion of the Supreme Court, Du Pont is liable for the economic loss suffered on the basis that it is wrongful to put a product into circulation which causes damage when it is used in a normal way and in accordance with its purpose. The Dutch Supreme Court did, however, refer to the standard of the *Halcion* case and stated that they correspond. Based on the aforementioned cases, it seems that there is more than one standard to measure the wrongfulness of the conduct of a producer, depending on the type of user and the type of damage. The test of *Du Pont/Hermans* slightly deviates from the defectiveness standard of the Directive as used in the *Halcion* case, as the Directive proceeds on the basis of a reasonably expected use of a product. This may include more than just normal use, whereas the *Du Pont/Hermans* wording imply that a producer cannot be held liable under Article 6:162 DCC for damage which is the result of use other than the normal use intended by the producer. The *Du Pont/Hermans* decision has been criticized in the academic literature for its narrow scope (Dommering-Van Rongen 2000, p. 33). Scholars have suggested using one uniform test for assessing the wrongful conduct of producers. The standard of the Directive should also be used with respect to producer's liability under Article 6:162 DCC in the event that the victim is a professional party. Perhaps the Supreme Court took account of the criticism. In HR 29 november 2002, *NJ* 2003, 50 (*Onkruidverdelger Thyram*), the Supreme Court linked the two criterions together by stating that in the case at hand (which was based on Article 6:162 DCC), the product did not offer the safety that the lettuce growers were entitled to expect, taking into account the expected use.

44 Dommering-Van Rongen 2000, p. 32; Bloembergen's note at HR 22 oktober 1999, *NJ* 2000, 159 (*Koolhaas/Rockwool*); Franken 2009. In HR 29 november 2002, *NJ* 2003, 50 (*Onkruidverdelger Thyram*), the Supreme Court also held that this yardstick for assessing the wrongfulness of the conduct of manufacturers/producers should also be applicable to suppliers. In a previous decision, the Supreme Court ruled the opposite and used a different yardstick to assess the wrongfulness of the conduct of a supplier in the distribution chain, see HR 22 september 2000, *NJ* 2000, 644 (*Vladeko/VSCI*).

45 HR 25 november 2005, *NJ* 2009, 103 (*Eternit*). See also HR 17 december 2004, *NJ* 2006, 147 (*Hertel/Van der Lugt*). See for a discussion of the decision on appeal of Hof Arnhem 13 april 2004, *NJ* 2004, 612: Meijer 2004.

> Eternit had knowledge or ought to have had knowledge of these dangers at that time.
> The Dutch Supreme Court held that the wrongfulness of the conduct of the producer/supplier of asbestos boards needs to be determined in light of the societal views at the time when the producer/distributor's conduct took place. From the time that within the social circle, which included the producer/distributor, it should have been common knowledge that there are health dangers attached to working with asbestos, there was an increased duty of due care in view of the interests of those who are in direct vicinity of a place where people work with asbestos. It depends on the circumstances of the case and the existing knowledge and insights at the time of the conduct, what safety measures could have been expected of the producer/distributor. Factors that need to be included in the considerations are the degree of certainty with respect to the health dangers associated with working with asbestos and the nature and seriousness of the risks. The Supreme Court noted that the Court of Appeal was correct in its assessment that the producer/supplier is seriously to blame, since he/she failed to warn the public against the health risks associated with asbestos of which the producer/supplier had knowledge.[46]

Several general unwritten obligations regarding producers' conduct have been formulated by the Dutch Supreme Court. Wrongfulness can be established if the damage is a consequence of the violation of a specific duty which rests upon the producer, such as a producer's obligation to have his/her product examined before it is put into circulation, the duty to inform, the duty to warn, the duty to take account of a certain amount of careless and inadvertent behaviour of potential victims and the duty to sufficiently check products before putting them into circulation.

In the *Koolhaas/Rockwool* judgement the Dutch Supreme Court formulated an important yardstick, which can be applied to both producers of finished and semi-finished products.[47] According to the Dutch Supreme Court, a producer is obliged to take those measures which can reasonably be required from him in his capacity of a careful producer in order to prevent damages caused by his product. In addition, a producer must ascertain which implications a new or renewed finished or semi-finished product may bring.[48] It

46 HR 25 november 2005, *NJ* 2009, 103 (*Eternit*). See for a discussion of the decision on appeal of Hof Arnhem 13 april 2004, *NJ* 2004, 612: Meijer 2004.

47 HR 22 oktober 1999, *NJ* 2000, 159 (*Koolhaas/Rockwool*).

48 Rockwool is the manufacturer of *steenwol, i.e.* an insulating material used to improve the structure of potting compost. Rockwool supplied the above-mentioned insulating material as a component part to manufacturers of potting compost, who then supplied the potting compost to growers of plants, such as the injured party Koolhaas. Rockwool directly addressed these growers in advertisements. At some time Rockwool altered the composition of the insulating material from type 2 to type 6. During the time Koolhaas was using potting

has been noted in the literature that this decision implies that a producer has a duty to investigate what effects new or renewed products can have, which will often concern conducting research into the potentially negative effects on the health of others. This duty will generally entail that a producer must carry out tests before putting the product on the market.[49]

Another duty formulated in case law is the producers of semi-finished products' duty to inform. The Dutch Supreme Court also held in the *Koolhaas/Rockwool* case that, in the event that the producer of a semi-finished product who puts the semi-finished product with a modified version into circulation, without having it tested thoroughly and without having disclosed the test results to the public or without having provided a precaution regarding the modified use of the product, cannot merely inform the direct buyers of the component part. The producer of a semi-finished product must also ensure that the ultimate buyers of the end product are informed about the modification before the modified version is put into circulation.[50]

Another unwritten duty following from Dutch case law is the producers' responsibility to warn against dangers. The *Halcion* case has drawn the attention in this field. In this ruling, the Dutch Supreme Court acknowledges that producers have a responsibility to warn consumers of dangers attached to their products.[51] It listed several circumstances. Although in this case it concerned the liability of a producer of a medicine, the factors may be interpreted in a general way so that they can be applicable to other cases. These factors concern the nature and seriousness of the injury, the probability of injury, the utility of the product, product information and the cost of an alternative design of the product. As has been remarked in the literature, these factors resemble the *Kelderluik* factors and fit in a risk-utility analysis.[52]

Dutch *Halcion* case
Several consumers suffered serious injuries as a result of the side effects of the sleeping drug *Halcion*, such as suicide attempts and anxiety attacks. These side effects were not described in the package insert. The consumers instituted their

compost with type 6, he suffered losses to his plants. Grower Koolhaas holds the view that the damage is a consequence of the unfitness of the compost with type 6 for his plants. According to him, Rockwool acted wrongfully because Rockwool had omitted to investigate the consequences of the added type 6 to plants and because he had not informed the growers of the effects of the altered insulation material when he put it into circulation. In lower instances, the claim was rejected. The Supreme Court reversed the ruling of the Court of Appeal.

49 Bloembergen's note at HR 22 oktober 1999, *NJ* 2000, 159 (*Koolhaas/Rockwool*); Dommering-Van Rongen 2000, p. 34.

50 HR 22 oktober 1999, *NJ* 2000, 159 (*Koolhaas/Rockwool*).

51 HR 30 juni 1989, *NJ* 1990, 652 (*Halcion*).

52 Dommering-Van Rongen 2000, p. 45, 53; Stolker (Onrechtmatige daad I), Art. 186, aant. 14.

claim for damages against the producer Upjohn. Upjohn advanced the argument that there was a low probability that the side effects would occur.
The District Court rejects liability. According to the District Court, it was known and to a certain extent accepted in society that medicines may cause harmful side effects. The Court of Appeal nevertheless held Upjohn liable.
The Supreme Court quashed this decision by considering that a medicine is defective if the nature and the seriousness of possible harmful side effects and the chance thereof is such that these side effects are not in proportion to the seriousness of the disease or ailment which the remedy is meant to heal or suppress, given all other circumstances of the case such as the remedy's efficiency, product information and the potential damaging natures of substitutes. Moreover, according to the Supreme Court the user does not have to anticipate side effects of which he is not warned.[53]

An important obligation for producers that is formulated by the Dutch Supreme Court and that needs to be mentioned here relates to anticipating a certain amount of careless behaviour of potential victims in order to properly assess the likelihood that accidents will occur. As was touched upon in paragraph 2.2.3.2, the *Kelderluik* ruling has in general stipulated the significance of this factor. As for product liability, the importance of taking into account the probability that product users may not be cautious and attentive all the time is demonstrated in the *Warmwaterkruik* case.[54] The Dutch Supreme Court held that producers must not only take into account the users who take all necessary precautionary measures during product use to prevent potential damage, but also those people for whom the product is intended but who fail to take the appropriate safety measures. The duty to carefully monitor the absence of possible (manufacturing) defects in the products prior to marketing them can also be deduced from this decision.[55]

Dutch *Warmwaterkruik* case
A maternity assistant put a hot-water bottle (*i.e. warmwaterkruik)* in the cradle of a new born baby. Due to a defective screw cap the bottle leaked and severely scalded the baby. An action for damages was brought against the producer of the bottle under Article 6:162 DCC. The producer called Jumbo argued that the risk of a leaking hot-water bottle cannot be entirely excluded and that the damage would not have occurred if the nurse had been more careful.

53 HR 30 juni 1989, *NJ* 1990, 652 (*Halcion*).
54 HR 2 februari 1973, *NJ* 1973, 315 (*Warmwaterkruik*).
55 Dommering-Van Rongen 2000, p. 35.

At first instance and on appeal the courts rejected the claim. The Court of Appeal rejected liability, because the producer proved that having an effective controlled system to prevent hot-water bottles from leaking was impossible. Furthermore, the Court of Appeal assumed that the defect was an accidental occurrence as the findings of a research report showed that a number of factors were needed to bring about the defect and these factors should all occur at the same time and in the right order.

In contrast, the Supreme Court decided that, in general, a producer cannot invoke the defence that the accident would not have occurred in the event that all precautionary measures had been taken by the user that would have prevented the injurious effect of that defect and that the defendant was entitled to expect that the precautionary measures would be adopted. The proof of effective control of the hot-water bottles does not relieve the producer from liability. To determine the probability of accidents as a result of such defects, which factor must be taken into account by the producer, not only those users who take all precautions when using these products should be considered, but also the entire public for which the product is intended and of whom a part fails to adopt the precautions. As regards the Court of Appeal's judgement that fault on the part of the producer is lacking as a result of the consequence that sufficient monitoring of the absence of leaking of these types of bottles is not possible to achieve, is, having regard to the research report, an understandable ground according the Supreme Court. However, the Court of Appeal's ground does not make clear that there are no measures possible, which can prevent bottles with a concrete defect as is stated here from being manufactured and put into circulation. Consequently, the Court of Appeal should have investigated if this concrete bottle contained a danger of such a nature that this product should not have been put into circulation.[56]

2.2.5 Conclusion

This paragraph dealt with fault-based liability in tort. The cornerstone of fault-based liability in European tort law is the standard of reasonable conduct of the tortfeasor. The emphasis in this paragraph was its development under Dutch law.

In paragraph 2.2.2, the Dutch general rules of fault-based liability that are laid down in Article 6:162 DCC were described. Fault-based liability consists of two main elements: the wrongfulness of the act and the attributability/imputability of the act to the wrongdoer. The other requirements pertain to the damage sustained and the causal connection between

56 HR 2 februari 1973, *NJ* 1973, 315 (*Warmwaterkruik*).

the damage and the wrongful act. According to paragraph 2 of Article 6:162 DCC three sorts of wrongful acts can be distinguished: (1) conduct which infringes a person's right; (2) conduct contrary to a statutory duty; and (3) any act or omission which violates a rule of unwritten law regarding a proper social conduct. The latter category is the most popular one. Within this category, the most important subcategory concerns the unwritten safety standards that are applicable in situations of endangerment. Endangerment refers to creating a dangerous situation that may harm people or goods or allowing such a situation to continue.

Paragraph 2.2.3 described the assessment of wrongful endangerment in more detail. Of essence is the landmark *Kelderluik* decision of the Dutch Supreme Court in which it was held that the wrongfulness of breaching unwritten safety standards depends on a balancing test of the circumstances of the case. This decision represents the classical example of the balancing process that is needed to determine whether conduct, such as in case of endangerment, is wrongful. It concerns whether the level of precaution taken by the wrongdoer was sufficient in view of the level of risk. In this respect, it has been argued in the literature that the *Kelderluik* standard is the Dutch equivalent of the American risk-utility test of Judge Learned Hand to determine negligence. The Dutch Supreme Court has provided four circumstances that should be considered when assessing whether there is wrongful endangerment: the probability that potential victims are careless and inattentive; the likelihood that this leads to accidents; the gravity of the consequences of the accident; and the costs of taking precautionary measures. These are termed the *Kelderluik* factors and they have been cited by the Dutch Supreme Court and lower courts on a regular basis ever since. The factor of the cost of taking precaution relates to whether, in the absence of the precautionary measure, the measure was possible to take and customary as well as the inconvenience in terms of time, effort and financial costs of adopting the measure.

The duty to warn was addressed in paragraph 2.2.3.3. Providing a warning is one means of taking precaution and as a result the duty to warn adequately should be assessed on weighing the concrete circumstances of the case, such as the *Kelderluik* factors. Dutch case law indicates that for determining whether there is a failure to warn additional factors can be of relevance, such as the obviousness of the risk, knowledge of the risk, and the improbable behaviour of potential victims. The above-mentioned factor of the costs of providing a warning has also been shown to be of importance for assessing a failure to warn. It was noted that courts generally consider providing warnings a simple and easy way of exercising due care and therefore often not too high of a cost to adopt. In cases where a warning was given yet damage occurred, liability commonly revolves around answering the question whether the adopted measure of the warning can be considered sufficiently adequate. An important ruling of the Dutch Supreme Court governing the adequacy of warnings under Article 6:162 DCC is the recent *Jetblast* case, which holds that to answer the question whether a warning can be considered an adequate measure to protect against a certain danger, it is of decisive importance whether it is to be expected that the warning will lead to an act or omission as a result of which the danger is avoided. Warnings are often not the only measure that can be adopted to prevent a specific risk. In this regard, the Dutch Supreme Court ruled in the *Multivac-machine* case that the mere inclusion of a warning onto a dangerous machine will in principle not be sufficient to avoid liability. Employers should first investigate whether the safety of the machine's design can be improved by other more effective measures.

Following the discussion on wrongful conduct in general, paragraph 2.2.4 discussed the wrongful conduct of producers. The wrongfulness of the conduct of producers should be assessed on the basis of weighing the relevant circumstances, including the *Kelderluik* factors. Unwritten obligations following from Dutch case law that pertain to how reasonable producers should act when designing products were described. One of them concerns the *Halcion* case of 1989. In this case the Dutch Supreme Court provided guidance to courts by using the defectiveness standard of the EPLD to determine whether the producer of the medicine Halcion was liable under Article 6:162 DCC. Hence, it was concluded that the standard of defectiveness has been declared applicable to assess the conduct of producers under fault-based liability. Note, however, that fault-based liability also requires imputability of the act to the producer. The *Halcion* ruling has additional bearing. The Dutch Supreme Court established a producers' duty to warn. It has listed several circumstances which may be of importance when assessing the wrongfulness of a producer's conduct. These boil down to the utility of the product, the seriousness of the injury, the probability of injury and the cost of an alternative reasonable design of the product. Hence, these are of a similar nature as the *Kelderluik* factors. Another important decision in this field concerns the *Koolhaas/Rockwool* judgement in which the Dutch Supreme Court ruled that a producer is obliged to take those measures which can reasonably be required from him in his capacity of a careful producer in order to prevent damages caused by his product. Furthermore, the Dutch Supreme Court held in the *Warmwaterkruik* case that producers must not only take into account the users who take all necessary precautionary measures during product use to prevent potential damage, but also those people for whom the product is intended but who fail to take the appropriate precautionary measures.

2.3 PRODUCT LIABILITY WITHOUT FAULT: DIRECTIVE 85/374/EEC

2.3.1 *Introduction*

The previous paragraph generally described the requirements for determining the liability of producers based on fault. In addition to or along with a claim based on fault, persons who sustain personal injury or property damage as a result of a product can claim compensation on the basis of the provisions of the liability regime of the Directive. This liability regime can be considered strict, that is, it requires no proof of fault on the part of the producer. As a result of this, product liability claims in Europe are nowadays often primarily based on this regime and only alternatively on fault-based liability. The essential requirements of the Directive are briefly discussed in paragraph 2.3.2. The next paragraph pays specific attention to the important requirement of defectiveness. After that, paragraph 2.3.4 describes in more detail why it has often been said in the literature that the regime of the Directive is a strict liability regime with elements of fault. The final paragraph (2.3.5) provides a summary.

2.3.2 The Directive's Requirements in a Nutshell

The European Product Liability Directive 85/374/EEC on liability for defective products came into effect on 25 July 1985. The implementation of the Directive expired on 31 July 1988. A majority of the Member States, including The Netherlands, missed that deadline by some years. The Directive was adopted in the Netherlands by passing the Netherlands Product Liability Act of 13 September 1990, which came into effect on 1 November 1990. The provisions are laid down in Articles 185-193 of Book 6 DCC. Article 15 left a few options open to the Member States to derogate from the Directive. The Dutch legislator adopted the development risk defence of Article 7(e). The provision in Article 16(1) granting the possibility of providing a limit for the total liability of the producer was not used.

The implementation of the EPLD into the laws of Member States introduced a system of liability without fault. According to the Directive's first recital of the preamble, approximation of the laws of the Member States concerning the liability of the producer for damage caused by the defectiveness of his products is necessary, because the existing divergences may distort competition and affect the movement of goods within the common market and entail a differing degree of protection of the consumer against damage caused by a defective product to his health or property.

Under Article 267 of the Treaty on the Functioning of the European Union, courts of Member States can refer questions to the ECJ on the interpretation of the Directive.[57] In three judgements, the ECJ has confirmed that the Directive is a maximal harmonisation Directive.[58] This basically means that Member States do not have the power to provide a higher level of consumer protection in areas that are dealt with by the Directive. The ECJ

57 That is the former Article 234 EC Treaty.

58 ECJ 25 April 2002 (Case C-52/00), *ECR* 2002, p. I-3827 (*Commission v. France*), paras. 16, 19, 24; ECJ 25 April 2002 (Case C-154/00), *ECR* 2002, p. I-3879 (*Commission v. Greece*); ECJ 25 April 2002 (Case C-183/00), *ECR* 2002, p. I-3901 (*María Victoria González Sánchez v. Medicina Asturiana SA*). See for a discussion of this aspect Howells 2005c.

notes that the EPLD is an internal market measure as the Directive was adopted by the Council by unanimity under Article 100 of the EC Treaty[59] concerning the approximation of such laws, regulations or administrative provisions of the Member States as directly affect the establishment or functioning of the common market. The ECJ holds that the margin of discretion available to the Member States in order to make provision for product liability is entirely determined by the Directive and must be inferred from its wording, purpose and structure. The fact that the Directive provides for certain derogations or refers in certain cases to national law does not mean that in regard to the matters which it regulates harmonisation is not complete. It follows that the Directive seeks to achieve, in regard to those matters, complete harmonisation of the laws, regulations and administrative provisions of the Member States.

The Directive's first article provides that a producer shall be liable for damage caused by a defect in his product, irrespective of fault. The injured party has to prove the existence of the damage, the defectiveness of the product and the causal link between that defect and the damage suffered.[60] The cornerstone of the Directive's liability system is the requirement of defect. Article 6 reads as follows:

> 1. A product is defective when it does not provide the safety which a person is entitled to expect, taking all circumstances into account, including:
> (a) the presentation of the product;
> (b) the use to which it could reasonably be expected that the product would be put;
> (c) the time when the product was put into circulation.
> 2. A product shall not be considered defective for the sole reason that a better product is subsequently put into circulation.

Some key concepts that have been defined in the Directive are the meaning of 'product', 'producer' and 'damage'.[61] In addition, the ECJ has recently explained the meaning of the term 'putting into circulation'.[62] In short, a product must be considered as having been put into circulation when it leaves the production process operated by the producer and enters a marketing process in the form in which it is offered to the public in order to be used or consumed. Generally, it is not important in that regard that the product is sold

59 The article was amended to 94 EC Treaty and is again amended to Article 115 of the Treaty on the Functioning of the European Union.

60 Article 4 of the Directive.

61 Articles 2, 3 and 9 of the Directive, respectively.

62 ECJ 9 February 2006 (Case C-127/04), *ECR* 2006, p. I-1313 (*Declan O'Byrne v. Sanofi Pasteur MSD Ltd and Sanofi Pasteur SA*), paras. 27-30.

directly by the producer to the user or to the consumer or that that sale is carried out as part of a distribution process involving one or more operators.

Another important article of the Directive is Article 7 as it exhaustively lists the six defences on the basis of which a producer in terms of the Directive is freed from liability. Two defences that merit attention here are the development risk defence of Article 7(e) and the compliance defence of Article 7(d). The development risk defence entails that a producer is not liable if he proves that the state of scientific and technical knowledge at the time when he put the product into circulation was not such as to enable the existence of the defect to be discovered. The ECJ has explained the meaning of the defence. The ECJ first comments that in order for a producer to incur liability under the Directive, the victim does not have to prove that the producer was at fault. However, in accordance with the principle of fair apportionment of risk between the injured person and the producer set forth in the seventh recital in the preamble to the Directive, the producer has a number of defences. With regard to the development risk defence, the ECJ holds that the producer has to prove that the objective state of scientific and technical knowledge, including the most advanced level of such knowledge, without any restriction as to the industrial sector concerned, was not such as to enable the existence of the defect to be discovered. Nevertheless, the relevant knowledge must have been accessible at the time when the product in question was put into circulation.[63] Article 7(d) lays down that proof that the defect is due to compliance of the product with mandatory regulations issued by the public authorities bars liability under the Directive. This defence is discussed in more detail later on.

A producer's obligation to pay compensation can be reduced or even extinguished in the event the damage is caused by both a defect in the product and by the fault of the injured person or of a person for whom the injured person is liable.[64]

Article 13 stipulates that the Directive does not affect any rights which an injured person may have according to national liability rules, *i.e.* contractual liability, fault-based liability and other special liability systems, which existed in Member States prior to the implementation of the Directive. The ECJ has also explained the correct interpretation of Article 13. According to the ECJ, Article 13 cannot be interpreted as giving the Member States the possibility of maintaining a general system of product liability different from that provided for in the Directive. The rights conferred under the legislation of a Member State on the victims of damage caused by a defective product under a general system of liability having

63 ECJ 29 May 1997 (Case C-300/95), *ECR* 1997, p. I-2649 (*Commission v. United Kingdom*). See for a discussion of the defence *e.g.* Mildred 2005; Hodges 1998.

64 Article 8(2) of the Directive.

the same basis as that put in place by the Directive may be limited or restricted as a result of the Directive's transposition into the domestic law of that State.[65]

Note that despite the ECJ rulings and the text of the Directive, difficulties may still arise when key concepts need to be interpreted in a specific case by a national court. This may lead to different interpretations between Member States even though each court must in principle apply the provisions of the Directive in accordance with the wording, purpose and structure of the Directive. Another disparity in harmonisation may arise due to the fact that certain aspects of the Directive are left to domestic law, such as specific rules concerning the standard of proof and the assessment of damages.

2.3.3 The Defectiveness Standard

2.3.3.1 The Three Specified Circumstances

According to Article 6:186(1) DCC implementing Article 6(1) of the Directive, a product is defective if it does not provide the safety that a person is entitled to expect, taking into account all the circumstances of the case at hand and in particular the presentation of the product, the use to which it could reasonably be expected that the product would be put, and the time the product was put into circulation. The Dutch legislator noted in the Explanatory Memorandum to the Act implementing the Directive that Article 6 provides the court a broad margin of discretion in assessing whether or not a product can be regarded as defective, since there are countless circumstances that can be determining factors.[66]

The first factor that is explicitly mentioned is the presentation of the product. Member States interpret this factor broadly, suggesting that warning statements against a danger, instructions for use, the packaging, advertising and so forth that may influence the safe use of the product can contribute to the assessment of defectiveness. The Dutch Explanatory Memorandum provides in this context the example that a product can be insufficiently safe because the producer failed to provide certain directions for use or failed to warn against the risks attached to the use of the product.[67]

The second factor concerns the reasonably expected use of the product. Both producers and European courts may encounter difficulties when assessing whether or not a certain product use actions may be considered reasonably expected. Some guidance for the interpretation of this factor can be deduced from the sixth recital to the preamble that stipulates that the safety is assessed by excluding any misuse of the product not reasonable under the circumstances. The implication arising from this factor is that a producer should not

---

65 ECJ 25 April 2002 (Case C-183/00), *ECR* 2002, p. I-3901 (*María Victoria González Sánchez v. Medicina Asturiana SA*), paras. 30-34.

66 *Parliamentary Papers II* 1985/86, 19 636, no. 3, p. 9.

67 *Parliamentary Papers II* 1985/86, 19 636, no. 3, p. 9.

only take account of the intended use of a product, but should also consider the reasonably foreseeable use of a product.[68] The Dutch Explanatory Memorandum remarks that for the interpretation of this factor, one must not only take the use intended for the product into account, but also which persons will use the product.[69]

The latter factor reflects that the level of safety should be assessed on the basis of the time when the product was put into circulation and not the (higher) degree of safety that could be achieved at the time the claimant suffered injuries.

The second paragraph of Article 6:186 DCC (and of the Directive) elaborates further on this aspect by stating that a product shall not be considered defective for the sole reason that a better product is subsequently put into circulation. Hence, defectiveness must be assessed on the basis of generally accepted safety standards applicable at the time when the product was put into circulation.[70]

2.3.3.2 The Legitimate Safety Expectations of the Public at Large

Defectiveness must be measured objectively. It is neither the safety expectations of the particular injured person nor of the particular producer that is decisive, but the degree of safety persons generally are entitled to expect. This can be distilled from the recital to the preamble that states that to protect the physical well-being and property of the consumer, the defectiveness of a product should not be related to its fitness for use, but to the lack of the safety which the public at large is entitled to expect.[71] This also makes the test applicable to bystanders and other third parties that suffer injuries as a result of the product but who are not users or consumers.[72] According to the Dutch Explanatory Memorandum objective measurement also implies that in the event that the safety expectations for a specific type of product only exist in certain professional circles, these safety expectations should be taken as a basis.[73]

As Burton J pointed out in the *English Hepatitis C* decision, the safety is not what is actually expected by the public at large, but what they are entitled to expect. He favours the term 'legitimate expectations' to 'entitled expectations'. It may be that the actual expectations of persons generally are congruent with the objectively assessed legitimate expectations, but they may also differ from the legitimate expectations and impose a lower standard of safety.[74] It is thus important to keep this distinction in mind.

68 Cf. HR 2 februari 1973, *NJ* 1973, 315 (*Warmwaterkruik*).
69 *Parliamentary Papers II* 1985/86, 19 636, no. 3, p. 9.
70 *Parliamentary Papers II* 1985/86, 19 636, no 3, p. 9. Cf. HR 25 november 2005, *NJ* 2009, 103 (*Eternit*).
71 See *e.g.* Dommering-Van Rongen 2000, p. 43; Miller & Goldberg 2004, p. 367.
72 Grubb & Howells 2007, p. 351.
73 *Parliamentary Papers II* 1987/88, 19 636, no. 6, p. 22.
74 [2001] 3 *All ER* 289 (*A v. National Blood Authority*), para. 31; Grubb & Howells 2007, p. 42.

Dutch *Datafan* case
Take for example the Dutch case of the Court of Appeal of Arnhem. In this case, the producer Indolec relies upon the expertise of the victim. The victim is a mechanic who had hurt his right hand on a sharp edge on the underside of the *Datafan*, *i.e.* a ventilation cylinder, while installing it. According to the producer, it is reasonable to expect of an experienced mechanic, that he/she is aware of the sharp edges of the steel-plated ventilator and the risk of sustaining cuts during the use of it.
In contrast, the Court of Appeal first and foremost stated that defectiveness should be assessed on the basis of the safety expectations of the public at large. But in the event that for specific products in certain circles other or higher safety expectations exist, these expectations should be used as a starting point. The Appellate Court ruled that in the case at hand, there was no specific professional circle of people on the basis of which the producer should have assessed the level of safety, since on the one hand, the ventilation cylinder is used by mechanics who are charged with installing it and on the other hand, after installation, by persons who have purchased the product. It has neither been argued, nor has it become evident that during the installation of the product the claimant failed to act in a way that can be expected of installers. The way in which claimant used the product is a way of use that producer should have reasonably expected. The court also noted that despite the fact that this use can be reasonably expected, Indolec had not warned against the sharp edges underneath the ventilator, nor had Indolec advised users to wear gloves or take other precautions. This is of importance, because in view of the nature of the product, it cannot be argued that the sharp edge is an inherent danger which users should reckon with. Furthermore, having regard to the conclusion of the expert's report that the edges could have easily been finished off and the fact that Indolec failed to put forward the argument that the measures to reduce the sharpness are of such inconvenience that they cannot be reasonably requested of Indolec, the Court of Appeal concluded that the ventilator is a defective product.[75]

2.2.3.3 Weaknesses of the Defectiveness Standard

Academics have criticised the defect requirement for its flaws.[76] First of all, although it is generally agreed that the test of defectiveness does not impose a standard of absolute safety, it is uncertain what level of relative safety is appropriate. Instead of providing the answer

75 Hof Arnhem 14 oktober 2003, *NJF* 2004, 46 (*Datafan*).
76 *E.g.* Howells 1993, p. 11; Stapleton 1994, p. 234; Henderson & Twerski 1999.

to the question when does a product fail to meet the safety expectations of the public, the test raises the next question regarding how much safety is the general public entitled to expect of a product.[77]

Furthermore, the test provides scant guidance in the context of technologically complex products, as consumers have difficulty assessing the safety level of such products. They will be likely to have no expectations or no idea of how the product could have been designed in a safer way.[78]

In addition, the Directive leaves it to the discretion of the court of the Member State to decide which circumstances are significant in a particular case. Notwithstanding the reference to 3 circumstances in Article 6, it has generally been argued in the European literature that this list of factors is non-exhaustive and uncertainty remains with regard to how to determine what circumstances are of relevance, their relative weight and how they should be balanced against each other.[79]

Dealing with the factor of consumers' knowledge of the product risk or its obviousness has also been considered problematic with the defectiveness test.[80] A strict application of the test to a product with a generally known or obvious risk can lead to the conclusion that defectiveness cannot be established since consumers are aware or should have been aware of the risk and in consequence should have adapted their safety expectations and use to it. However, it can be argued that such an interpretation focuses too heavily on one single factor (*i.e.* consumers' knowledge or the obviousness of the risk), thereby taking insufficient consideration of other factors that may be of relevance for determining defectiveness.

2.2.3.4 Distinction in Product Defects

A common tool for assessing whether a product has a defect is by categorising the deficiencies of products. The most important categories are design defects, warning defects and manufacturing defects.[81] The US Restatement of the Law (Third), Torts on product liability also distinguishes between types of defects and additionally has set forth the legal meaning of design, manufacturing and warning defects.[82] Pursuant to these definitions, a manufacturing defect exists when the product departs from its intended design even though all possible care was exercised in the preparation and marketing of the product. A design

77 *E.g.* Stapleton 1994, p. 234; Whittaker 2005, p. 485; [2001] 3 *All ER* 289 (*A v. National Blood Authority*), para. 31.

78 Miller & Goldberg 2004, p. 360.

79 Miller & Goldberg 2004, p. 354; Dommering-Van Rongen 2000, p. 44; Stolker (Onrechtmatige daad I), Art. 186, aant. 3. In contrast: Taschner 2005; Deards & Twigg-Flesner 2001; Burton J in the *Hepatitis C* case.

80 Grubb & Howells 2007, p. 370, 349.

81 See *e.g.* Grubb & Howells 2007, p. 7. In contrast, Judge Burton introduces the distinction between 'standard' and 'non-standard' products.

82 US Restatement of the Law (Third), Torts: Products Liability, § 2a, b, c (American Law Institute 1998, p. 14).

defect occurs when the foreseeable risks of harm posed by the product could have been reduced or avoided by the adoption of a reasonable alternative design, and failure to use the alternative design renders the product not reasonably safe. A warning or instructions for use are defective when the foreseeable risks of harm posed by the product could have been reduced or avoided by reasonable instructions or warnings, and their omission renders the product not reasonably safe.

A consequence of this distinction in the US Restatement of the Law (Third), Torts is that it imposes different standards to the different types of product defects.[83] It is important to note that the Directive does not refer to types of defects. All product deficiencies are bound to the same set of rules. This traditional classification is nevertheless frequently mentioned in European literature because of its usefulness and some Member States have even made some type of distinction in their domestic legislation.[84] Moreover, the categorisation into three defect types has also been used by European judges to establish defectiveness, in particular by German courts.[85] Nevertheless, the usefulness of this approach can be weakened by the fact that the distinction between design and manufacturing defects is not always clear-cut.[86]

2.3.4 *The Directive: Strict Liability with Elements of Fault*

2.3.4.1 General

It is said that the strict liability regime of the Directive is a mixture of risk and fault elements. The question of interest here is the extent of this mixture. This has been a topic of debate during the development of the provisions as well as later on when courts needed to interpret and apply vague concepts of the Directive.

In principle, the Directive advocates that it imposes a liability system without fault. Fault on the part of the producer does not have to be proven by the injured party, which embodies an essential component of strict liability. On the other hand, the Directive's regime does not purport to impose an absolute form of strict liability; the mere fact that a product caused damage to a user is not sufficient for finding liability. There must be a defect in the product. Furthermore, producers have been given escape routes. Given that there should be a fair apportionment of risk between the injured person and the producer as made clear in the preamble, the Directive's liability system therefore provides producers

83 Owen 2008, p. 346.

84 Grubb & Howells 2007, p. 7.

85 See *e.g.* Campbell 2007 for how European countries have implemented the concept of defectiveness. See also Lenze 2005 and in particular the German decision of BGH 9 May 1995, *NJW* 1995, 2162 (*Exploded mineral bottle*).

86 Grubb & Howells 2007, p. 13. This problem also came up in the case concerning a 'may contain' warning against the risk of allergen cross-contamination in food products, see Pape 2009.

with the possibility to free themselves from liability if they furnish proof as to the existence of certain exonerating circumstances. One of these defences is the development risk defence.[87] Defectiveness and the development risk defence are important concepts that can introduce fault into the framework of the Directive. The degree in which they 'contaminate' the strict liability system depends on the interpretation of these concepts.[88]

2.3.4.2 The Interpretation of the Development Risk Defence

Liability for unknown and undiscoverable risks is an element that can be regarded as an element of a true strict liability system.[89] Under the principle of fault-based liability, producers cannot be held liable if they did not have to have knowledge of the risk.[90] Within the Directive's liability regime, the development risk defence protects the producer against liability for damage caused by a risk which is present, but not yet discoverable at the time the product was put into circulation. Because of disagreement among Member States about excluding or including the development risk defence, a compromise was reached which permitted Member States to choose the option of not implementing the defence into domestic law.[91] Not having the possibility to rely on the development risk defence is likely to benefit the level of consumer protection. On the other hand, the disadvantages pertaining to the absence of the defence are that its exclusion can negatively influence the innovation of products and in addition that it could create considerable problems with regard to how to deal with the insurability of such risks.[92]

After the implementation of the defence into national legislation, there has been uncertainty with regard to when a producer can succeed in invoking this defence.[93] Does undiscoverability mean that it is absolutely impossible to discover the defect or does it

87 Article 7(e) stipulates that the producer shall not be liable as a result of this Directive if he proves that the state of scientific and technical knowledge at the time when he put the product into circulation was not such as to enable the existence of the defect to be discovered.

88 Note that the burden of proof is a relevant material point of difference between fault-based liability and strict liability of producers.

89 Cf. Grubb & Howells 2007, p. 30; Stoppa 1992, p. 223.

90 Van Dam 2000, p. 208. Under Dutch fault-based liability, the duty to investigate whether there are risks attached to the dangerous conduct can be stringent. It seems to lean towards strict liability. The extent of this duty generally depends on the possible size of the risk and the cost of conducting research. Especially in the event that the risk may be severe, it is required to do research. It follows from case law that a producer must take those measures which can reasonably be required from him in his capacity of a careful producer in order to prevent damages caused by the product (HR 22 oktober 1999, *NJ* 2000, 159 (*Koolhaas/Rockwool*)). This can include the responsibility to keep abreast of the scientific developments in his area of expertise. Furthermore, the argument of the injuring party that a certain method of research could not be considered standard practice in the branch of industry has no overriding importance to the assessment of unlawfulness (see *e.g.* HR 6 april 1990, *NJ* 1990, 573 (*Janssen/Nefabas*); HR 2 oktober 1998, *NJ* 1999, 683 (*De Schelde/Erven Cijsouw*) pertaining to employer's liability based on fault).

91 Article 15(1) of the European Product Liability Directive.

92 See Lovells 2003, p. 49.

93 See *e.g.* Mildred 2005; Hodges 1998.

involve the question whether it cannot be reasonably required from a producer that he/she should have discovered the risk? In the *EC v. UK* case, the ECJ shed some light on this issue.[94] The ECJ held that the state of scientific and technical knowledge must be measured objectively. It referred to the most advanced level and is not specifically directed to the practices and safety standards in use in the industrial sector in which the producer is operating.[95] Consequently, the subjective level of knowledge of the producer is not of importance and producer's proof that he took reasonable steps to discover the defect will not excuse him.[96] The standard of this defence is set high and it will thus be a difficult task for producers to successfully prove that the defect was undiscoverable.[97] The ECJ, however, has reduced the exemption's effect and consequently introduced an element of fault by requiring that the knowledge must have been accessible at the time when the product in question was put into circulation.[98] In consequence, if the relevant knowledge about the defect was not accessible, the producer escapes liability. Unfortunately, the term 'accessible' was not defined by the ECJ. As a result, some uncertainty still remains regarding the exact scope of this defence: is the relevant knowledge accessible (and is the defence rejected) if it has been published, in scientific literature or in English? It still seems possible to interpret the meaning of this term in favour of the producer by exercising restraint in accepting accessibility. Nevertheless, as a result of the interpretation of the ECJ, it has generally been recognised that the scope of the defence is narrow; in practice producers will seldom succeed in invoking the defence.[99] Although producers have the possibility of escaping liability for undiscoverable risks, this element can be viewed as an important ingredient for strict liability under the framework of the Directive.

---

94 ECJ 29 May 1997 (Case C-300/95), *ECR* 1997, p. I-2649 (*Commission v. United Kingdom*).

95 In this respect, the decision in the Dutch case Rb. Amsterdam 3 februari 1999, *NJ* 1999, 621 (*HIV*) regarding blood infected with the HIV virus can be viewed as incorrect. In this case, the blood was tested two times using an ordinary screening test and with negative results. A third test, *i.e.* the *PCR test*, gave a questionable result, but this test was not approved at that time. The producer raised the defence that it was impossible to detect the virus with the regular tests in the window period. The court held that the Foundation was entitled to rely on this defence, because of the practical inability of the Foundation to employ the third experimental test at the time of the blood transfusion and the Foundation could not have been expected to use this test. The decision has received considerable criticism in the Netherlands: Dommering-Van Rongen 2000, p. 40; Van Boom & Van Doorn 2006. The defence should not have been allowed, since it was possible to discover the HIV virus with a test.

96 Hodges 1998, p. 565.

97 The opinion of the Dutch legislator is in line with this. The Memorandum of Reply implementing the Directive notes that a producer must fulfil high standards in order to successfully rely upon this defence. He has to furnish proof that the defect was undiscoverable, even with the most intensive and advanced control. According to the legislator, producer's argument that he had no knowledge of the defect as well that he had no means or time fails: *Parliamentary Papers II* 1987/88, 19 636, no. 6, p. 18.

98 ECJ 29 May 1997 (Case C-300/95), *ECR* 1997, p. I-2649 (*Commission v. United Kingdom*).

99 Lovells 2003, p. 50.

2.3.4.3 The Requirement of Defectiveness

Another important concept that tends to blur the boundaries between fault-based product liability and the strict liability regime of the Directive is the requirement of defectiveness. It has been argued in the literature that this requirement is of such a nature that it is problematic not to give in to adopting some form of a fault-based analysis to evaluate whether the product is defective.[100]

Defectiveness is an open standard that needs to be applied to concrete cases. The text of the Directive explicitly refers to three circumstances. With respect to determining the reasonably expected use of the product, it is difficult not to take the reasonableness of the conduct of the producer into account.[101] In addition, the article states that all circumstances should be taken into account. As has been mentioned earlier, this gives courts the discretion to evaluate what other circumstances are of relevance for determining whether the product is defective or not. Although three factors have been specified, these alone will not help a judge to make a sound judicial analysis of the product's defectiveness. Clearly, there are more factors that carry weight and that can thus be of relevance. Circumstances such as the likelihood that dangers associated with the use of the product materialise, the degree of harm arising from that and the burden of taking extra precautionary measures to avoid or reduce the harm can generally be considered essential in a case. Especially the burden of taking precaution, which refers to the availability, practicability, and feasibility, also in terms of cost, of designing an alternative product that is (presumed to be) safer. These factors are originally from fault-based liability, where the use of a risk-utility analysis determines whether the producer has taken sufficient precautionary measures in respect of the risks involved.

By contrast, other scholars have contended that the Directive does not permit the use of factors of reasonableness as it would mean reintroducing fault-based liability by way of the back door.[102] Especially as regards the question whether the harm was avoidable and if so, the practicability and cost of it, commentators have asserted that such considerations do not belong here. Hence, it has been argued by Judge Burton in the English *Hepatitis C* case that the expression 'all circumstances' in Article 6 of the Directive should be read as being limited to 'all relevant circumstances'.[103] In consequence, the above-mentioned circumstances cannot be included as 'relevant'.

To tackle this, Justice Burton proposed a new approach to defectiveness in the English *Hepatitis C* case that, in his eyes, serves the purpose of the Directive and differs from fault-based liability. It entails a distinction between 'standard' and 'non-standard' products. A 'standard' product is one which performed as the producer intended. A 'non-standard'

---

100 *E.g.* Miller & Goldberg 2004, p. 354, 383; Grubb & Howells 2007, p. 341; Stapleton 1999.

101 Dommering-Van Rongen 2000, p. 36.

102 Burton J in the *Hepatitis C* case; Taschner 2005; Deards & Twigg-Flesner 2001.

103 [2001] 3 All ER 289 (*A v. National Blood Authority*), para. 57.

product is one which is deficient or inferior in terms of safety from the 'standard' product, and whose harmful characteristic, which is not present in the 'standard' product, has caused the material injury or damage.[104] Having characterised a product as 'non-standard', the next step is whether the public at large accepts the 'non-standard' nature of the product, by taking into account the relevant circumstances. Whether it would have been possible, practicable or costly to avoid the defect is not relevant to consumers' expectations. However, a risk-utility balancing can be applied in the limited circumstance regarding whether with full information and proper knowledge the public does and ought to accept the risk.[105] Unfortunately, it seems that applying this approach also raises further questions, especially relating to those products that have inherent risks.[106]

In the Netherlands, it has been accepted that risk-utility factors can form part of the assessment of product defectiveness. The legislator has noted that Article 6:186 DCC stipulates that all circumstances should be taken into account. Additional factors that have been mentioned in Dutch case law and literature to be of relevance for determining defectiveness are the seriousness of the damage, the probability of damage, the benefit(s) or utility of the product, whether the danger is known to users, the availability of a safer product, the feasibility (in terms of financial cost and technology) of a safer design of the product, the product's price, the nature of the product, and compliance with public safety provisions.[107]

2.3.5 Conclusion

Paragraph 2.3 dealt with the strict liability regime of the European Directive, especially its main liability requirement of defectiveness. After describing the basics of the EPLD in paragraph 2.3.2, paragraph 2.3.3 further explained the requirement of defectiveness, laid down in Article 6 of the Directive. It stipulates that there is a product defect when the product does not provide the safety which a person is entitled to expect. All circumstances need to be taken into account. Three are mentioned specifically: (a) the presentation of the product; (b) the use to which it could reasonably be expected that the product would be put; (c) the time when the product was put into circulation. These were discussed in paragraph 2.3.3.1. In addition, the paragraph explained that the standard of defectiveness must be measured objectively. It is neither the safety expectations of the particular injured person nor of the particular producer that is decisive, but the degree of safety persons generally are entitled to expect. Another important remark made was that the test concerns

104 [2001] 3 *All ER* 289 (*A v. National Blood Authority*), paras. 36 and 67.
105 [2001] 3 All ER 289 (*A v. National Blood Authority*), para. 68.
106 Howells & Mildred 2002, p. 99. See *e.g.* the case of allergen cross-contamination in Pape 2009.
107 Dommering-Van Rongen 2000, p. 45; Stolker (Onrechtmatige daad I), Art. 186, aant. 3.

a level of product safety that the public at large are entitled to expect, rather than what they actually expect. This paragraph paid attention to the flaws of the defectiveness test that have been brought up in the legal literature as well. An important flaw is that the standard of defectiveness does not answer the question on the subject of how much safety the general public is entitled to expect. Furthermore, the test can be problematic to utilise with regard to products that have obvious risks or generally known risks, because such risks influence the safety expectations of people and can lead to the conclusion that persons are not entitled to expect a higher level of safety.

After that, paragraph 2.3.4 described in more detail why it has often been said in the literature that the regime of the Directive is a strict liability regime with elements of fault. Important elements that can introduce fault into the framework of the Directive are the requirement of defectiveness and the optional development risk defence. The degree in which they 'contaminate' the strict liability system depends on the interpretation of these concepts. It was shown that the development risk defence was interpreted narrowly by the ECJ as a result of which the possibility to escape liability for undiscoverable risks has become small. It is worthy to note though, that there is still some room left for a fault-based type interpretation of the defence as a result of the interpretation of the added requirement by the ECJ that the scientific and technical knowledge to discover the defect must have been accessible to be liable. As regards the concept of defect, it is topic of debate whether and to what extent the application of the defectiveness requirement diverges from applying the requirement of wrongfulness under fault-based liability. It has been argued in the literature that the defect requirement is of such a nature that it is problematic not to give in to adopting some form of a fault-based analysis to evaluate whether the product is defective and to include factors such as the avoidability of the risk by adopting precautionary measures and the burden of doing that. Nevertheless, there is a minority that considers the use of the latter factors as not permitted, because they inevitably refer to the reasonableness of the conduct of producers. As for utilising the defectiveness test under Dutch law, case law and literature suggest that the tests of wrongfulness and defectiveness are practically identical.

2.4 WARNING REQUIREMENTS UNDER EUROPEAN PRODUCT LIABILITY LAW

2.4.1 Introduction

Following the discussion on the requirements of the EPLD and the difficulties associated with interpreting concepts of the Directive, paragraph 2.4 discusses the key part of this chapter: warnings and product liability. A remark that is needed here before the following paragraph begins concerns the use of the term 'warning'. It has been noted in the European literature that warnings can be distinguished from instructions.[108] Warnings allow persons to be aware of a risk, whereas instructions describe how to use the product safely. Hereafter, I shall treat warnings and instructions for use in this chapter together under the joint name of warnings or duty to warn, unless specific reference is made to the one or the other.[109]

This paragraph is structured as follows: paragraph 2.4.2 further elaborates on the interpretation of the defectiveness test in relation to warning claims. As for claims in product warning cases, two types of actions are popular in court: claims that centre on (1) the inadequacy of the given warning and (2) the absence of a warning. Because of the wealth of product safety legislation under public law that the European legislator has enacted, paragraph 2.4.3 first provides a peek into the general product safety provisions and into the warning requirements for certain consumer products under Community public law. Secondly, it discusses the influence these requirements can have on the assessment of the defectiveness requirement under the Directive. The next paragraphs address the warning claim pertaining to the absence of a warning under the Directive (paragraph 2.4.4) and the warning claim pertaining to the adequacy of the warning under the Directive (paragraph 2.4.5). The final paragraph provides an outline of this paragraph (2.4.6).

2.4.2 The Standard of Defectiveness for Warning Claims

Scholars in Europe, including in the Netherlands, have argued that the introduction of the liability system of the Directive does not really change the way in which product liability is established in practice.[110] It has been asserted that this is especially true with respect to design and warning cases.[111] In theory, the liability standards diverge. On the one hand, fault-based product liability assesses the negligent behaviour of the producer by applying a risk-utility test. In warning cases, this comes down to answering the question whether the producer has acted wrongfully by violating the duty to warn adequately, or by failing to warn at all when he could and should have acted so in a proper way. On the other hand, the Directive's liability regime is based on the product itself. It analyses whether the

108 Grubb & Howells 2007, p. 13; Miller & Goldberg 2004, p. 430; cf. Hodges 1993b, p. 103.

109 The issue of the legal interpretation of the term 'warning' is examined more closely in chapter 4.

110 *E.g.* Stapleton 1994; Stoppa 1992; Lenze 2003c; Miller & Goldberg 2004, p. 354, 417; Lord Griffiths, De Val & Dormer 1987-1988. See for the Netherlands: Snijders 1984; Dommering-Van Rongen 2000, p. 32; Van Dam 2000, p. 289; Stolker & Westerdijk 1984; Stolker (Onrechtmatige daad I), Art. 186, aant. 8, 9.

111 Henderson & Twerski 1990, p. 275; Miller & Goldberg 2004, p. 433. The US Restatement of the Law (Third), Torts: Products Liability has accepted the view that warning and design defects need to be assessed on the basis of reasonableness. In contrast, for manufacturing defects it is generally agreed that strict liability is well suited. Since the product has failed to meet the manufacturer's own standard, it will no doubt be considered to have failed to meet consumer expectations, see American Law Institute 1998, p. 14. See also (Owen 2008, p. 330); Stoppa 1992, p. 211; Van Dam 2006, p. 377.

product's dangerous characteristics such as the absence or the inadequacy of a warning failed to meet the safety expectations of the general public. Notwithstanding this theoretical distinction, the circumstance that the Directive does not hold a producer liable for the absence of a warning of an unknown or undiscoverable risk questions whether there is really a difference between establishing defectiveness and wrongfulness on this point. Furthermore, the determination of producer's liability in product warning cases under both legal theories is accepted by virtue of a legal appraisal of the circumstances of the case. Even though the Directive prescribes a test of consumer expectations, the factors involved in a risk-utility test, such as the size of the risk (probability and seriousness of consequences) and the avoidability of risk such as the cost and practicability of a safer design, can hardly be discounted in warning cases.[112]

In view of the legal literature and case law, it seems to be permitted that risk-utility factors can play a role and can be taken into consideration under the defectiveness test. Dutch case law is illustrative in this respect.[113] As shown earlier, the Dutch Supreme Court linked the defectiveness criterion to the standard of fault-based liability of the producer, and as a result the *Kelderluik* factors for determining wrongfulness are also of importance for determining defectiveness.[114] As noted above, some scholars believe that the Directive does not permit the use of these factors of reasonableness as it would mean reintroducing fault-based liability by way of the back door.[115] Nonetheless, this still seems to be the minority view. After the decision of Judge Burton, a number of European courts have continued to use the language of fault in their judgement of defectiveness or have expressly stated that these tests merge.[116]

2.4.3 *Compliance with European Public Product Safety Provisions*

2.4.3.1 General

A factor that can be regarded as relevant in determining defectiveness is whether the product meets the requirements of the European public safety provisions.[117] Consumers are entitled to expect that the product conforms to its applicable product safety legislation under EU public law.

---

112 Miller & Goldberg 2004, p. 354, 383, 417; Howells & Borghetti 2010, p. 466.

113 Relevant case law is summarised in paragraphs 2.4.4 and 2.4.5.

114 HR 30 juni 1989, *NJ* 1990, 652 (*Halcion*); HR 22 oktober 1999, *NJ* 2000, 159 (*Koolhaas/Rockwool*); HR 29 november 2002, *NJ* 2003, 50 (*Onkruidverdelger Thyram*).

115 Burton J in the *Hepatitis C* case; Taschner 2005; Deards & Twigg-Flesner 2001.

116 See *e.g.* [2002] *EWHC* 490 (*Bogle v. Mc Donald's Restaurants*); [2000] *PIQR* 95 (*Worsely v. Tambrands Ltd*); Rb. Zwolle 24 april 2002, *Praktijkgids* 2002, 5921 (*Mini-tampon*); Hof 's-Hertogenbosch 15 mei 2007, *LJN* 2007, BA6838 (*Nagelstyling*); Hof Arnhem 14 oktober 2003, *NJF* 2004, 46 (*Datafan*).

117 See *e.g.* Miller & Goldberg 2004, p. 380.

Over the past 20 years, the European legislator has been increasingly active in the field of product safety. Many European directives, regulations and standards have been developed and adopted.[118] These also encompass product information requirements. It falls outside the scope of this thesis to provide an extensive overview of the public product safety requirements – including product warning requirements – that are laid down for various (categories of) consumer products. This thesis centres on the product warning requirements under European product civil liability law. Providing an outline of the general EU product safety and warning requirements and providing a peek into warning requirements for certain consumer products suffices here to point out the influence they may have on the assessment of the defectiveness requirement under European product liability law.

Before discussing the scope of the Directive's defence that pertains to compliance with EU public law in paragraph 2.4.3.5, paragraph 2.4.3.2 starts with an outline of the general safety provisions for non-food products and foodstuffs, followed by a sketch of specific warning requirements related to toys and chemicals in paragraph 2.4.3.3. Paragraph 2.4.3.4 briefly deals with the concept of voluntary standards that can pertain to the quality and safety of products next to mandatory provisions. These can also play a role, albeit a minor one compared to mandatory legislation, in the discussion of defectiveness.

2.4.3.2 General Safety Provisions for Non-Food and Food Products

General Product Safety Directive

An important EU product safety instrument is the General Product Safety Directive (GPSD) of 2001 which provides obligations for producers and distributors with regard to safe products.[119] It replaces the Directive of 1992 and is applicable as from January 2004.[120] The objective of the revised GPSD is to ensure a high level of product safety throughout the EU for consumer products that are not covered by specific sector legislation (*e.g.* toys, chemicals, cosmetics, machinery). The Directive also complements the provisions of sector legislation which do not cover certain matters, for instance in relation to producers' obligations and the authorities' powers and tasks.[121]

---

118 Dutch product safety legislation depends for a large part on the European developments in this area. EU Directives need to be implemented in national law by the national authorities and EU Regulations are directly applicable under Article 288 of the Treaty on the Functioning of the European Union (249 EC Treaty old). Important Dutch legislation in the field of product safety is the Commodities Act (*Warenwet*). This Act is a framework Act, which means that on the basis hereof Decrees and Regulations can be adopted.

119 Directive 2001/95/EC of the European Parliament and the Council of 3 December 2001 on general product safety (*OJ* 2002, L 11/4).

120 Directive 92/59/EEC of the European Parliament and the Council of 29 June 1992 on general product safety (*OJ* 1992, L 228/24).

121 Article 1(2) GPSD.

The GPSD imposes the general requirement on producers of putting only safe products onto the market.[122] Article 2(b) GPSD defines a 'safe product' as follows:

> 'Safe product' shall mean any product which, under normal or reasonably foreseeable conditions of use including duration and, where applicable, putting into service, installation and maintenance requirements, does not present any risk or only the minimum risks compatible with the product's use, considered to be acceptable and consistent with a high level of protection for the safety and health of persons, taking into account the following points in particular:
> (i) the characteristics of the product, including its composition, packaging, instructions for assembly and, where applicable, for installation and maintenance;
> (ii) the effect on other products, where it is reasonably foreseeable that it will be used with other products;
> (iii) the presentation of the product, the labelling, any warnings and instructions for its use and disposal and any other indication or information regarding the product;
> (iv) the categories of consumers at risk when using the product, in particular children and the elderly.
> The feasibility of obtaining higher levels of safety or the availability of other products presenting a lesser degree of risk shall not constitute grounds for considering a product to be 'dangerous'.

Article 2(c) defines a 'dangerous product' as any product which does not meet the definition of 'safe product'. And a 'serious risk' means any serious risk, including those whose effects are not immediate, requiring rapid intervention by the public authorities.[123]

As shown above, the relevance of product information to the safety of non-food products is reflected in the GPSD too. Any type of information regarding the product must be taken into account when determining whether the consumer product is safe. In addition, Article 5(1) GPSD formulates an information obligation for producers:

> Within the limits of their activities, producers must provide consumers with the relevant information to enable consumers to assess the risks inherent in a product throughout the normal or reasonably foreseeable period of its use, where such risks are not immediately obvious without adequate warnings and to enable them to take precautions against those risks.

122 Article 3 GPSD.
123 Article 2(d) GPSD.

Article 5(1) GPSD stipulates that the presence of warnings does not exempt any person from compliance with the other requirements set out in the GPSD.

General Food Law

The General Food Law (GFL) of 2002 lays down general principles and requirements as regards the safety of food and feed. It also establishes the European Food Safety Authority and it lays down procedures for matters with an impact on food and feed safety.[124] The central objective of the European Commission's food safety policy is to assure a high level of protection of human life and health and the protection of consumers' interests in relation to food, whilst ensuring the effective functioning of the internal market.[125] The Commission's guiding principle to ensure the safety of food and feed is to apply an integrated 'farm to fork' approach covering all stages of the food/feed chain, including feed production, primary production, food processing, storage, transport and retail sale. The primary responsibility for ensuring compliance with food law, and in particular the safety of the food, rests with the food business. Similarly, this principle is applied to the feed business. In this regard, Article 17 GFL determines that food and feed business operators at all stages of production, processing and distribution within the businesses under their control shall ensure that foods or feeds satisfy the requirements of food law which are relevant to their activities and shall verify that such requirements are met.[126]

The central article of food law of the GFL is the general food safety requirement, which is laid down in Article 14.[127] This requirement also takes into account the influence that information can have on safety. The first 2 paragraphs of a total of 9 read as follows:

> 1. Food shall not be placed on the market if it is unsafe.
> 2. Food shall be deemed to be unsafe if it is considered to be:
> (a) injurious to health;
> (b) unfit for human consumption.
> 3. In determining whether any food is unsafe, regard shall be had:
> (a) to the normal conditions of use of the food by the consumer and at each stage of production, processing and distribution, and
> (b) to the information provided to the consumer, including information on the label, or other information generally available to the consumer concerning

124 Regulation No 178/2002 of the European Parliament and of the Council of 28 January 2002 laying down the general principles and requirements of food law, establishing the European Food Safety Authority and laying down procedures in matters of food safety (*OJ* 2002, L 31/1).

125 Articles 1, 5 and 7 GFL and recitals 1-3.

126 Other responsibilities of food and feed business operators are laid down in Article 18 (track and trace responsibility) and Articles 19 and 20 (responsibility to withdraw unsafe food from the market and inform the authorities and possibly inform consumers and ultimately recall the products).

127 The feed safety requirement is laid down in Article 15 GFL.

> the avoidance of specific adverse health effects from a particular food or category of foods.
> 4. In determining whether any food is injurious to health, regard shall be had:
> (a) not only to the probable immediate and/or short-term and/or long-term effects of that food on the health of a person consuming it, but also on subsequent generations;
> (b) to the probable cumulative toxic effects;
> (c) to the particular health sensitivities of a specific category of consumers where the food is intended for that category of consumers.

As regards information, the GFL provides the additional obligation that the labelling, advertising, and presentation of food or feed, shall not mislead consumers.[128] Moreover, the proposal for a new Regulation on the provision of food information to consumers has formulated new mandatory labelling requirements with regard to the content and format of information on food labels.[129] Recital 25 holds that food labels should be clear and understandable to assist consumers wanting to make better-informed food and dietary choices. Articles 7(2) and 14 elaborate this. The first article states that food information must be accurate, clear and easy to understand. In addition, Article 14 stipulates that consumers need to be given information that is marked in a conspicuous place in such a way as to be easily visible, clearly legible and, where appropriate, indelible. It shall not in any way be hidden, obscured, detracted from or interrupted by any other written or pictorial matter or any other intervening material.

2.4.3.3 Specific Product Safety and Informational Provisions

Product safety law under EU public law also lays down safety requirements for specific types of products, such as toys, cosmetics, personal protective equipment, chemicals, pharmaceuticals, machinery, products with low voltage limits and recreational craft. This is sector-specific product legislation.[130] The recent provisions governing toy safety and chemicals are dealt with here briefly to illustrate the EU requirements set on warnings.

128 Article 16 GFL.

129 Proposal of the Commission of 30 January 2008 for a Regulation of the European Parliament and of the Council on the provision of food information to consumers 2008. It consolidates and updates the important Directive 2000/13/EC on the labelling, presentation and advertising of foodstuffs and Directive 90/496/EEC on nutrition labelling for foodstuffs. It also recasts six other Directives that pertain to the labelling of certain categories of foods.

130 The Commission department has set up guidance documents for Member States on the relationship between these sector-specific Directives and the GPSD.

Toy Safety

Product legislation can impose mandatory warning requirements, sometimes even of such a nature that the warning has already been defined by law. This is the case for the new Toy Safety Directive (TSD). The new TSD was adopted in 2009 and replaces the Directive 88/378/EEC of 1988 that fell short in dealing with new technological developments in the toys market and ensuring a high level of protection of children.[131] The TSD lays down the essential safety requirements with regard to toys and specific safety provisions regarding the physical and mechanical properties, flammability, chemical properties, electrical properties, hygiene and radioactivity which toys must meet during manufacture and before being placed on the market. It also brings about changes and new requirements with regard to toy warnings. The European Commission considers them necessary in order to further promote safe conditions of use of toys.[132]

Article 11(1) in conjunction with part A of Annex V of the TSD states that where appropriate for safe use, general warnings must appear on toys specifying the appropriate conditions and limitations of use. User limitations include at least minimum or maximum age of the user and where appropriate, the abilities of the user, the maximum or minimum weight of the user and the need to ensure that the toy is used only under adult supervision.

Part B of Annex V stipulates that for certain categories of toys specific warnings must be used, and precautions and/or instructions must be taken or used.[133] The warning sentences that need to be used are laid down explicitly.[134] Illustrative are the specific warning requirements as regards toys that are not intended for use by children under 36 months. Toys which might be dangerous for children under 36 months of age must bear a warning such as: 'Not suitable for children under 36 months' or 'Not suitable for children under three years' or a warning symbol in the form of the generally known graphic:

131 Directive 2009/48/EC of the European Parliament and of the Council of 18 June 2009 on the safety of toys (*OJ* 2009, L 170/1).

132 Recital 30.

133 These toys are (1) toys not intended for use by children under 36 months; (2) activity toys; (3) functional toys; (4) chemical toys; (5) skates, roller skates, online skates, skateboards, scooters and toy bicycles for children; (6) aquatic toys; (7) toys in food; (8) imitations of protective masks and helmets; (9) toys intended to be strung across a cradle, cot or perambulator by means of strings, cords, elastics or straps; (10) packaging for fragrances in olfactory board games, cosmetic kits and gustative games.

134 For instance, part B of the annex provides that toys contained in food or co-mingled with food shall bear the following warning: 'Toy inside. Adult supervision recommended'.

The Annex states that these warnings shall be accompanied by a brief indication, which may appear in the instructions for use, of the specific hazard calling for this precaution. Furthermore, it rules that this point shall not apply to toys which, on account of their function, dimensions, characteristics or properties, or on other cogent grounds, are 'manifestly unsuitable' for children under 36 months. Related to this is the general obligation of Article 11(1) TSD which determines that toys shall not bear one or more of the specific warnings where that warning conflicts with the intended use of the toy, as determined by virtue of its function, dimension and characteristics. Pursuant to recital 30 to the Directive's preamble, the main reason for introducing this obligation was to prevent the misuse of warnings to circumvent the applicable safety requirements. This has occurred particularly in case of the warning stating that the toy is not suitable for children under 36 months.

Paragraph 2 of Article 11 provides additional warning design requirements. The manufacturer is obliged to mark the warnings in a clearly visible, easily legible and understandable and accurate manner. They must be marked on the toy, on an affixed label or on the packaging. If appropriate, warnings should also be included in the instructions for use which accompany the toy. Small toys which are sold without packaging shall have appropriate warnings affixed to them. Another new warning requirement stipulates that the warnings shall be preceded by the words 'Warning' or 'Warnings', as the case may be. Lastly, paragraph 2 obliges that warnings which determine the decision to purchase the toy, such as those specifying the minimum and maximum ages for users and the other applicable warnings set out in Annex V, shall appear on the consumer packaging or be otherwise clearly visible to the consumer before the purchase, including in cases where the purchase is made on-line.

The final paragraph of Article 11 notes that in accordance with Article 4(7), a Member State may, within its territory, stipulate that those warnings and the safety instructions shall be written in a language or languages easily understood by consumers.

Chemical Safety

Mandatory requirements with regard to warning related information are also laid down in legislation pertaining to products that contain chemical substances. The developments concerning the safety of chemical substances are relevant for consumer products as hazardous chemicals can be found in a range of consumer products such as detergents, cleansing agents, plastic toys, cosmetics and so forth.

Due to the inadequacy of existing legislation, the EU has modernised its framework on chemicals by introducing the REACH (Registration, Evaluation and Authorization of Chemicals) system and by setting up a European Chemicals Agency. The new rules are laid down in the European REACH Regulation 1907/2006/EC (REACH Regulation), which entered into force in stages as from 1 June 2007.[135]

Article 1(3) states that the REACH Regulation is based on the principle that it is for manufacturers, importers and downstream users to ensure that they manufacture, place on the market or use such substances that do not adversely affect human health or the environment. Hence, the obligation for safety rests with the industry and not the authorities. Furthermore, the Regulation lays down specific duties and obligations on manufacturers, importers and downstream users. In short, Article 6 REACH Regulation determines that manufacturers and importers of chemical substances in quantities of 1 tonne or more per year have the obligation to submit a registration dossier with relevant information to the European Chemicals Agency. This obligation applies to substances as such and in preparations.[136] The registration dossier contains data on the characteristics of the substances they import or manufacture.[137] Only registered substances are allowed to circulate on the internal market.[138] Substances in quantities of 10 tonnes or more per year that are subject to registration must also be accompanied by a chemical safety report with information on the assessment of the health and environmental risks related to the substance and the applied risk management measures to ensure that any risks from the uses of a substance are adequately controlled.[139] The Agency will perform different types of evaluations, including checking that the registration dossiers comply with the Regulation's requirements and evaluating the dangerous characteristics of registered substances in more detail.[140]

The REACH system is complemented by Regulation 1272/2008/EC on the Classification, Labelling and Packaging of substances and mixtures (CLP Regulation).[141] This Regulation integrates the classification criteria and rules on labelling of the United Nations' Globally

---

135 Regulation (EC) No 1907/2006 of the European Parliament and of the Council of 18 December 2006 concerning the Registration, Evaluation, Authorisation and Restriction of Chemicals (REACH), establishing a European Chemicals Agency, amending Directive 1999/45/EC and repealing Council Regulation (EEC) No 793/93 and Commission Regulation (EC) No 1488/94 as well as Council Directive 76/769/EEC and Commission Directives 91/155/EEC, 93/67/EEC, 93/105/EC and 2000/21/EC (*OJ* 2006, L 396/2). The REACH Regulation will replace an old set of Directives and Regulations. According to Article 1 of the Regulation, the main purposes of the REACH Regulation is to improve the protection of human health and the environment whilst maintaining competitiveness and strengthening the spirit of innovation in Europe's chemicals industry. It lays down provisions on substances and preparations within the meaning of Article 3 and these provisions are applicable to the manufacture, placing on the market or use of such substances on their own, in preparations or in articles and to the placing on the market of preparations.

136 A special registration regime applies for substances in articles (*e.g.* manufactured goods such as cars and textiles). Certain substances are exempted from registration.

137 See Articles 10 and 12 REACH Regulation.

138 Article 5 REACH Regulation.

139 Article 14 REACH Regulation.

140 Article 20 REACH Regulation.

141 Regulation (EC) No 1272/2008 of the European Parliament and of the Council of 16 December 2008 on classification, labelling and packaging of substances and mixtures, amending and repealing Directives 67/548/EEC and 1999/45/EC, and amending Regulation (EC) No 1907/2006 (*OJ* 2008, L 353/1).

Harmonized (GHS) System with EU legislation. The GHS system provides a single system to identify hazards and to communicate them in transporting and supplying chemicals across the world. Hence, the GHS system and its labels will be used and recognised worldwide. The CLP Regulation entered into force on 20 January 2009. The introduction of the CLP Regulation and its new labels, especially the symbols, will also take place in stages. This process of harmonisation enhances protection of human health and the environment, and improves the free circulation of chemical substances and mixtures.[142] Enterprises must classify, label and pack their substances and mixtures in line with the provisions of this Regulation before putting them on the market. They must determine which properties of substances and mixtures should lead to a classification as hazardous, in order for the hazards of substances and mixtures to be properly identified and communicated. These properties include physical hazards as well as hazards to human health and to the environment, including hazards to the ozone layer. Annex I of the CLP Regulation establishes the criteria for the classification and labelling of hazardous substances and mixtures. The Annexes of the Regulation also include a mandatory list of hazard statements (Annex III), a list of precautionary statements (Annex IV), pictograms for each hazard class (Annex V) and the lists of classifications and labelling harmonised at Community level.

Article 17 of the CLP Regulation provides general rules for the content of labels of hazardous substances or mixtures. In accordance with the classification of the hazardous substance or mixture, the label should contain the following elements: contact information about the supplier; the quantity of the substance or mixture; product identifier(s); hazard pictogram(s); signal word(s); hazard statement(s); and supplemental information. The articles that follow provide additional information about these informational elements. Article 32 of the CLP Regulation provides labelling requirements with regard to its format, such as the requirement that the pictograms, signal word, hazard and precautionary statements should be located together on the label. Furthermore, Annex I contains additional labelling requirements, for example with regard to the dimensions of the label.

> Examples of hazard statements are H251: 'Self-heating: may catch fire.' or H304: 'May be fatal if swallowed and enters airways.' Examples of precautionary statements are P202: 'Do not handle until all safety precautions have been read and understood.' or P234: 'Keep only in original container.'
> The first pictogram shown below is new and is the GHS08 symbol relating to long-term health hazards, such as respiratory sensitisation, aspiration hazard, reproductive toxicity. The second symbol (GHS06) that expresses the acute toxic nature of the hazard and the third symbol (GHS09) that represents a

142 Article 1 CLP Regulation.

hazard to the aquatic environment are not new in the sense that the former legislation contained a symbol to express these hazardous properties.

2.4.3.4 Product Standards

In addition to public product safety provisions, there are also standards that can be of relevance to products. They provide guidance with regard to how to comply with essential requirements of EU pubic law. Standards are shaped by consensus among enterprises, public authorities, consumers, and trade unions, through a consultation process organised by independent, recognised standardisation bodies at national, European and international level.

These standards are distinct from public law, because they do not have any binding power. The use of European standards is on a voluntary basis. A number of product standards have been harmonised, which means that these have been published in the 'Official Journal of the European Union'. Article 4 GPSD lays down a procedure for standards to be adopted and provides that compliance with them will provide a presumption of conformity. As regards specific sector legislation, toys are for example subject to a presumption of conformity if they comply with harmonised standards published in the Official Journal of the European Union. Hence, producers can use these harmonised standards to demonstrate that their products comply with EU legislation.[143] However, European product liability law can require a higher level of safety than that that is laid down in product standards. Furthermore, the defence discussed below is only applicable to mandatory provisions and is not applicable in cases where the defectiveness is a result of compliance with a voluntary standard.

2.4.3.5 The Scope of the Defence of Article 7(d)

Compliance with a public product safety provision can be a relevant indication of non-defectiveness, given that public law sets a minimum standard of safety that must be achieved before the product is placed on the market.[144] Consumers are entitled to expect at least that such products on the market meet those requirements. One of the defences of the EPLD relates to compliance with public law:

143 See <http://ec.europa.eu/enterprise/policies/european-standards/index_en.htm#top> on this subject.
144 See Van Boom 2007, p. 424.

> Article 7 stipulates that the producer shall not be liable as a result of this Directive if he proves:
> (d) that the defect is due to compliance of the product with mandatory regulations issued by the public authorities.

This exemption has a narrow scope and can seldom be invoked successfully. One reason for this is that most of the above-mentioned safety requirements are of a general and open nature. Producers and courts must give meaning to these provisions. For example, they are obliged to apply the general criterion of 'safe product' of the GPSD or GFL, or more specifically the general warning requirement of 'understandable information' to their product and consequently investigate what hazards are attached to their products, when consumers need to be warned, and how to present the information. Because these yardsticks are open, it requires a thorough assessment of whether the concrete provision has been met or violated. In this respect, the warning requirements arising from public law are quite similar to those of liability law. Both stipulate that a reasonably foreseeable product use by consumers must be taken into account and that consumers need adequate instructions about and warnings of the dangers attached to the product. The violation of such general requirements renders the product defective if under the circumstances of the case the producer has failed to implement the accurate level of safety. Nonetheless, some of the warning requirements arising from public law have been laid down in a detailed way by law. For instance, the hazard and precautionary statements and the symbols for consumer products that contain chemicals are explicitly formulated by the CLP Regulation. It is likely that the content of these statements cannot be tested under the Directive because of the aforementioned defence, even if one questions the understandability of the content of a specific statement or symbol.[145]

In addition to the warning requirements that arise from public law, liability law imposes unwritten obligations on producers as well. These warning requirements that can be deduced from case law and the academic literature are dealt with in the following paragraphs.

2.4.4 *Defectiveness: The Absence of a Warning*

Following the discussion on the relevance of EU product safety regulations, the upcoming two paragraphs go deeper into the assessment of the common warning claims in European product liability law, starting with the claim that alleges that a warning should have been present to render the product non-defective.

145 It may be possible to contend that there is too little information available to be understandable since producers carry the responsibility to determine which statements are given.

The Directive emphasises the relevance of warning information and its effect on the safety expectations of consumers by specifically referring to the presentation of a product as a factor for establishing defectiveness. Because of the specific reference to this factor, it follows that the Directive supports the view that the absence or presence of warning information can influence the safety expectations of product users and that it subsequently is an important issue for determining defectiveness under the Directive.

Various circumstances can be of significance when determining whether the product is defective as a result of the absence of a warning. As already touched upon earlier, in the Netherlands but also in other European countries, it seems that the way of assessing whether a product is defective as a result of the absence of a warning does not differ from the fault-based assessment of whether there rests a duty to warn upon the producer.[146] Those factors that are used by courts in the process of determining defectiveness are in essence identical in a failure-to-warn case under fault-based liability.[147] So, even though the liability theories diverge, it would be difficult to imagine that the outcome under the two standards is different. The assessment under fault-based liability entails weighing the relevant factors in order to determine whether the safety benefits of providing a warning are outweighed by its costs. It is generally agreed that the more practicable and less costly the precautionary measure (the warning) to protect against the risk, the greater the likelihood that it will be required under law.[148] In the Netherlands, the *Kelderluik* factors and the *Halcion* factors in particular are of relevance for these types of product warning claims. It can thus be said that – in terms of the absence of a product warning – these can also be considered valuable for assessing defectiveness under the liability regime of the Directive.

Applicable cases from Dutch case law and a number of European cases that have been described in European literature were examined. An outline of the cases is provided in the next paragraph. The cases suggest that the most important factors in this regard are:

- the intended and reasonably expected use of the product;[149]
- the time that the product was put into circulation;[150]
- probability that a product danger emerges;[151]
- the degree of harm arising from that;[152]

---

146 See paragraphs 2.3.4.3 and 2.4.2.

147 Stoppa 1992, p. 221; Miller & Goldberg 2004, p. 433.

148 See paragraphs 2.2.2.1 and 2.2.2.3.

149 See *e.g.* Rb. Maastricht 21 maart 2002, *LJN* 2002, AE0776 (*Versgeperste jus d'orange*); Rb. Amsterdam 17 december 2008, *NJ* 2009, 311 (*Rokersclaim*); OLG Düsseldorf 20 December 2002, 14 U 99/02, *VersR* 2003, 912 (*Chocolate bar*); Hof 's-Hertogenbosch 15 mei 2007, *LJN* 2007, BA6838 (*Nagelstyling*); Hof Arnhem 14 oktober 2003, *NJF* 2004, 46 (*Datafan*).

150 Cf. HR 25 november 2005, *NJ* 2009, 103 (*Eternit*).

151 See *e.g.* HR 30 juni 1989, *NJ* 1990, 652 (*Halcion*); Hof 's-Hertogenbosch 15 mei 2007, *LJN* 2007, BA6838 (*Nagelstyling*).

152 See *e.g.* Rb. Amsterdam 17 december 2008, *NJ* 2009, 311 (*Rokersclaim*); [2002] *EWHC* 490 (*Bogle v. Mc Donald's Restaurants*); HR 25 november 2005, *NJ* 2009, 103 (*Eternit*); HR 30 juni 1989, *NJ* 1990, 652 (*Halcion*)

- the nature of the product hazard;[153]
- the burden/costs of implementing a warning or a design change;[154]
- a hidden or obvious risk;[155]
- general knowledge of the risk;[156]
- the degree of certainty with respect to the health hazards associated with product use;[157]
- (non)compliance with public product safety provisions or voluntary standards;[158]
- the utility of the product.[159]

Due to the casuistic nature of the cases presented to the courts, it is difficult to provide general statements as to whether a warning or its absence results in a defective product. As noted above, assessing defectiveness involves a weighing of several relevant circumstances and their value differs per case. The value of the circumstances is left for assessment of the court. A number of guidelines can be deduced from legal literature and case law.

Circumstances such as the presence of a hidden danger to consumers or the seriousness of the injury weigh in favour of providing a warning message.[160] But, if the probability of damage is trivial, demanding a warning under product liability becomes less necessary.[161]

Whether there should be a warning also generally depends on the way in which consumers use the product. Article 6 of the Directive makes reference to the reasonably expected use of the product. Similarly, the GPSD expressly refers to the normal or reasonably foreseeable conditions of use in the definition of safe product. It stipulates that the product use should be interpreted widely: it takes account of the duration of use, and the different phases of use such as putting the product into service, installation and maintenance. It follows that the wording of Articles 2 GPSD and 6 EPLD recognise that merely anticipating and warning of risks that can arise from product use that is normal and intended by the producer is a too narrow viewpoint. Consumers do not behave cautiously and attentively every minute during product use and they can use it in other ways that are

153 See *e.g.* Hof Arnhem 14 oktober 2003, *NJF* 2004, 46 (*Datafan*); Hof 's-Hertogenbosch 15 mei 2007, *LJN* 2007, BA6838 (*Nagelstyling*); HR 25 november 2005, *NJ* 2009, 103 (*Eternit*); HR 30 juni 1989, *NJ* 1990, 652 (*Halcion*).

154 See *e.g.* HR 30 juni 1989, *NJ* 1990, 652 (*Halcion*); Hof 's-Hertogenbosch 15 mei 2007, *LJN* 2007, BA6838 (*Nagelstyling*); Hof Arnhem 14 oktober 2003, *NJF* 2004, 46 (*Datafan*).

155 See *e.g.* Rb. Amsterdam 17 december 2008, *NJ* 2009, 311 (*Rokersclaim*).

156 See *e.g.* Rb. Maastricht 21 maart 2002, *LJN* 2002, AE0776 (*Versgeperste jus d'orange*); Rb. Amsterdam 17 december 2008, *NJ* 2009, 311 (*Rokersclaim*); Hof Arnhem 14 oktober 2003, *NJF* 2004, 46 (*Datafan*); [2002] *EWHC* 490 (*Bogle v. Mc Donald's Restaurants*); OLG Düsseldorf 20 December 2002, 14 U 99/02, *VersR* 2003, 912 (*Chocolate bar*); OGH 5 December 2002, 2 Ob 249/02k (*Stepladder*).

157 See *e.g.* HR 25 november 2005, *NJ* 2009, 103 (*Eternit*).

158 See *e.g.* Rb. Maastricht 21 maart 2002, *LJN* 2002, AE0776 (*Versgeperste jus d'orange*).

159 See *e.g.* HR 30 juni 1989, *NJ* 1990, 652 (*Halcion*).

160 See *e.g.* HR 30 juni 1989, *NJ* 1990, 652 (*Halcion*); Rb. Amsterdam 17 december 2008, *NJ* 2009, 311 (*Rokersclaim*).

161 Grubb & Howells 2007, p. 375. This issue was raised in the context of a 'may contain' warning against the risk of allergen cross-contamination in food products, see Pape 2009.

more satisfactory for the user. As stipulated by the wording of these articles governing product safety, uses that are not intended, but reasonably expected or foreseeable should also be considered when designing a product[162] – for example, by providing a warning. This resembles the unwritten obligation laid down in the *Warmwaterkruik* case in which the Dutch Supreme Court ruled that producers are obliged to take into account the circumstance that a part of the users will omit to take all necessary precautions.[163] Hence, it is argued here that this decision made under fault-based liability continues to be of value within the Directive's liability regime.[164] In general, warnings cannot be considered needed if the risk that materialised was caused by product use that falls outside the scope of what use can reasonably be expected.[165] As noted in paragraph 2.3.3.1, the sixth recital to the Directive's preamble mentions that any misuse of a product cannot be regarded reasonable. Under such circumstances, the responsibility of potential victims to use the product safely precludes a successful claim for compensation. The limit of unreasonable misuse does not seem to be very helpful in assessing what usage must be taken into account by producers, as it does not tell when reasonably expected use results into unreasonable misuse.

Another consideration for determining whether a warning can be regarded as necessary is the characteristics of the users, such as their level of knowledge.[166] Circumstances such as the presence of a generally known risk and/or an obvious risk imply that warning information is not required.[167] Especially the circumstance that this risk is known and/or obvious due to the fact that it is inherent to a normal use of the product can absolve the need for a warning. One can think of damage resulting from the use of alcohol and tobacco.[168] This type of damage, *i.e.* risks that are inherent within a product that nevertheless is justifiable to market, is usually referred to as *system damage*.[169] The circumstance that obvious inherent risks need no warning also follows from public law. Nevertheless, the GPSD specifies the requirement by stating that the risk must have been immediately obvious.[170] Moreover, it could follow from the application of the test of defectiveness that products with such a generally known or obvious risk need no warning information since

162 Miller & Goldberg 2004, p. 373, 424.

163 HR 2 februari 1973, *NJ* 1973, 315 (*Warmwaterkruik*).

164 Dommering-Van Rongen 2000, p. 36; Spier e.a. 2009, p. 136; Pape 2006, p. 375.

165 See *e.g.* Rb. Maastricht 21 maart 2002, *LJN* 2002, AE0776 (*Versgeperste jus d'orange*).

166 Grubb & Howells 2007, p. 372.

167 Miller & Goldberg 2004, p. 436; Grubb & Howells 2007, p. 373, 375; Spier e.a. 2009, p. 135. See *e.g.* [2002] *EWHC* 490 (*Bogle v. McDonald's Restaurants*); Rb. Maastricht 21 maart 2002, *LJN* 2002, AE0776 (*Versgeperste jus d'orange*).

168 Rb. Amsterdam 17 december 2008, *NJ* 2009, 311 (*Rokersclaim*). See on this latter decision: Franken 2009; Grubb & Howells 2007, p. 368.

169 The term was introduced by B. Dahl. Exposure to the risk can be justified because of the benefits of the product and the lack of a direct safer substitute product, see Grubb & Howells 2007, p. 12, 368; Miller & Goldberg 2004, p. 419.

170 Article 5(1) GPSD.

consumers are already aware of it and as a consequence should adapt their safety expectations and behaviour accordingly. However, such interpretation of the test may not always be valid and should be considered too strict. Even though a risk is known or obvious, other circumstances such as the seriousness of the injury, the probability of harm and the expected inadvertent or careless moments of product users may yet tip in favour of a warning or another design measure.[171] Lastly, there will usually not be a need to warn of risks that are considered socially acceptable in view of the utility of the product, as these are also often generally known and/or inherent.[172]

Dutch *Versgeperste jus d'orange* case

This case of the Dutch District Court of Maastricht concerns a victim who suffered serious injury to her eyes as a result of an exploding plastic bottle filled with 'versgeperste jus d'orange', *i.e.* freshly squeezed orange juice. She had not stored the bottle in a cool place for a few days. As a result of the warm temperature, the content fermented. When she opened the bottle to pour its content into the sink, the juice spurted out and the bottle cap hit her eye.

The District Court of Maastricht ruled that the producer may have reasonably expected that the average consumer knows that freshly squeezed juices prepared without any preservatives can ferment and explode if not stored in a cool place for a number of days. A warning against such a danger resulting from storage outside the fridge is neither required by public law nor needed by reason of unwritten safety standards. The District Court also noted that a producer does not have to take into account that consumers will use the freshly squeezed orange juice, meant to be consumed soon after purchase, in such a way.[173]

Dutch *Rokersclaim* case

The recent Dutch *Rokersclaim* case, *i.e.* tobacco claim, also pertains to the question whether a warning was needed under civil liability law. The claim is instituted under fault liability (Article 6:162 DCC). From 1957-1983, claimant smoked cigarettes of three brands A, B, and C. Defendant produced brand B. Claimant smoked brand A in the period 1957-1963, B in the period of 1963-1980 and brand C in 1980-1983. As of 1982, a health warning was required by Dutch law. After several attempts, claimant successfully quit smoking in 1983, mainly because of health problems (shortness of breath). After he had stopped, the symptoms did not disappear. In January 1996, claimant went to see a lung

171 Miller & Goldberg 2004, p. 377.

172 Grubb & Howells 2007, p. 368; Howells 2005b, p. 165.

173 Rb. Maastricht 21 maart 2002, *LJN* 2002, AE0776 (*Versgeperste jus d'orange*).

specialist for the first time. He was diagnosed with chronic obstructive pulmonary disease (COPD) that was caused by his smoking. In 2002 en 2004, claimant had a cerebral infarction, as a result of which claimant became partly paralysed.

Claimant instituted a claim against the producers A, B and C based on fault liability and alleged that he suffered health injury as a result of the cigarettes that were made by defendant. The cigarettes should be considered defective because they have serious health threatening characteristics that defendant knew or should have known about. Furthermore, claimant alleged that defendant failed to warn him in time and failed to warn adequately against the health risks attached to smoking cigarettes. Instead, defendant used advertising to promote smoking and tried to disguise the harmful nature of cigarettes. Claimant argued that he did not have knowledge of the serious health effects prior to 1981.

The District Court rejected the claim against A on the basis of the limitation period. As regards the rejection of the claim against producer C, the court holds that the cigarettes were produced and distributed by others than defendant C. As regards the claim against defendant of brand B the District Court ruled as follows.

First, the District Court assumed that the facts given by claimant mentioned above concerning the time when he smoked the brands and the health consequences he suffered thereof are correct. The District Court decided that a producer cannot be held liable for dangers that are inherent with regard to a normal use of the product in so far as the dangers are generally known. According to the District Court, at the least since 1963 (when claimant started smoking brand B) it has been generally known to the average consumer that smoking cigarettes causes serious life-threatening risks to one's health and that if one starts smoking, it will be difficult to quit. The District Court based its conclusion that the inherent dangers should have been of general knowledge to the general public since 1963 on the following circumstances. Since 1950 there have been scientific studies and reports that addressed the relationship between smoking and lung diseases. In 1962, the UK Royal College of Physicians issued a report concluding that smoking causes the life-threatening disease lung cancer. During this period, the media also reported extensively on this matter. Furthermore, claimant stated that he read the newspaper almost on a daily basis during that period. The court also referred to claimant's (good) level of education. The court does not consider it a requisite that all forms of lung diseases that can result from smoking cigarettes need to be brought to the attention of potential users. It is only required that it is sufficiently clear to the average consumer

what type of dangers is attached to the product. Hence, if it is known that smoking can have life-threatening consequences like lung cancer, a producer is not obliged to warn against all other harmful diseases such as COPD and hart and vascular diseases as well. The mere fact that nicotine generally has an addictive effect is insufficient for accepting liability. It depends on whether in this particular case, the addiction is of such a nature that subsequently claimant was unable to quit smoking at the time that the health dangers were becoming generally known.
As regards the claim concerning the failure to warn (*i.e.* providing misleading information and failing to warn of the dangers), the court dismissed claimant's argument that the general knowledge of the dangers associated with smoking were downplayed and undermined by the defendant's advertisements and one report in 1988 that questions the causal relationship between cancer and smoking. The court argued that the advertisements shown by claimant neither contain any claims that smoking is good for your health nor a denial of the dangers. Even though they do arouse positive associations, they do not mislead the average consumer with regard to the health risks of smoking. Furthermore, the advertisements are not from the defendant's brand. As regards the statements in the report of 1988, the court considered them reprehensible. It is a manufacturer's duty to warn product users of dangers that are inherent, as far as these dangers are not of general knowledge, and also not to downplay them. The public manifestations alone are no ground for liability. There was information present in the media that adequately informed the average consumer of the dangers and that enabled the average consumer to make an informed decision with regard to the dangers of smoking. Consequently, the responsibility of claimant to choose whether or not to smoke has not been nullified by these manifestations. The court decided that as regards the existence of a duty to warn prior to a mandatory warning, warnings are required only if the risks are hidden. A producer is not obliged to warn against dangers if these dangers are generally known to the average consumer.[174]

English *McDonald's* case
In this case, the claimants argued that the tea and coffee served by McDonald's were defective because there should have been warnings of the risk of burning. The warnings should have appeared at the point of service and alerted the customers to how hot the drinks were and how serious the injury could be.

174 Rb. Amsterdam 17 december 2008, *NJ* 2009, 311 (*Rokersclaim*).

The judge denied liability by arguing that the safety of hot coffee and tea meet the expectations of consumers generally since consumers expect the purchased drinks to be hot and consumers know that if a hot drink is spilled onto someone a serious scalding injury can result. The judge also took into consideration that the staff were trained to cap the drinks securely before they were handed to customers. Consequently, a warning is not required and the safety of the hot drinks in a cup of which the lid can be removed is such as persons generally are entitled to expect.[175]

German *Chocolate bar* case

In 1998, the victim was diagnosed with diabetes. The victim had regularly consumed the defendant's chocolate bars since childhood. At least four years prior to the diagnosis he had been eating a minimum of two bars of chocolate per working day.

The victim instituted a legal action against Masterfood and alleged that the bars were defective under the provisions of the German Product Liability Act 1989, because of a construction defect and an instruction defect. He argued that the chocolate bars 'Mars' and 'Snickers' of the producer Masterfood had caused his diabetes. As to the construction defect, the claimant argued that the chocolate bars contained an intolerable amount of sugar, that is to say almost 50% refined sugar. Furthermore, he argued that the bars contained addictive ingredients. As to the instruction defect, the claimant alleged that the producer was obliged to issue warning notices informing consumers that eating chocolate bars might lead to diabetes or to cavities or other tooth problems.

The Higher Regional Court of Düsseldorf ruled that the product cannot be considered defective neither in construction nor instruction. As regards the alleged instruction defect, the court held that the defendant was not obliged to provide warning notices of the heath dangers of consumption on the packaging. The court noted that the utility of a warning lies in informing consumers about the risks connected with the use of that product of which the consumer is not yet aware. The Higher Regional Court of Düsseldorf held that this ruling does not apply here, since the ordinary consumer has general knowledge of the risks connected with the excessive consumption of refined sugar and is also aware that chocolate bars consist in large part of refined sugar. Given the sugary taste, the chocolate and caramel texture and the list of ingredients the ordinary con-

175 [2002] EWHC 490 (*Bogle v. Mc Donald's Restaurants*).

sumer could not help but realise that 'Mars' and 'Snickers' are strongly sweetened foods.[176]

Austrian *Hammock* case
In this case, the claimant bought a set of hammock hooks, screws and dowels to construct his hammock. The packaging contained no instructions about the way in which to hang up a hammock. Whilst he first lay down on it, the hammock collapsed and he badly injured his eyes. The hook had loosened, came out of the wall and sprang into the claimant's eye.
The victim instituted a legal action against the supplier of hammocks and alleged that the hammock was defective under the provisions of the Austrian Product Liability Act because of an instruction defect. It was established at trial that the victim had used the wrong dowels and that he had fixed the hooks in a layer of mortar instead of drilling into the actual bricks. He argued that this would not have happened had the defendant supplied the hooks with adequate instructions. There should have been an instruction leaflet explaining that the solidity of the fastening of the hammock depended on the composition of the wall and the type of dowel being used to affix the hooks. Furthermore, the claimant argued that a warning was required stating not to drill into a wall with unknown composition or to leave this to a professional.
In contrast to the Court of First Instance, the Vienna Court of Appeal dismissed the claim. The Austrian Supreme Court upheld the latter decision. The Supreme Court agreed with the Court of Appeal that it is obvious for everyone who intends to hang up a hammock that an improperly fixed hook could loosen and come off the wall. The correct dowel depends on the composition of the wall. Everyone who has even drilled holes in a wall knows that the first step is to check the wall for any pipes or wires and also for any layers of mortar in which screws might not hold. Associated instructions or warnings of this nature are common knowledge and would seem absurd.[177]

2.4.5 *Defectiveness: Inadequacy of a Warning*

In addition to warning cases that deal with the absence of a warning, cases that centre on the warning that has been given by the producer are popular as well in product liability

176 OLG Düsseldorf 20 December 2002, 14 U 99/02, *VersR* 2003, 912 (*Chocolate bar*). Reported in *EPLR*, with note by M. Hannes, see Hannes 2003; Howells & Borghetti 2010, p. 470.
177 13 November 2002, 7 Ob 245/02h (*Hammock*). Derived from Lenze 2003b with note by S. Lenze.

litigation. Victims claim compensation by alleging that the product is defective as a result of the inadequacy of the product warning that is present.

The Directive is also silent on the matter of when warning information including the safety instructions for use is sufficiently adequate to render a product safe. It follows that a warning can be considered inadequate under the EPLD if its presence affects the product in such a way that it does not provide the safety that the public at large is entitled to expect. It has been noted in the legal literature that adequate warnings allow consumers to assess the product's risks and to avoid them.[178] This resembles the definition that is set out in Article 5(1) of the GPSD.[179] In addition, Article 6(2) of the Directive stipulates that a product shall not be considered defective for the sole reason that a better product has subsequently been put into circulation. Applying this rule to product warnings, it indicates that the Directive does not require a warning to be designed in the best possible way. Liability is not imposed only because a better alternative is available. In line with this is the consideration that the defectiveness of a product should be judged according to a minimum standard.[180]

As noted earlier, it can be said that there is no or no meaningful difference between the determination of warning inadequacy claims brought under the Directive and under fault-based liability. When answering the question whether a given warning renders the product non-defective, fault-based considerations such as whether the producer exercised reasonable care in designing a warning for a specific risk or adopted insufficient care because the warning could and should have been designed in another, better way, continue to be of relevance I believe.[181]

A non-exhaustive list of factors for assessing the adequacy of a warning can be derived from various Member State court decisions and European academic literature. An outline of the cases is provided in the next paragraph. These circumstances are:

- probability that a product danger emerges;[182]
- the severity of harm arising from that;[183]
- nature of the product hazard;[184]
- noticeability of the information;[185]

---

178 Howells 2005b; Hodges 1993b, p. 100.

179 See paragraph 2.4.3.2.

180 Miller & Goldberg 2004, p. 464; Grubb & Howells 2007, p. 377. See [2000] *PIQR* 95 (*Worsely v. Tambrands Ltd*).

181 See paragraphs 2.3.4.3 and 2.4.2.

182 See *e.g.* Hof 's-Hertogenbosch 15 mei 2007, *LJN* 2007, BA6838 (*Nagelstyling*); [2000] *PIQR* 95 (*Worsely v. Tambrands Ltd*).

183 See *e.g.* Hof 's-Hertogenbosch 15 mei 2007, *LJN* 2007, BA6838 (*Nagelstyling*); [2000] *PIQR* 95 (*Worsely v. Tambrands Ltd*).

184 See *e.g.* Hof 's-Hertogenbosch 15 mei 2007, *LJN* 2007, BA6838 (*Nagelstyling*).

185 See *e.g.* OLG Celle 29 January 2003, 9 U 176/02, *VersR* 2004, 964 (*German mixed concrete*); [2000] *PIQR* 95 (*Worsely v. Tambrands Ltd*).

- legibility of the information;[186]
- unambiguous, understandable language;[187]
- complete information (*e.g.* about hazard, consequences, precautions);[188]
- location;[189]
- language;[190]
- knowledge/expertise of the product user.[191]

A number of guidelines can be deduced from legal literature and case law. It is generally agreed in the literature that this adequacy depends on factors that influence the form and content of a warning. A warning should be specific about the risk(s) involved so that the user is aware of the extent and nature of the risk he is exposed to.[192] In fault-based liability, it is generally agreed that the greater the likelihood of damage, the more comprehensive the warning that will be required.[193] This guideline is also instructive for the assessment of warning adequacy under the Directive. Hence, risks that differ in size need different warnings. Besides, it can be said that warning messages must be conspicuously designed so that they have enough impact on users.[194] Furthermore, warning messages need to be legible and comprehensible to be considered adequate. This latter requirement entails among other things that in the event that the provided warning information is too vague for product users to understand how to use the product safely, the warning renders the product defective.[195] In addition, the location of the warning on the product might play a role in the balancing process of whether the warning renders the product defective.[196] Also, the absence of symbols or other representations and the inadequacy of their content can

186 See *e.g.* [2000] *PIQR* 95 (*Worsely v. Tambrands Ltd*); OLG Celle 29 January 2003, 9 U 176/02, *VersR* 2004, 964 (*German mixed concrete*).

187 See *e.g.* Rb. Zwolle 24 april 2002, *Praktijkgids* 2002, 5921 (*Mini-tampon*); [2000] *PIQR* 95 (*Worsely v. Tambrands Ltd*); Hof 's-Hertogenbosch 15 mei 2007, *LJN* 2007, BA6838 (*Nagelstyling*); OLG Celle 29 January 2003, 9 U 176/02, *VersR* 2004, 964 (*German mixed concrete*); [2006] *EWHC* 1284 (QB) *All ER* (D) 86 (Jun) (*Klunk Klip device*); OGH 5 December 2002, 2 Ob 249/02k (*Stepladder*).

188 See *e.g.* Rb. Middelburg 13 juli 2005, *JA* 2005, 104 (*Betonmortel*); Rb. Zwolle 24 april 2002, *Praktijkgids* 2002, 5921 (*Mini-tampon*); 5 OGH December 2002, 2 Ob 249/02k (*Stepladder*); OLG Celle 29 January 2003, 9 U 176/02, *VersR* 2004, 964 (*German mixed concrete*).

189 See *e.g.* [2000] *PIQR* 95 (*Worsely v. Tambrands Ltd*); OLG Celle 29 January 2003, 9 U 176/02, *VersR* 2004, 964 (*German mixed concrete*).

190 See *e.g.* Hof 's-Hertogenbosch 15 mei 2007, *LJN* 2007, BA6838 (*Nagelstyling*).

191 See *e.g.* Rb. Middelburg 13 juli 2005, *JA* 2005, 104 (*Betonmortel*); Rb. Zwolle 24 april 2002, *Praktijkgids* 2002, 5921 (*Mini-tampon*); OLG Celle 29 January 2003, 9 U 176/02, *VersR* 2004, 964 (*German mixed concrete*).

192 Howells 2005b, p. 161.

193 Miller & Goldberg 2004, p. 463; Hodges 1993b, p. 108. See *e.g.* Hof 's-Hertogenbosch 15 mei 2007, *LJN* 2007, BA6838 (*Nagelstyling*).

194 Grubb & Howells 2007, p. 375. See *e.g.* [2000] *PIQR* 95 (*Worsely v. Tambrands Ltd*).

195 Grubb & Howells 2007, p. 374. See *e.g.* Rb. Zwolle 24 april 2002, *Praktijkgids* 2002, 5921 (*Mini-tampon*) and [2006] *EWHC* 1284 (QB) *All ER* (D) 86 (Jun) (*Klunk Klip device*).

196 Grubb & Howells 2007, p. 376. See *e.g.* [2000] *PIQR* 95 (*Worsely v. Tambrands Ltd*).

be brought forward as a circumstance that affects defectiveness.[197] The type of consumers who is likely to use the product and their characteristics, such as knowledge level and language skills, can be of relevance for the adequacy of the content and format of a warning.[198] Finally, as was previously discussed, (non)compliance with public product safety provisions or voluntary standards can influence the judgement of defectiveness.

The question whether there is a safer alternative available in terms of a change of the intrinsic design of the product is a factor that should also be considered when assessing whether the product can be considered defective, even if there is a warning present. Warnings only may not always be sufficient to reduce the risk. This relates to the viewpoint that warnings should not be used to preclude liability for flawed design. If the risk could easily be avoided by adopting a safer alternative design, then the failure to do that can render the product defective.[199] This notion could be derived from the GPSD which stipulates that the presence of warnings does not exempt a person from compliance with the other product safety requirements. Moreover, Burton J argued in the *Hepatitis C* case that this could also follow from Article 12 EPLD which states that product liability cannot be limited or excluded by producers.[200] Hence, having provided a warning is not sufficient to prevent liability. It is decisive whether the quality of the warning itself is sufficient in respect of the risk.[201]

> Dutch *Betonmortel* case
> The Dutch claimant owns an agricultural business and ordered 'betonmortel', *i.e.* concrete fluid, from defendant to concrete a part of his yard. Claimant severely scalded his knees and legs after he had used the liquid concrete. The victim underwent several skin transplantations. There was a warning that consisted of a symbol of a cross with the word 'irritating' beneath it.
> The claimant argued that the defendant is liable on the basis of Article 6:185 DCC for the damages he suffered to his knees and legs, because the scalded knees and legs are a consequence of using the liquid concrete. Claimant advanced the argument that defendant failed to warn against the irritating characteristics of concrete. This was not known to him. He argues that if the use of concrete fluid entails that certain precautions need to be taken, the defendant should have informed him of this. The defendant put forward in his defence that it is an inherent, generally known and accepted fact that concrete

197 Grubb & Howells 2007, p. 375.

198 Grubb & Howells 2007, p. 375, 377; Hodges 1993b, p. 108. See *e.g.* Rb. Middelburg 13 juli 2005, *JA* 2005, 104 (*Betonmortel*); [2000] *PIQR* 95 (*Worsely v. Tambrands Ltd*); Hof 's-Hertogenbosch 15 mei 2007, *LJN* 2007, BA6838 (*Nagelstyling*).

199 Grubb & Howells 2007, p. 476. See *e.g.* the German *Floor panel stripper* case below.

200 [2001] 3 All ER 289 (*A v. National Blood Authority*), para. 65.

201 Grubb & Howells 2007, p. 379.

has inherently dangerous characteristics such as skin irritation and burns. Hence, according to the defendant, claimant should have known that it is necessary to wear special, protective clothing. Instead, the victim wore just jeans and gloves. In the opinion of the producer the warning was indeed sufficient.

The Dutch District Court held to the contrary and ruled that defendant had not adequately fulfilled his duty to warn under Article 6:186 DCC (defectiveness standard). The District Court found that a user of concrete, especially an unprofessional one like the victim, should not have to conclude from the warning that special, protective clothing is required.[202]

Dutch *Mini-tampon* case

An inexperienced sixteen-year-old girl mistakenly inserted a mini-tampon into the urethra after she applied some Vaseline on the tampon to facilitate the insertion. The directions for use described among other things that the vagina lies between the urethra and the anus and that the tampon is adequately inserted 'when you do not feel it anymore'. There were also pictures included of the female genital area and of the lower part of the body showing a correctly inserted tampon. An action for damages is instituted against the manufacturer.

The parents of the girl held the producer liable on the basis of fault-based liability or the provisions implementing the EPLD. Claimant based the claim that the product is defective on the assertion that there were no specific instructions for use that contained warnings or directions for the intended target population which has no or little experience with inserting tampons. Claimant also alleged that the user instructions did not describe that in the event that there is pain, it is possible that the tampon has not been inserted in the vagina. The claim was also based on the allegation that the tampon is unsafe and defective because its pointed shape and small size make it possible to insert the tampon in the urethra.

The Dutch District Court of Zwolle dismissed the claim. In the opinion of the court, the producer cannot be blamed with respect to the content of the instructions for use and there is thus no defect in the presentation of the product, as referred to in Article 6:186(1a) DCC. The District Court ruled that the first assertion does not hold, because these instructions for use are preeminently addressed to women who have no or almost no experience with inserting tampons. The instructions are clear with respect to the opening where the tampon needs to be placed. Furthermore, it is unlikely that even inexperi-

202 Rb. Middelburg 13 juli 2005, *JA* 2005, 104 (*Betonmortel*).

enced women need more specific instructions, because apart from the pain, it becomes immediately apparent that a tampon is wrongly inserted as it does not achieve its intended effect. The court also ruled that providing the information, that in case of pain, the tampon may not be inserted into the vagina, is useless. Once the pain occurs, the tampon has already been wrongly inserted. It follows from the printed information that the tampon is adequately inserted 'when you do not feel it anymore', that in case of pain the tampon is wrongly inserted. As regards the claim of a defective design, the court concluded that the tampon cannot be regarded as defective since the way in which the tampon was inserted is not reasonably to be expected.[203]

Dutch *Nagelstyling* case

The *Nagelstyling* case of the Court of Appeal of 's-Hertogenbosch also deals with the adequacy of a warning label. During March 1999-July 1999, claimant followed the course 'Creative Nail Design Acryl' offered by the defendant called the Beauty Company once or twice a week. The defendant distributes hand and nail care products and provides courses in 'nagelstyling', *i.e.* nail styling. The products that contain acryl are accompanied by a warning that makes mention of the irritating characteristic of acryl and associated safety precautions. After that period claimant opened a nail studio. During October-November 1999, claimant suffered injury to her hands and was later diagnosed with having eczema caused by an allergic reaction to acryl. In consequence, claimant closed her studio down.

The victim alleged that her allergy for acryl is caused by the products of the Beauty Company and she held the defendant liable for the damages suffered by her as a result thereof, because the products are defective and because the Beauty Company had acted wrongfully by failing to warn or failing to warn adequately against the risks associated with the use of the products.

In its interlocutory judgement, the District Court concluded that the Beauty Company has a duty to warn and that defendant failed to warn adequately for the risk of developing an allergic reaction to acryl. Furthermore, the District Court instructed the defendant to prove the statement that the causal connection between the absence of an adequate warning and the damage suffered by the victim was missing. In a second interlocutory ruling, the District Court reached the decision that the defendant failed to comply with the order to produce the evidence.

203 Rb. Zwolle 24 april 2002, *Praktijkgids* 2002, 5921 (*Mini-tampon*).

The Court of Appeal squashed the decision of the District Court and rejected the victim's claim. The Court of Appeal first stated that for both legal bases, that is to say product defectiveness and wrongful behaviour, it revolves around the question whether, taking all relevant circumstances into account, defendant's products cause an unsafe situation/danger during normal use and if so, whether there is adequate warning against the unsafe situation/danger.

The Court of Appeal ruled as follows. It dismissed defendant's ground that there is no duty to warn. The court agreed with the District Court's decision that claimant furnished sufficient evidence that acryl is a substance that has the risk of causing an allergic reaction to users who use it frequently. According to the Court of Appeal, the fact that the nature and/or concentration of the amount of acryl in the defendant's products pose a lesser risk and the circumstance that there is a small number of cases in which the risk will manifest during product use do not lead to the conclusion that no warning against the risks associated with the use of acryl should be provided at all. The Court of Appeal also took into consideration that a warning is not a major measure and that such a measure can prevent far-reaching adverse consequences in a fairly simply way.

Furthermore, it ruled that the nature and extent of the risks attached to the defendant's products are of relevance for determining the nature of the warning that can be expected of the defendant. On the basis of the evidence produced by the claimant, the Court of Appeal ruled that it cannot be concluded that the risk of developing an allergy associated with the use of defendant's products is such that there should be a specific warning against this on the label of the products and that it was not sufficient to only mention the irritating characteristic of the chemical and the related precaution statements of 'Avoid skin contact' and to 'Discontinue use immediately if redness or other allergic symptoms occur'. As a result, the Court of Appeal ruled that there is insufficient reason to consider the given warnings inadequate. The fact that a number of the labels on the defendant's products were not in Dutch text is not relevant for the judgement that the warnings are inadequate, since claimant had in any case taken note of the warnings when the text of the label was written in Dutch. The Court of Appeal also noted that the District Court's conclusion with regard to the burden of proof of the causal connection must be reversed, since it is in principle the claimant that has the obligation to furnish facts and to produce evidence on both the defectiveness of the products and the causal connection.[204]

204 Hof 's-Hertogenbosch 15 mei 2007, *LJN* 2007, BA6838 (*Nagelstyling*).

English *Tampon* case

This English case of *Worsley v. Tambrands* also concerns the adequacy of warnings on a tampon box and in an enclosed leaflet. Here, the claimant suffered toxic shock syndrome due to the use of the tampons. The woman lost the leaflet and thought the symptoms were the result of food poisoning. The package contained a warning of the risk of toxic shock syndrome. The warning on the box stated that there is an association between toxic shock syndrome and tampon use and directs the menstruating woman to the internal leaflet for full details. The leaflet described the syndrome's symptoms and advised users to contact a doctor in the event of experiencing any of these symptoms. The United Kingdom leaflet was multilingual. The information was set out in columns of relatively small print in blue type on white. There were four columns on the page which contained the English language warning. The symptoms of TSS were in bold type.

The claim was brought under negligence and under the strict liability regime of the Directive. No distinction was made between negligence and strict liability. The victim argued that the tampon was defective because there should have been printed full details on the packaging rather than on the enclosed leaflet, as the defendant ought to have foreseen that the internal leaflet might not be kept and/or read. Worsley also alleged that the warning in the leaflet was not designed in such a way as to have a sufficient impact on her. She referred to American warnings of toxic shock syndrome that were more prominent, fuller and that would have such an impact.

The claim failed. Judge Ebsworth held that the warning on the box and in the leaflet are adequate because the defendant had done what a menstruating woman was, in all the circumstances, entitled to expect. The way in which the warnings were given, *i.e.* one on the box and a full explanation in the leaflet, was sufficient in view of the seriousness of TSS and its small probability. Furthermore, the defendant had provided a clearly legible warning on the box and the leaflet was legible, literate and unambiguous. It described the symptoms and advised the action to be taken should they occur. The judge also ruled that even though the design of the American warnings is better than the version of the United Kingdom, that is not the issue here.[205]

English *Klunk Klip device* case

In this English case, the claimant brought her claim under the Consumer Protection Act 1987, which implements the Directive in the United Kingdom. The

205 [2000] PIQR 95 (*Worsely v. Tambrands Ltd*).

injuries were suffered when a product was used incorrectly by a consumer. The consumer was driving a vehicle and lost control after suffering an epileptic fit. The car crashed and the driver was killed. The claimant was a six-year-old child at that time and she sustained severe brain damage in that accident. The claimant had been sitting in the front seat of the car and her head sustained a major impact with the side panelling of the car. She had been wearing a seat belt with a 'Klunk Klip' device attached which had been designed to introduce slack into the belt in order to maximise comfort to the wearer. The device was produced by the defendant and was sold with instructions for use. In this case, however, the users did not have access to the instructions because they had purchased the vehicle second-hand, and the device was already installed.
This English High Court accepted liability and held that the 'Klunk Klip' has a tendency to induce some members of the public to introduce excessive slack. The High Court found it likely that in using the device, a member of the public would introduce more slack than appropriate, thus converting the seat belt from an inertia reel seat belt into a potentially dangerous static seat belt. The court held that the device was liable to misuse even if the instructions were available since the device compromised the actual operation of the seat belt. The instructions were regarded as incomplete as they failed to notify the user to disengage the 'Klunk Klip' at the outset of attaching the device. Although the instructions warned against 'excessive slack', the term was not further explained which gives rise to life threatening situations.[206]

Austrian *Stepladder* case

In this Austrian case, the claimant argued that his stepladder was defective under the Austrian Product Liability Act because it was poorly designed and not supplied with sufficient warnings. To position the ladder, the legs needed to be pulled apart at their lower ends so that the stepladder formed a triangle. The front and the rear section of the stepladder were connected with nylon cords which, once the legs were pulled apart, secured the correct positioning of the ladder and helped stabilise it. The stepladder was also accompanied by an instruction leaflet that contained a picture displaying that the nylon belts should be fully extended before the ladder was used. The claimant suffered several injuries falling down the ladder when the ladder collapsed during his climb up. Before claimant climbed onto the ladder, he had failed to check if the two nylon cords were fully extended.

206 [2006] *EWHC* 1284 (QB) *All ER* (D) 86 (Jun) (*Klunk Klip device*). Reported in *EPLR*, with note by R. Freeman, see Freeman 2006.

Both the Court of First Instance and the Court of Appeal found that the stepladder was not defective. As regards the design of the product, the Austrian Court of Appeal reasoned that the ladder was safe enough, even though nylon belts were not the perfect accessories to guarantee the ladder's stability. The mere fact that there were safer products on the market did not necessarily lead to a finding of defectiveness. The price of the product was an important factor to take into account and this stepladder was a product in the lower price range. As regards the warning claim, the appellate court rejected the argument that the instruction leaflet had failed to adequately warn consumers of the risks incumbent in the wrong positioning of the ladder. It held that there is no obligation to instruct and warn if consumers generally know of the risk. The leaflet was sufficient enough. It was not necessary to describe all possible risks in great detail especially since the ladder was a standard (common) product. Consumers were, according to the appellate court, able to draw the necessary conclusions from the picture.

The Austrian Supreme Court allowed the claimant's appeal and reversed the judgement. The Supreme Court did not address the matter of the design defect. It refrained from referring the case back for further taking of evidence as it held that the stepladder in any event contained a warning defect. The Supreme Court held that even if consumers could not expect a safer design of the product, they should at least be told what consequences there might be if the nylon cords were not fully extended. The Supreme Court argued that the picture did not display the full magnitude of the specific risks as expressly as it possibly could, because although consumers may recognise the instability of the ladder if the belts were not fully extended, they would not necessarily know that the nylon cords could rip out of their fastening, or that the ladder could collapse. The Supreme Court also ruled that there was a quarter of contributory negligence on the part of the claimant.[207]

German *Mixed concrete* case

This German case deals with the extent of manufacturers' obligation to warn under tort law. The circumstances of this German case have some similarity with those in the Dutch *Betonmortel* case. A do-it-yourself (DIY) worker had sustained knee injuries from using ready mixed concrete from the defendant who delivered to commercial and private customers. The following information was present on the delivery note that was given to the victim: 'Ready mixed

207 OGH 5 December 2002, 2 Ob 249/02k (*Stepladder*). Reported in *EPLR* with note by S. Lenze, see Lenze 2003a.

concrete is alkaline, therefore protect skin and eyes! On contact wash thoroughly with water! On contact with eyes call doctor immediately!'. An orange coloured warning sign appeared next to the word 'irritant'. The claimant wore work clothing in order to prepare and even out the concrete with a straight edge. Within a few minutes he suffered severe pain in the knee area and therefore took off his clothing. Due to the irritant and corrosive effect of the ready mixed concrete he sustained severe skin lesions. The claimant instituted proceedings and alleged that the instruction was not adequate. The letters of the word 'irritant' were too small and the word was not distinguishable from the remainder of the text. The victim also argued that the warning sign was similar to the defendant's company logo and the instructions needed to be on a separate sheet.

The Court of First Instance dismissed the claim on the ground that the defendant fulfilled its warning obligation with the printed warning instruction on the delivery note. On appeal, the Higher Regional Court also found the defendant not liable. The defendant had warned of the existing dangers associated with the product and had made it sufficiently clear how the user needed to protect himself. The Higher Regional Court ruled that the graphical design of the warning instruction was adequate. It did not need to be on a separate sheet. Because the warning sign and instructions were on the delivery note, it gave an increased assurance that users would notice it. The court judged that if the delivery note and the instruction were separated this would pose an increased risk that the claimant would not read the instruction, particularly in view of the fact that ready mixed concrete requires quick processing. The delivery note contained only the most vital information and the instruction was so clearly highlighted that it immediately caught one's eye. It did not disappear in a multitude of handling instructions or advertisements and due to the three succinct sentences each ending with an exclamation mark was easily legible. The other allegations of the claimant also failed. In addition, the court found the content of the warning adequate. The terms 'alkaline' and 'irritant' made clear that the concrete can be dangerous. The instruction 'skin and eyes to be protected' also clearly explains why the user should avoid contact with the product. The instruction did not need to expressly require the use of impermeable protective clothing. It was evident that clothing would only provide protection if it were impermeable. A warning of the consequences of failing to use the concrete properly was not required according to the court. The court stated that the product is not a standard consumer product, but rather a material used by commercial consumers and DIY workers and this target group must have had some previous knowledge of the material they were using.[208]

German *Floor panel stripper* case

The claimant who was an employee at a furnishing company had brought his action under the German Product Liability Act implementing the Directive. He suffered injuries to his fingers when he was using the floor panel stripper produced by the defendant during work. The floor panel stripper was designed to remove an old layer of double coated floor panels. The panel needed to be inserted into a tray, where it would be taken up by a roll guiding the panel inside the machine. The double coated floor panel was subsequently transported to a blade which would shave off the upper of the two floor panels. The manual instructed the operator to press the emergency button whenever irregularities would occur and never to put a hand inside the machine whilst in operation. The manual also warned against the danger of crushing one's fingers under the moving roll if the instructions were not followed. The claimant got hurt when he tried to press a panel under the roll with his fingers. He pressed the panel because the roll did not take up the panel. At the time of the accident the machine was not equipped with a protection device to prevent the user from getting his hands under the roll. After the accident, the defendant added such a protective device to the stripper.

The District Court of Düsseldorf found the defendant liable on the grounds of a defective design. The court held that the safety expectations that a consumer is entitled to expect are defined by the relevant product safety regulations, here the European Machinery Directive. Article 1.3.7 of the Annex of the Machinery Directive lays down that moving parts of machinery must be designed to prevent all risk of contact which could lead to accidents.[209] The court stated that the producer had breached this article, because the roll can be considered a hazardous moving part in terms of the Machinery Directive. Furthermore, the court held that the product is also defective in design because the risk could have been reduced by adding a simple protection device. The court went on to hold that warnings against dangers of misuse do not clear the product of its design defect, since warnings do not replace necessary design features. The

208 OLG Celle 29 January 2003, 9 U 176/02, *VersR* 2004, 964 (*German mixed concrete*). Reported in *EPLR* with note by M. Hannes, see Hannes 2004; Grubb & Howells 2007, p. 376.

209 Directive 2006/42/EC of the European Parliament and of the Council of 17 May 2006 on machinery, and amending Directive 95/16/EC (*OJ* 2006, L 157/24).

court reduced the damages on the basis of contributory negligence, since the claimant failed to follow the instructions in the manual.[210]

2.4.6 Conclusion

The final paragraph 2.4 aimed at presenting a consensus on product warning requirements for accepting liability under European product liability law. This was done on the basis of studying case law and legal literature. It started with the interpretation of the defectiveness test for warning claims. For both claims, it was argued that it can be questioned whether there is really a difference between establishing defectiveness and wrongfulness on the issue of the absence or inadequacy of warnings. Relevant circumstances, guidelines and case law pertaining to the claim were discussed accordingly.

Because of the wealth of product safety legislation that the European legislator has enacted, paragraph 2.4.3 first provided a short introduction into the general public law product safety provisions with regard to food and non-food products. Secondly, it touched upon the nature of the specific warning requirements for toys and products that contain chemicals. Thirdly, this paragraph discussed the influence that these mandatory requirements and non-binding standards can have on the assessment of the defectiveness requirement. Whether the product is in accordance with European public safety provisions can be viewed as a relevant factor. The applicable obligations are usually described in general terms and less often in detail. Consumers are entitled to expect that the product conforms to applicable public product safety legislation. However, the mere fact that a producer complied with public law warning requirements is not an automatic defence against product liability; this requires a balancing of all the relevant circumstances. Moreover, it must be borne in mind that the provisions of public law are minimum requirements of safety. The obligations of producers under civil liability can require the adoption of a higher level of safety.

One of the defences of the EPLD relates to compliance with public law. It basically states that in the event that the finding of product defectiveness can be attributed to this compliance, a producer is freed from liability. Because this defence has a narrow scope, it will rarely be invoked successfully.

The next paragraphs addressed the warning claim centred on the absence of the warning under the Directive (paragraph 2.4.4) and the warning claim centred on the adequacy of the warning under the Directive (paragraph 2.4.5). A non-exhaustive list of factors for assessing the absence as well as adequacy of a warning was made up on the basis of an inventory of Dutch case law and a number of European cases. To determine whether a

210 Landgericht Düsseldorf 30 November 2005, 10 O 144/04, *NJW-RR* 2006, 1033 ff (*Floor panel stripper*). Derived from and reported in *EPLR*, with note by S. Lenze, see Lenze 2006.

product is defective as a result of the absence of a warning, circumstances that have shown to be of relevance are: the intended and reasonably expected use of the product; the time that the product was put into circulation; probability that a product danger emerges; the degree of harm arising from that; the nature of the product hazard; the burden/costs of implementing a warning or a design change; a hidden or obvious risk; general knowledge of the risk; the degree of certainty with respect to the health hazards associated with product use; (non)compliance with public product safety provisions or voluntary standards; and the utility of the product. The circumstances that can be of relevance to assess whether an inadequate warning renders the product defective are: probability that a product danger emerges; the severity of harm arising from that; the nature of the product hazard; noticeability of the information; legibility of the information; unambiguous, understandable language; complete information (*e.g.* about hazard, consequences, precautions); location; language; and knowledge/expertise of the product user. The final paragraph also noted that the presence of a warning is not always sufficient to reduce the risk. A product can still be considered defective, even if there is a warning of the risk, when the intrinsic design of the product could have easily avoided the risk. This was demonstrated in the German *Floor panel stripper* case. This view that warnings must not be used to cover up a design flaw is also supported in the legal literature.

2.5 SUMMARY

In this closing paragraph, the conclusions of the previous paragraphs are grouped together to give an outline.

Product Liability Based on Fault

Paragraph 2.2 first explained the general ground of fault-based liability of Article 6:162 DCC, which basically consists of the element of wrongful conduct of the wrongdoer (*e.g.* the producer) and imputability of the act to the wrongdoer. The other requirements pertain to the damage sustained and the causal connection between the damage and the wrongful act. According to paragraph 2 of Article 6:162 DCC three sorts of wrongful acts can be distinguished: (1) conduct which infringes a person's right; (2) conduct contrary to a statutory duty; and (3) any act or omission which violates a rule of unwritten law regarding a proper social conduct. The latter category is the most popular one. Within this category, the most important subcategory of wrongfulness concerns the unwritten safety standards that are applicable in situations of endangerment. Endangerment refers to creating a dangerous situation that may harm people or goods or allowing such a situation to continue.

Paragraph 2.2.2 described the assessment of wrongful endangerment in more detail. Of essence is the landmark *Kelderluik* decision of the Dutch Supreme Court in which it was held that the wrongfulness of breaching safety standards depends on a balancing test of the circumstances of the case. This decision represents the classical example of the balancing process that is needed to determine whether conduct, such as in case of endangerment, is wrongful. It concerns whether the level of precaution taken by the wrongdoer was sufficient in view of the level of risk. In this respect, it has been argued in the literature that the *Kelderluik* standard is the Dutch equivalent of the American risk-utility test of Judge Learned Hand to determine negligence. The Dutch Supreme Court provided four circumstances that should be considered when assessing whether there is wrongful endangerment: the probability that potential victims are careless and inattentive; the likelihood that this leads to accidents; the gravity of the consequences of the accident; and the costs of taking precautionary measures. These are termed the *Kelderluik* factors and they have been cited by the Dutch Supreme Court and lower courts on a regular basis ever since. The factor of the cost of taking precaution relates to whether, in the absence of the precautionary measure, the measure was possible to take and customary as well as the inconvenience in terms of time, effort and financial costs of adopting the measure.

There are several ways to exercise care and providing a warning is one means of taking precaution. As a result, whether there is a duty to warn should be assessed on weighing the concrete circumstances of the case, such as those mentioned above. The duty to warn was addressed in paragraph 2.2.2.3. Dutch case law indicates that for determining whether there is a failure to warn additional factors can be of relevance, such as the obviousness of the risk, knowledge of the risk, and the improbable behaviour of potential victims. In addition, the above-mentioned factor of the cost of providing a warning has also been shown to be of importance for assessing a failure to warn. It was noted that courts generally consider providing warnings a simple and easy way of exercising due care and therefore often not too high of a cost to adopt. In cases where a warning was given yet damage occurred, liability commonly revolves around answering the question whether the adopted measure of the warning can be considered sufficiently adequate. An important ruling of the Dutch Supreme Court governing the adequacy of warnings under Article 6:162 DCC is the recent *Jetblast* case, which holds that to answer the question whether a warning can be considered an adequate measure to protect against a certain danger, it is of decisive importance whether it is to be expected that the warning will lead to an act or omission as a result of which the danger is avoided. Warnings are often not the only measure that can be adopted to prevent a specific risk. In this regard, the Dutch Supreme Court has ruled in the *Multivac-machine* case that the mere inclusion of a warning onto a dangerous machine will in principle not be sufficient to avoid liability. Employers should first investigate whether the safety of the machine's design can be improved by other more effective measures.

Following the discussion on wrongful conduct in general, paragraph 2.2.4 discussed the wrongful conduct of producers. The wrongfulness of the conduct of producers should be assessed on the basis of weighing the relevant circumstances, including the *Kelderluik* factors. Unwritten obligations following from Dutch case law that pertain to how reasonable producers should act when designing products were described. One of them concerns the *Halcion* case of 1989. In this case the Dutch Supreme Court provided guidance to courts by using the defectiveness standard of the EPLD to determine whether the producer of the medicine Halcion was liable under Article 6:162 DCC. Hence, it was concluded that the standard of defectiveness has been declared applicable to assess the conduct of producers under fault-based liability. Note, however, that fault-based liability also requires imputability of the act to the producer. The *Halcion* ruling has additional bearing. The Dutch Supreme Court established a producer's duty to warn. It listed several circumstances which may be of importance when assessing the wrongfulness of a producer's conduct. These boil down to the utility of the product, the seriousness of the injury, the probability of injury and the cost of an alternative reasonable design of the product. Hence, these are of a similar nature as the *Kelderluik* factors. Another important decision in this field concerns the *Koolhaas/Rockwool* judgement in which the Dutch Supreme Court ruled that a producer is obliged to take those measures which can reasonably be required from him in his capacity of a careful producer in order to prevent damages caused by his product. Furthermore, the Dutch Supreme Court held in the *Warmwaterkruik* case that producers must not only take into account the users who take all necessary precautionary measures during product use to prevent potential damage, but also those people for whom the product is intended but who fail to take the appropriate precautionary measures.

Product Liability without Fault: Directive 85/374/EEC

Paragraph 2.3 dealt with the strict liability regime of the EPLD, especially its main liability requirement of defectiveness. After describing the basics of the EPLD in paragraph 2.3.2, 2.3.3 further explained the important requirement of defectiveness (Article 6). It stipulates that there is a product defect when the product does not provide the safety which a person is entitled to expect. All circumstances need to be taken into account. Three are mentioned specifically: (a) the presentation of the product; (b) the use to which it could reasonably be expected that the product would be put; and (c) the time when the product was put into circulation. These were discussed in paragraph 2.3.3.1. In addition, the paragraph explained that the standard of defectiveness must be measured objectively. It is neither the safety expectations of the particular injured person nor of the particular producer that is decisive, but the degree of safety persons generally are entitled to expect. Another important remark made is that the test concerns a level of product safety that the public at large are entitled to expect, rather than what they actually expect. This paragraph also paid attention to the flaws of the defectiveness test that were brought up in the legal literature. One important flaw is that the standard of defectiveness does not answer the question in regard to how much safety the general public is entitled to expect. Furthermore, the test can be problematic to utilise with regard to products that have obvious risks or generally known risks, because such risks influence the safety expectations of people and can lead to the conclusion that persons are not entitled to expect a higher level of safety.

After that, paragraph 2.3.4 described in more detail why it has often been said in the literature that the Directive's regime is a strict liability regime with elements of fault. Important elements that are prone to introducing fault are the defectiveness requirement and the optional development risk defence. The degree in which they 'contaminate' the strict liability system depends on their interpretation. It was shown that the development risk defence has been interpreted narrowly by the ECJ as a result of which the possibility to escape liability for undiscoverable risks has become small. It is worthy to note that there is still some room left for a fault-based type of interpretation of the defence as a result of the requirement added by the ECJ that the scientific and technical knowledge to discover the defect must have been accessible to be liable. As regards the concept of defect, it is topic of debate whether and to what extent the application of the defectiveness requirement diverges from applying the requirement of wrongfulness under fault-based liability. It has been argued in the literature that the defect requirement is of such a nature that it is problematic not to give in to adopting some form of a fault-based analysis to evaluate whether the product is defective and to include factors such as the avoidability of the risk by adopting precautionary measures and the burden of doing that. Nevertheless, there is a minority that considers the use of the latter factors as not permitted, because they inevitably refer to the reasonableness of the conduct of producers. As for the utilisation of the defectiveness test under Dutch law, case law and literature suggest that the tests of wrongfulness and defectiveness are practically identical.

Warning Requirements under European Product Liability Law

Paragraph 2.4 aimed at presenting the consensus on product warning requirements for determining defectiveness in European product liability law on the basis of studying case law and legal literature. Because of the wealth of product safety legislation under public law that the European legislator has enacted, paragraph 2.4.3 first provided a short introduction into the general public law product safety provisions with regard to food and non-food products. Secondly, it touched upon the nature of the specific warning requirements for toys and products that contain chemicals. Thirdly, this paragraph briefly discussed the influence that these mandatory requirements and non-binding standards can have on the assessment of defectiveness. Whether the product is in accordance with European public safety provisions can be viewed as a relevant factor. The applicable obligations are usually described in general terms and less often in detail. Consumers are entitled to expect that the product conforms to applicable public product safety legislation. However, the mere fact that a producer complied with public law warning requirements is not an automatic defence against product liability; this requires a balancing of all the relevant circumstances. Moreover, it must be borne in mind that the provisions of public law are minimum requirements of safety. Civil liability can require the adoption of a higher level of safety. One of the defences of the EPLD relates to compliance with public law which basically states that in the event that the finding of product defectiveness can be attributed to this compliance, a producer is freed from liability. Because this defence has a narrow scope, it will rarely be invoked successfully.

The next paragraphs addressed the warning claim centred on the absence of the warning under the Directive (paragraph 2.4.4) and the warning claim centred on the adequacy of the warning under the Directive (paragraph 2.4.5). For both claims, it was argued that it can be seriously questioned whether there is really a difference between establishing defectiveness and wrongfulness on the issue of the absence or inadequacy of warnings. In addition, relevant circumstances, guidelines and case law pertaining to the claim were discussed accordingly.

A non-exhaustive list of factors for assessing the absence as well as adequacy of a warning was made up on the basis of an inventory of Dutch case law and a number of European cases. To determine whether a product is defective as a result of the absence of a warning, circumstances that have shown to be of relevance are: the intended and reasonably expected use of the product; the time that the product was put into circulation; probability that a product danger emerges; the degree of harm arising from that; the nature of the product hazard; the burden/costs of implementing a warning or a design change; a hidden or obvious risk; general knowledge of the risk; the degree of certainty with respect to the health hazards associated with product use; (non)compliance with public product safety provisions or voluntary standards; and the utility of the product. The circumstances that can be of relevance to assess whether an inadequate warning renders the product defective are: probability that a product danger emerges; the severity of harm arising from that; the nature of the product hazard; noticeability of the information; legibility of the information; unambiguous, understandable language; complete information (*e.g.* about hazard, consequences, precautions); location; language; and knowledge/expertise of the product user. The final paragraph also noted that the presence of a warning is not always sufficient to reduce the risk. A product can still be considered defective, even if there is a warning of the risk, when the intrinsic design of the product could have easily avoided the risk. This was demonstrated in the German *Floor panel stripper* case. This view that warnings must not be used to cover up a design flaw is also supported in the legal literature.

3 Product Warnings and Human Behaviour

3.1 Introduction

Having concentrated on the law in relation to product warnings, the second step that needs to be taken before these two worlds merge in chapter 4 involves discussing the interplay between product warnings and psychology. Hence, as the title shows, this chapter only deals with the theory and research on how humans interact with warnings in real life. No comparisons and conclusions are made with regard to the law. This is done in chapter 4.

Over the past 30 years, a tremendous body of empirical research literature in the design and effectiveness of product warnings has been developed. It largely involves American literature.[1] The main focus of these empirical studies has been on identifying the factors that influence the effectiveness of processing warnings and on providing design implications and guidelines on the basis of the findings. Several articles and books have been published that provide overviews and reviews of the academic literature and research.[2]

As has become clear from the legal chapter on product warnings and European product liability law, warnings play a role in determining product liability. The research results and the literature on warnings is a valuable source of information for framing producers' legal obligation to warn adequately, as it provides explanatory information about important associated warning issues, such as the functions of warnings, the positive and negative effects warnings can really have on humans and the factors that influence whether warnings are effectively processed by humans. These findings can, in turn, offer recommendations with regard to the (re)appraisal of warnings under European product liability law.

In view of this, the objective of this chapter is to present an analysis of the main topics that have received attention in the warning research and literature. To achieve this, the structure of this chapter is built around answering the following questions:

---

1 Three reasons have been brought forward by the warning research literature to explain the increasing interest in research on the topic of warnings, mainly in the United States. First, there has been a growing concern in safety and health, which in part relates to the rising costs of health care. Furthermore, the growth in warning research can also be explained by legal concerns. There has been a growth in mandatory legislation and standards that deals with safety, such as occupational and consumer safety. Moreover, the product liability litigation in the United States that deal with the inadequacy of warnings has been a major spur or impetus for research on warnings. Human factor specialists who are experts on the topic of warnings have been increasingly involved in American litigation in the role of an expert witness, as a result of which interesting researchable questions were identified that led to research in this field, see Wogalter, DeJoy & Laughery 1999b, p. xiii.

2 Important and comprehensive books are Wogalter 2006; Wogalter, DeJoy & Laughery 1999b and the earlier work of Lehto & Miller 1986.

- Which disciplines have mainly contributed to the growing body of empirical studies on product warnings and to our understanding of the warning process? (paragraph 3.2);
- What is a product warning and what is it intended to achieve? (paragraph 3.3);
- How can the warning process be modelled? (paragraph 3.4);
- How has the warning research been conducted? (paragraph 3.5);
- Which factors influence the effective communication and processing of warnings? (paragraph 3.6);
- Which characteristics of warning recipients affect the processing of warning information? (paragraph 3.7);
- What hazards need a warning? (paragraph 3.8);
- When should there be a warning in relation to other design methods? (paragraph 3.9).

Paragraphs 3.3, 3.6, 3.8 and 3.9 discuss key warning issues that can be considered relevant for European product liability law. The other paragraphs fulfil a supportive role to better understand the warning research and the warning process.

The first warning issue is dealt with in paragraph 3.3. To answer the question why warn, this paragraph first explains how warnings can be defined in order to discuss next what purposes warnings intend to have according to the warning research literature.

The fundamental warning issue concerns the factors that underlie the effective processing of warnings by warning receivers and the implications that follow from the warning research literature to design effective warnings. This issue is addressed in paragraph 3.6. Various empirical warning studies are described in detail to show the reader what they actually entail. When reading this paragraph, it is beneficial to have some foreknowledge so that this lengthy discourse on this warning issue can be better understood. First of all, it is desirable to have some general knowledge of how people process information. Several theoretical models have been proposed by researchers to describe and explain the warning process. These are dealt with in paragraph 3.4. Furthermore, background information on the main sciences that are involved in the warning research (paragraph 3.2) and on the methods that warning researchers have employed to achieve empirical findings (paragraph 3.5) are desirable when reading about the warning studies that have been done. Paragraph 3.7 is related to paragraph 3.6 in that it takes a closer look at the characteristics of warning receivers that affect the effective processing of warnings.

Paragraph 3.8 pays attention to the warning issue with respect to what hazards to warn about and the considerations that have been made in this regard by warning researchers.

Paragraph 3.9 discusses another warning issue that has been given attention in the warning research literature, that is the role of warnings compared to other design methods that deal with product hazards. The final paragraph (paragraph 3.10) provides a summary of the previous paragraphs.

A final comment is that many of the paragraphs that follow are organised around the C-HIP model, a popular theoretical model of the warning process.[3] This model is described in more detail in paragraph 3.4. Although the model is a simplified version of how humans process warnings in reality, the model is considered a refined model of the warning process. It has been designed by leading scholars in the field and they have repeatedly used the model to organise the considerable body of warning-related research.

3.2 Main Disciplines of the Warning Research

3.2.1 *Cognitive Psychology*

Research on the design and the effectiveness of warnings has been carried out by various disciplines such as cognitive psychology, ergonomics (human factors), safety engineering, consumer behaviour and marketing, communications, social psychology and so on.[4] As regards product warnings, two of them warrant special attention here.

The first and foremost discipline that can explain how people interact with product warnings is cognitive psychology. Cognitive psychology is an empirical research science that has as its primary goal understanding the basic cognitive processes that underlie the workings of the human mind. The internal processes are attention, perception, learning, memory, language, problem solving, reasoning and thinking.[5]

To study the human mind, cognitive psychologists make use of theories, which can be viewed as a set of related statements proposed to explain the researcher's phenomena of interest. Subsequently, research experiments are set up that allow the researcher to test his theoretical questions. If the results of the research are consistent with the predictions of the theory, the confidence in the theory increases. Contradictory results indicate that the theory needs to be modified or even abandoned.

Cognitive psychologists vary in their approach to studying the internal processes of the mind and the theories they embrace. Nonetheless, they share three basic assumptions. The first assumption is that mental processes exist. Secondly, humans can be viewed as active information processors and thirdly, mental processes can be scientifically investigated through the use of measures.[6]

3 Wogalter 2006b.
4 Stewart & Martin 1994, p. 1; Laughery 2006, p. 467.
5 Eysenck & Keane 2000, p. 1; Payne & Wenger 1998, p. 3.
6 Payne & Wenger 1998, p. 22.

Cognitive psychologists have employed many theoretical models as a tool to try to explain the human mind.[7] One dominant model within cognitive psychology stems from the information processing approach, which proposes that information is processed through a series of cognitive stages, with each stage performing unique operations on the information.[8] Such a model can be used to account for how people take in information, store it, and then use the stored information to make responses.[9] An early version of the information processing model included the stages of attention to a stimulus, perception of the stimulus,[10] thought processes, decision and finally response or action. It assumed that only one internal process occurs before the next begins (serial processing). This assumption has been abandoned because is it grossly oversimplified. Nowadays, it is recognised that human information processing is more complex. Human cognition not only involves bottom-up processing, but also top-down processing, which means that processing is affected by what the individual contributes, *e.g.* his knowledge and expectations, rather than the stimulus itself.[11]

3.2.2 *Human Factors/Ergonomics: An Applied Discipline*

Ergonomics is the second meaningful discipline in relation to product warnings. Ergonomists or human factor specialists have played a crucial role in the development of the research on warnings as a substantial portion of the research has been carried out by them and published in the ergonomics literature.[12] Its main purpose is to design systems, jobs, products, environments that (better) match to the psychical and mental abilities and limitations of people.[13] Since warnings form part of the interface between people using and maintaining a technological product and these interfaces are the domain of ergonomics, its relevance to the design and evaluation of product warnings is not surprising.[14]

Ergonomics has an interdisciplinary nature. Ergonomic professionals come from a variety of academic fields, such as engineering, biomechanics, medicine, physiology, physiotherapy and psychology. These sciences are used to support the ergonomics discipline.

7 A model is a description or an analogy that is used to account for a limited set of phenomena. Models generally account for a rather small range of observations, see Payne & Wenger 1998, p. G-8.

8 Payne & Wenger 1998, p. G-6.

9 Payne & Wenger 1998, p. 23.

10 Perception means the acquisition and processing of sensory information in order to see, hear, taste or feel objects in the world, see Eysenck & Keane 2005, p. 31.

11 Eysenck & Keane 2005, p. 2, cf. Payne & Wenger 1998, p. 24. Bottom-up processing is processing that is directly influenced by the environmental stimuli, see Eysenck & Keane 2005, p. 556.

12 Laughery 2006, p. 467; Laughery & Wogalter 2008.

13 Helander 2005, p. 1-2; McCormick & Sanders 1992, p. 4.

14 Laughery 2006, p. 467.

In other words, interdisciplinary knowledge is applied in the design activities of ergonomists. This includes knowledge from cognitive psychology.[15]

An important approach within this discipline is the *systems approach*, emphasising the interaction between operator and machine in their environment. A system is the combination of human, technological and environmental components relevant to system performance; the interactions among the human, technological and environmental components; and the interfaces that facilitate or hinder the various interactions.[16] The approach focuses on the human component. Important areas of consideration within the human component entail the anatomical, physiological and psychological capabilities of humans.[17] These need to be accommodated by the technological component. The anatomical characteristics of humans include for example the dimensions of the body, such as the hand size grip for chain saw handles or children's' height, weight, reach and grip for playground equipment, but also strength and application of forces, such as the pulling strength for the starting of mowers. Physiological capabilities refer to how a human's body functions and involves for example our sensory systems, the neurological system, and the immune system. Some example applications for this area are fatigue effects or noise interference during the operation of power saws. Our psychological capabilities concern areas such as information processing, decision making, attitudes and behaviour. An ergonomic application of this is for instance the size, shape and legibility of warning labels or the degree of risk-taking behaviour in the use of roller skates.[18] To obtain system optimisation, the environmental component also needs to be addressed in the design of the system. For example, the climate conditions and the environmental lighting are essential: humidity, high or low temperature or too little (artificial) light can cause the design to malfunction.[19]

According to Sanders and McCormick, human factors/ergonomics has two major objectives. The first objective concerns enhancing the effectiveness and efficiency with which work is carried out. This includes the correlated objectives of increased convenience of use, reduced errors and increased productivity. The second objective concerns enhancing certain desirable human values, such as increased safety, reduced fatigue and stress, increased comfort, greater user acceptance, increased job satisfaction and improved quality of life.[20]

Helander provides a less extensive list of similar objectives and argues that ergonomics is a design methodology that is used to arrive at the objectives of safety, productivity and operator satisfaction. The objective of safety can be assessed by comparing the performance

15 Helander 2005, p. 1-4; Payne & Wenger 1998, p. 16; see also Dirken 2004, p. 27.
16 Bakken 2005, p. 28-1.
17 Ramsey 1985, p. 115; Bakken 2005, p. 28-3ff.
18 Ramsey 1985.
19 Bakken 2005, p. 28-11.
20 McCormick & Sanders 1992, p. 4.

requirements of the task and the environment in which it is performed with the performance limitations of the operator. For example, when designing a safe system, it is important to take account of the limitations attached to the information processing capabilities of operators during the performance of a certain task or else unsafe product use and consequent injuries are to be expected. Productivity, quality and time to perform a task also influence a system's design. Work with industrial machines that can be done more quickly and with fewer quality errors represents an important characteristic of machinery used in the work place. Operator satisfaction is a goal that can be widely interpreted, such as what makes a task performance comfortable and convenient for operators. The relative importance of these goals varies depending on the system. It is possible to improve safety as well as the quality of production at the same time. Nevertheless, in many cases trade-offs have to be made: improving production could mean comprising safety.[21]

3.3 DEFINITIONS AND PURPOSES OF PRODUCT WARNINGS

3.3.1 *Warnings Are Safety Communications*

Product warnings come in all shapes and sizes. Most are visual and generally contain information about the safety aspects of a product, such as details about the nature of the hazard, the safety instructions on how to avoid the danger, or information about the harmful consequences. The information can be presented visually in text and in the form of pictorial symbols, but it can also be auditory (a sound or a voice warning) or olfactory (in the sense of an odour cue). Warning information can be on the label of a product's package or in a user manual. This chapter will mainly deal with the effects of visual warning messages.

Regardless of the quantity of information provided, the way in which the information is expressed and presented, its location or how it communicates to our senses, it can be said that all warnings are in essence communications about safety. A more elaborate definition of product warnings provided by Wogalter and Laughery is that of a form of safety communications used to inform people about product hazards and to provide safety instructions so that undesirable consequences are avoided or minimised.[22] Wogalter notes that sometimes a distinction is made between warnings and instructions. The reason for this is that there are instances in which usage instructions are not related to safety: the usage instruction 'Use a plastic spoon' is not a safety instruction, contrary to 'Wear rubber

21 Helander 2006, p. 14-16. See also McCormick & Sanders 1992, p. 4, 661; Dirken 2004.
22 Wogalter 2006a, p. 3; Wogalter & Laughery 2006, p. 889, 891.

gloves' which is. Thus, warning messages usually include instructions, but not all instructions are part of a warning.[23]

The warning literature also provides other definitions of warnings. A related definition considers warnings as any information that has the potential to change behaviour and prevent accidents.[24] Similarly, warnings are information that attempts to influence user behaviour through the information presented.[25] Warnings have also been described as specific stimuli that alert a user to the presence of a hazard, thereby triggering the processing of additional information regarding the nature, probability and magnitude of the hazard.[26] Other researchers have stated that warnings are artefacts produced by a designer in relation to a situation or product which has some associated level of risk additional to that provided by – or which the user could be expected to bring to – the situation or product itself. In this respect, a warning is an artefact and a representation of the situation to which they refer.[27]

3.3.2 *Purposes of Warnings*

By drawing a parallel between the various definitions, it follows that their differences and similarities lie in the purposes or functions that have been assigned to warnings. Some put more weight on the ability of a warning to persuade warning users to produce the behaviour intended by the warning, whilst others put the emphasis on the alerting function and the informing function of warnings.

The first chapter of the 'Handbook of Warnings', a chapter that introduces important warning issues that form key themes throughout the various chapters of the book, tells us that warnings have four purposes.[28]

At a general level, warnings are intended to reduce or prevent accidents, health problems, personal injuries and property damage. It is generally agreed in the warning literature that this is the ultimate purpose of a warning.

The other three purposes can be distinguished at a more concrete level and concern the functions of informing consumers, influencing behaviour and reminding consumers. The first function follows from the ultimate goal of reducing or preventing injury. Warnings can be seen as to steer or influence the behaviour of the recipients of the warning in ways that will improve safety. Secondly, warnings are a method for communicating (new) important information about hazards so that people are able to make better, more informed

23 Wogalter & Laughery 2006, p. 900.
24 Ayres e.a. 1989, p. 426.
25 McCarthy e.a. 1982.
26 Lehto & Miller 1986, p. 14, 16.
27 Edworthy & Adams 1996, p. 1-3.
28 Wogalter 2006.

decisions on safety issues. A third purpose of warnings is that they can serve as a reminder to individuals who may already know about the hazard, but are unaware of it when the information is necessary. For example, the warning symbol in vehicles that reminds people to fasten seat belts. By calling into awareness information that may be latent in long-term memory, consumers can use the product safely at the right time.[29]

3.4 Theoretical Models of the Warning Process

3.4.1 Introduction

Throughout the years, a number of different theoretical models of how people interact with product warnings have been proposed. Theoretical models are valuable in more than one way. Using a theoretical framework helps to structure the rapidly growing body of warning research. Moreover, it provides a method to help uncover why a warning message is not effective in changing behaviour, to evaluate the effectiveness of warnings and to help designers develop better warnings.[30]

In this paragraph, three types of models that have gained prominence in the warning literature and that have been applied to warnings are presented, including the warning design implications that follow from these models. The structure of this paragraph is as follows. In paragraph 3.4.2, I deal with several versions of human information processing models. A recent model that merits attention here is the Communication Human Information Processing (C-HIP) model. This model provides a detailed model of how warnings are being processed in combination with a communication model. Other types of models that have gained the interest of warning researchers are human decision making models. This because behavioural (non)compliance with the directions of a warning can be viewed as a decision making process involving a weighing of the costs and benefits of exhibiting some type of behaviour. These are discussed in paragraph 3.4.3. Lastly, paragraph 3.4.4 provides information about a specific model about human behaviour that has originally been used to analyse the role of human error in accidents related to occupational safety. Lehto applied this model of Rasmussen's to the design of warnings. The closing paragraph 3.4.5 provides a conclusion.

29 Wogalter 2006a, p. 4; Laughery 2006, p. 468, Kalsher & Williams 2006, p. 313.

30 Lehto 2006, p. 63; Wogalter 2006b, p. 51.

3.4.2 *Human Information Processing Models*

3.4.2.1 Versions of Information Processing Models

Human information processing models are popular models in the warning literature. An early model that emerged in the warning literature involves the general warning tree information processing model of Lehto and Miller. According to their model, an effective warning must trigger a sequence of eight events. There must be: exposure to the warning, attention to the warning, active processing of the warning, comprehension and agreement with the warning, storage in memory, response selection, and response performance. Intervening events that must not occur if the warning is to be effective include filtering of the message, overload of the sensory channel, disagreeing with the message, forgetting the message, selecting an alternative action, or inability of performing the required response.[31]

Another information processing model that is worth mentioning is the model of Rogers, Lamson and Rousseau, who provide a more general model of the warning process that includes the broad components of noticing, encoding, comprehending and complying. They developed this framework in 2000 to review the warning research literature that had been conducted in the period 1980-1998.[32]

Ramsey has also offered a conceptualisation with regard to the effectiveness of warnings in relation to the prevention of accidents. His accident sequence model is based on an ergonomic approach and depicts the various stages encountered by humans in attempting to avoid an accident with a consumer product. The model begins with the assumption of exposure to some type of product hazard which has the potential to lead to an accident, followed by the stages of perception of the hazard by the human senses, cognition of the hazard, decision to avoid the hazard and finally the ability to avoid the hazard. If these stages are successfully processed, no accident will occur according to the model. This model emphasises the importance of ergonomic principles for the design of safe consumer products. If in each stage of the model, ergonomic areas of consideration are taken into account when designing a consumer product an accident is not likely to happen. To be able to avoid the hazard by carrying out the behaviour that is required by the warning, the consumer product must accommodate various characteristics of humans, such as their anthropometric, biomechanical and motor skill abilities.[33] For example, the warning instruction 'Open Valve First' should be consistent with the strength requirements of the persons required to perform this action.[34]

31 Lehto & Miller 1986; Lehto 2006, p. 67; Lehto & Papastavrou 1993, p. 570.

32 Rogers, Lamson & Rousseau 2000, p. 103.

33 Ramsey 1989; Ramsey 1985.

34 Ramsey 1989, p. 198.

3.4.2.2 The C-HIP Model of Wogalter, DeJoy and Laughery

A model of the warning process that has gained popularity in recent years is the C-HIP model devised by Wogalter, DeJoy, and Laughery.[35] It has recently been modified by Wogalter and is shown in figure 1.[36] The C-HIP model is a social-cognitive model of the warning process that combines a basic communication model with an information processing model.

The C-HIP model provides a framework for showing the stages of information flow from a source to a receiver, whereby the receiver successfully processes the warning information to produce subsequent compliant behaviour. A source (*e.g.* the producer) encodes a message (*i.e.* the product warning) into a channel (*e.g.* visual or auditory) that is transmitted and delivered to a receiver (*i.e.* the product user). As the effectiveness of a warning may also be influenced by characteristics of the communication source and the channel through which the warning is sent, it is of importance to include these components in the theoretical model.[37] Similarly, environmental stimuli, such as low illumination, background noise or the presence of others, can also affect the processing of a warning. If the warning is transmitted and delivered to the receiver, the information must then be mentally processed within the receiver. The receiver stage includes human information processing substages prior to carrying out the desired behavioural response to a warning. These substages are attention switch, attention maintenance, comprehension, beliefs and attitudes, and motivation. In the model, behavioural compliance with the warning is the culmination of these subsequent stages.

At each receiver stage, the warning information needs to flow through to the next stage, but it may produce a bottleneck that blocks the flow before the process ends in behaviour. For example, a warning that is not noticed cannot be read and comprehended and have an impact on behaviour. It must be noted that although the model represents a linear process, the stages interact with each other. Later stages can affect the processing in earlier stages. For example, the belief people hold with regard to the hazardousness of the product's risks can have an impact on whether they will attend to a warning.[38]

The devisers of the C-HIP model have emphasised the benefits of this particular model. First of all, it can function as a framework to organise the substantial body of diverse research findings of the last 30 years. The model can also aid in making predictions about the effectiveness of warnings and guide the warning design process. Moreover, the model

35 Wogalter, DeJoy & Laughery 1999a; Wogalter 2006.

36 Wogalter 2006b, p. 52. The model has been updated in four ways. The receiver stage of attention is now split up in the two stages of attention switch and attention maintenance. The stage of delivery is added, which shows that it is important that a warning reaches the target audience. Furthermore, more emphasis is put on the influence of environmental stimuli on warning effectiveness. The fourth difference is that the new model puts greater emphasis on the receiver's personal characteristics and task involvement.

37 Wogalter 2006b, p. 51; Cox III & Wogalter 2006; Cohen e.a. 2006.

38 Wogalter 2006b, p. 52.

can be useful as an investigative tool, as it can help pinpoint the stages that caused a bottleneck and thus form reasons for the failure of a warning to be processed adequately.[39]

Figure 1 The C-HIP model, originating from Wogalter 2006b

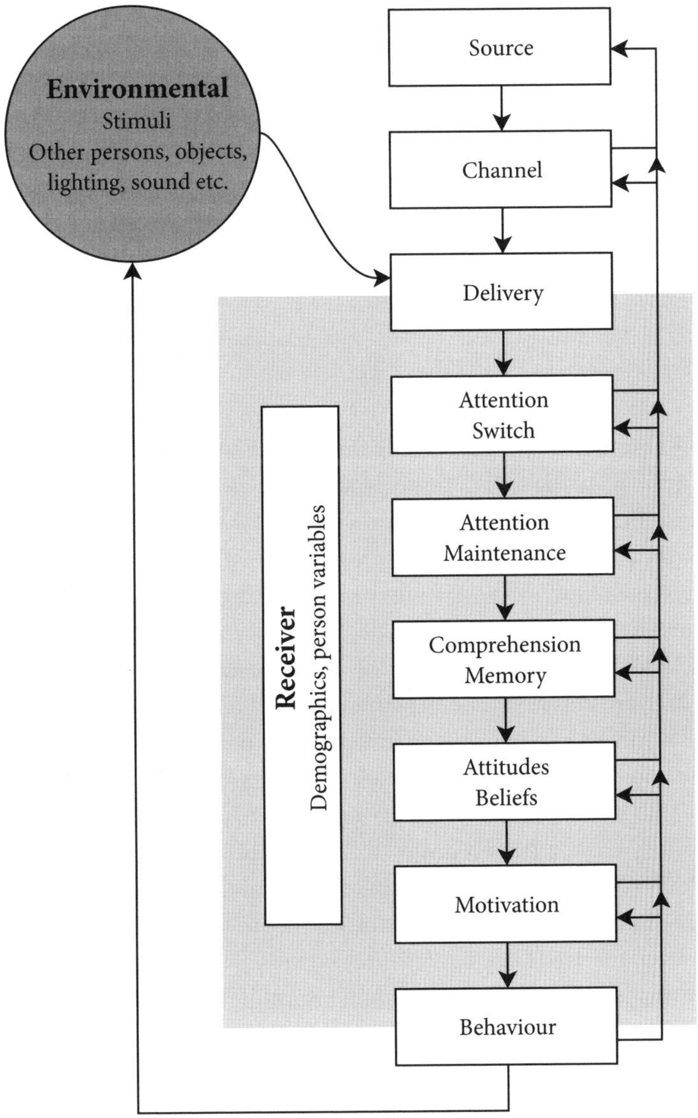

39 Wogalter 2006b, p. 59.

3.4.2.3 Implications

The information processing models that have been applied to warnings, including the C-HIP model, are quite similar to each other, in that they all describe the warning process as a sequence of several information processing stages. They imply that all the information processing stages must be successfully completed for the warning process to end in behaviour that corresponds with the warning. Hence, they predict that the behavioural effectiveness of a warning is determined by the success at each single stage of the model. Moreover, the effectiveness will not be greater than the least effective stage in the sequence.[40]

Although these models suggest that warning processing is linear, it is recognised that feedback from later stages can influence processing at earlier stages.[41] The models vary from each other in the degree of detail in which the warning process is described: some are divided in more information processing stages than others and some put additional emphasis on external factors that can influence the processing of warnings.

3.4.3 Human Decision Making Models

3.4.3.1 Value-Expectancy Models from Social Psychology

The second type of models that deserves attention here are human decision making models, since human decision making is related to information processing.[42] Behavioural compliance with warnings can be seen as the result of a decision making process: whether warning receivers comply with warnings or not will depend on whether they judge the benefits of compliance to outweigh the costs of compliance.

Theoretical perspectives and models of decision making that can be of use to the warning research are those from social psychology, particularly the persuasion literature.[43] A persuasive communication can be defined as any message that is intended to shape, reinforce or change the response of another, or others.[44]*Persuasion* is a significant topic for warnings. Since the ultimate purpose of warnings is to produce the intended behavioural effect, persuading users to exhibit protective or precautionary behaviours when encountering potentially dangerous products or situations forms a necessary function of warnings. The persuasive function of warnings addresses the motivational aspect of the message. This has, however, received fairly limited direct attention by warning researchers. Research focused more on investigating the alerting and informing functions of warnings. However, one should look beyond the functions of alerting and informing consumers of the unknown

40 Lehto 2006, p. 66, 68.
41 Rogers, Lamson & Rousseau 2000, p. 104; Wogalter 2006b, p. 52.
42 Lehto 2006, p. 73.
43 Cameron & DeJoy 2006.
44 Miller 1980; Cameron & DeJoy 2006, p. 302.

risks associated with a product. Merely informing consumers about new risks should not be considered sufficient enough. Consumers who may already be aware of the hazard associated with a product, yet have chosen to ignore the warning should also be addressees of the warning. Hence, warnings should also focus on these people by persuading them to change their belief, attitude, intention or behaviour. Theories and models that have been developed to explain health-related behaviour and that aim at improving the persuasive power of communication messages can therefore offer helpful implications for warning effectiveness and design.[45]

Value-expectancy models offer such a theoretical perspective that can be applied to the study of warnings. Value-expectancy models come in various forms and they differ from each other in view of their focus.[46] But they all are predicated on the idea that individuals will engage in a cost-benefit analysis, whilst taking into account the severity of the hazard, and then select an action that will maximise the expected outcome.[47] The various value-expectancy models highlight the personal and individual factors that can affect the decision whether to engage in precautionary behaviour or not. They share the assumptions that: (a) an individual's motivation for self-protection is created through one's desire to avoid a negative health outcome; (b) that one's belief about the likelihood of occurrence of a particular negative health outcome determines the effect of the negative outcome on one's motivation; (c) that if an individual expects that a suggested action will reduce either the likelihood or the severity of harm, then one will be more motivated to act; and (d) that one weighs the expected costs against the expected benefits, which determines the likelihood that one will engage in an action.[48]

3.4.3.2 Implications

The insights from these value-expectancy models are first that threat-related beliefs, which concerns the beliefs about personal severity (the perceived severity of a negative consequence or outcome that could result from assuming no behavioural action) and personal susceptibility (the perceived probability that the particular negative consequence could occur) are beliefs that can affect the motivation to act and should be taken into account for the design of effective warning messages. The importance of these beliefs has already been evidenced in the warning research. It consistently shows that message effectiveness increases with the perceived hazardousness of the product or activity. Although perceived

45 Cameron & DeJoy 2006, p. 301.

46 These value-expectancy models are the health belief model (Becker 1974), the theory of reasoned action (Ajzen & Fishbein 1980) and of planned behaviour (Ajzen 1991), protection motivation theory (Rogers 1975) and the extended parallel process model (Witte 1992).

47 DeJoy 1991.

48 Cameron & DeJoy 2006, p. 302; Weinstein 1993.

susceptibility has been less investigated in the warning-related studies, the perceived severity seems to be closely related to the perceived hazardousness of a product.

This is the same for the component of barriers or costs that form a part of the models. These costs can include factors such as financial cost, or personal factors such as the effort or inconvenience to perform an action. Warning studies clearly show that an increase in the costs of compliance decrease warning compliance.

Third, *self efficacy*, which concerns the belief about one's own ability to follow safety instructions successfully in order to avoid the danger, is a critical component for behaviour change in the models and may also be of relevance to warning compliance behaviour. If one does not believe that he/she can perform the warning's behavioural recommendations, then this belief can prevent compliance with the warning. Consequently, warning messages must communicate to receivers the fact that they are able to adopt the behaviour prescribed by the warning, because the action is easy to perform. Nevertheless, it is noted in the literature that because the instructions in warnings are usually simple, it will be less likely that a warning speaks to self-efficacy.

The models also highlight the potential importance of *response efficacy*. The belief refers to the utility of the recommended behaviour to prevent the threat. Hence, for warning messages to be persuasive, the recommendations for avoiding injury must be perceived as ones that will be efficacious.[49] This may be established by including specific information in the warning that emphasises how many serious injuries are prevented by the simple behavioural recommendations of the warning.[50]

Finally, the value-expectancy models (in)directly address the normative expectations of people. This concerns warning receivers' perceptions of the social environment that can have an impact on their beliefs, intentions and behaviour, *e.g.* the influence that the expectations of friends can have on someone's safety.[51]

3.4.3.3 Edworthy's Cost-Benefit Model of Warning Compliance

Another human decision making model that approaches behaviour from a cost-benefit analysis and that aims at understanding and predicting warning-related behaviour is that of Edworthy. Edworthy developed a framework derived from a decision making model to help understand the role of the decision making process in the warning process.[52] The model is based on a traditional cost-benefit utility model of decision making, which is the subjective expected utility model.[53] This model is widely used to predict different types of

49 Cameron & DeJoy 2006, p. 304, 305; DeTurck 2002, p. 348.
50 DeTurck 2002, p. 361.
51 Cameron & DeJoy 2006, p. 305.
52 Edworthy 2000; Edworthy 1998; Edworthy & Adams 1996, p. 20-24.
53 Edwards 1954.

behaviours that are the result of some sort of decision making. Edworthy applies this model to warning-related behaviour.

The subjective expected utility equation for modelling the behaviour of warning compliance can be expressed in the following way. The subjective expected utility of an action (whether to show compliant behaviour) is the product of the probability that injury will occur when compliance does not take place, minus the probability that injury will occur when compliance does take place times the seriousness of the risk, minus the cost of compliance. Thus, people will only comply with a warning when the result of the calculation produces a positive result. That is, where the likelihood of injury when complying, plus the cost of complying, is less in total than the likelihood of injury if non-compliant.[54]

When using this model, it is important to know the social cognitive variables or cues that are available to a person when confronted with a risk and that will influence the person's decision to behave safely. When people are confronted with a danger, cues or factors from the environment, the hazard and personal factors modify their behaviour towards that risk (whether a warning is present or not). To date, many warning studies have identified a wide range of variables that can influence behavioural warning compliance. In a specific case, their position in the equation will need to be determined.

However, the warning design variables should not be part of the equation. In Edworthy's framework, the role of the warning is that it can be viewed as a decision making aid; it can be an indicator of the level of risk or hazard involved in the situation. If a warning is present, the warning itself may provide cues that will allow a person to judge whether or not safety behaviour is necessary. When the design of a warning contains variables (such as colour) that can give an indication of the level of risk involved, the warning can help a person in deciding whether to act or not. According to Edworthy, the risk assessment that occurs on the basis of the warning design should be viewed separate from the cost-benefit analysis when actually deciding whether or not to exhibit safety behaviour. The warning complements the cost-benefit analysis. The argument is that a warning should be viewed as an artefact of the risk associated with its referent (in other words, the product or situation). As the warning is separate from its referent, the cost-benefit analysis should be based on the environment, not on the artefact provided. Furthermore, it is better to leave warning design considerations out of the equation until the effects of the variables are clearly known.[55] Also, design variables and social cognitive variables have different origins. One cannot really compare the effect of font size with a subjective belief.

Another important element of the model is to know the relative weights of the variables, so that each variable can be weighed accordingly. This may be problematic. It may not be easy to compare and generalise the findings of different studies that investigated the effect

54 Edworthy 2000, p. 772; Edworthy & Adams 1996, p. 20.
55 Edworthy & Adams 1996, p. 21, 22.

of a certain variable on warning compliance due to methodological differences between the studies.[56]

3.4.4 *Rasmussen's Levels of Performance Model*

3.4.4.1 Skill-Based, Rule-Based and Knowledge-Based Behaviour

The third type of model presented here is the *levels of performance model*, applied to warnings. Rasmussen formulated a model of naturalistic decision making by classifying tasks.[57] This model was originally developed for analysing the role of human error in accidents involving complex systems in the working environment. His model distinguishes between different forms of behaviour from the perspective of levels of task performance, namely skill-based, rule-based and knowledge-based levels of decision making and task performance. The hierarchical levels of performance correspond to a particular path between initial exposure to a stimulus and emitting a response. Each path differs in the depth to which information is processed.

Skill-based behaviour occurs at the bottom of the hierarchy and involves more or less automatic behaviour, in which there is little need for conscious decision making. For example, in manufacturing, an operator on the assembly line picks different parts from bins, assembles them, and puts the finished product on a conveyer belt. The operator could perform this task in his sleep, as it were. This mode is not very prone to error because of a greater skill base at this level.

Rule-based behaviour constitutes behaviour that represents a more conscious effort to follow rules. It is similar to skill-based behaviour in that it is procedurally oriented and generally routine in nature. The principal difference is that skill-based behaviour is a continuous procedure in which each step is automatically executed, whereas rule-based behaviour is a consciously executed step-by-step sequence. In the case of rule-based behaviour, the operator has several well understood stored rules of decision making. For example, if situation A occurs, I do X: if the floor is wet, then do not use an electric tool.

Task performance on a higher level is knowledge-based behaviour. Behaviour at the knowledge-based level involves conscious problem solving directed towards attaining a goal: the operator may first have to think about the purpose of the task before deciding what actions to take to solve the problem. Knowledge-based behaviour is typical for unfamiliar tasks: it occurs when people are not familiar with performing the task at hand and they have to find a solution based on their understanding and their previous knowledge or when experienced operators work in a complex or novel setting. Behaviour at this level is, therefore, most prone to error.

---

56 Edworthy 2000.

57 Rasmussen 1986.

Another important part of the model is that shifts between the different levels of performance can be expected. Transitions up the hierarchy from skill-based to knowledge-based processing levels occur as part of task performance, whilst transitions down correspond to longer term effects associated with learning or increased skill. The amount of conscious processing increases as operators move up the hierarchy. With increased experience and training, many knowledge-based tasks become rule-based tasks and rule-based tasks turn into skill-based behaviour.[58] Failure to make the needed shift up or down the hierarchy can cause human errors resulting into accidents. At each level, certain types of human error are prevalent.[59] By identifying these errors, design solutions aimed at eliminating or reducing these errors can be adopted.

3.4.4.2 Lehto's Extended Version of Rasmussen's Model

Lehto proposed a conceptual model of human behaviour that builds upon the model of Rasmussen and that provides a warning design methodology: it relates the levels of performance to the design of warnings.[60]

To better describe safety-related behaviour, Lehto added a fourth level of human performance to the framework, called judgement-based behaviour. Belief or judgement-based behaviour is placed at the top level, with knowledge, rule and skill-based behaviour at the lower levels of performance. At the judgement-based level, an affective reaction such as fear or pleasure is triggered by the stimulus (*e.g.* the warning) and changes the goal priorities set by the operator, such as a safe or convenient task performance. By performing value judgements, people assign priorities to goals which ultimately influence task performance. For example, if a warning message can elicit the belief that the product is dangerous, then it may trigger safe behaviour.

The model provides a fundamental way of classifying forms of warning information (signals, signs, symbols, and values) in terms of the depth of processing.[61] Each form of warning information is related to a specific level. At the skill-based level, information is not consciously processed. Information serves as signals that trigger automated behaviour. The rule-based level corresponds to 'reading to do'. The warning information needs to be interpreted in the form of nonverbal signs to trigger the conscious application of rules. At the knowledge-based level, warning information serves as symbols and is more deeply processed. This level corresponds to 'reading to learn'. The judgement-based performance corresponds to 'reading to evaluate'. At the judgement-based level, the operator develops

58 Lehto 1991, Helander 2006, p. 81-83; Lehto 2006, p. 78-81.

59 Reason 1990.

60 Lehto 1991.

61 Signals correspond to sensory data, signs indicate perceived or named states, symbols are abstract constructs which can be formally processed in mental models, and values represent deeply processed concepts associated with opinions, attitudes and beliefs, see Lehto 1991, p. 599.

goal priorities on the basis of agreement or disagreement with presented values and makes trade-offs between the goal priorities. Warnings at this level are often oriented toward persuading product users to form or change their beliefs and attitudes towards safety.[62]

3.4.4.3 Implications

A number of implications for the design of effective warnings can be drawn from this model. The primary insight is that a warning should be processable in a way that matches the level of task performance, namely, as signals when task performance is at the skill-based level, as signs at the rule-based level, as symbols at the knowledge-based level, and as values at the judgement-based level. It is thus critical to identify and determine the level of performance at which errors occur. Since the nature of a human error is mapped to a specific level of performance, the mapping can guide the selection of an appropriate remedy to reduce error. In other words, warnings should be designed to provide information in a form compatible with the level of performance at which human errors can occur.

Prevalent errors in skill-based behaviour include scenarios where people follow a routine sequence of actions (a script) which is inappropriate because of changes in familiar settings or differences between products (*e.g.* a person who previously used a nonflammable adhesive without ventilating their work-area might do the same when using a flammable adhesive). Errors at this level can also occur when scripts are incomplete or contain unsafe steps which have been reinforced by past experience (*i.e.* entering, starting, and driving a vehicle without automatically fastening a seat belt). Before the design of a warning can successfully modify skill-based behaviour, the warning receiver must make a transition upward to a rule-based or higher level. Creating awareness by increasing the conspicuousness of the warning or interrupting the automated behaviour by means of an interactive warning that requires interaction with the warning during task performance may be helpful. Another approach to reduce error at the skill-based level is to design the product in such a way that it provides signals and cues via multiple sensory channels when the product is in a hazardous state.

Rule-based errors occur when people omit to take a step/action in some essential procedure, or fail to shift up to a knowledge-based level in unusual situations where the rules they normally use are no longer appropriate.[63] When people are operating at a rule-based level, it is preferable to integrate the warning information into the task so that it can be easily mapped to immediately relevant actions. The reason for this is that due to the limitations of the short-term memory, information can only be retained in consciousness for a short time interval. In addition, when people operate at a rule-based level, they are trying to get the task completed. Therefore, the warning's design is likely to have the greatest

62 Lehto 1991, p. 599; Lehto & Papastavrou 1993, p. 583.
63 Helander 2006, p. 81-83; Lehto 2006, p. 78-81; see also Van Duijne 2005, p. 21.

influence on people at this level if the warning contains text that is brief or has concrete symbols that convey an action and that are perceived at a glance rather than a content that requires effort to process and that is likely to produce an information overload. Furthermore, because of the short-term memory limitations or distraction, the information must be given at the time needed. Product characteristics that nonverbally inform or remind will often be the most effective means of providing the information at critical times. A final note is that people at a rule-based level may use incorrect rules which have been developed on the basis of previous experience and which have moved down the hierarchy from knowledge or judgement to rule-based behaviour (*e.g.* failing to wear protective glasses). To remedy this, the rule-based behaviour must be modified into a higher level of behaviour. The routine rules of the operator must be changed and new ones must be learned. However, such a shift upward is a difficult task for a warning to accomplish.

At the knowledge-based level, it is expected that warnings have the greatest influence on behaving safely since people performing at this level may have a tendency to be seeking out information to solve the problem and may not yet have developed conflicting patterns of skill- or rule-based behaviour which would have to be modified to achieve behavioural compliance. Nevertheless, this does not suggest that simply providing information to inexperienced people is sufficient at this level. Errors at this level usually concern the initiation of inappropriate actions. Care should be taken to assure that product users are able to correctly interpret the information. Providing step-by-step information about how the product must be used seems especially useful as it may trigger the transition down the hierarchy in the future by teaching a rule.

At a judgement-based level, people may have inappropriate affective reactions that can frustrate safe behaviour, such as stress, annoyance or a sense of urgency. However, inappropriate judgements (*e.g.* about the products' dangers) are difficult to modify. Consequently, warnings focusing on influencing behaviour at this level are unlikely to be effective. Providing information that emphasises the dangerousness of the product hazards or the inconvenience of certain unsafe actions that a product user is likely to perform may change goal priorities and ultimately behaviour in a positive way.[64]

In sum, using the levels of performance model to analyse the behaviour of product users may help us understand why certain unsafe actions occurred during usage of a product. In addition, it may provide guidance as to how these errors could be reduced by changing the design of the product and/or the warning information.[65] An illustrative example in this regard is the study of Edworthy e.a. 2004 that used Rasmussen's taxonomy to fit the research findings concerning the measuring of behavioural compliance with the

64 Lehto 1991; Lehto & Papastavrou 1993, p. 581-591; Lehto 1992, p. 123-127; Lehto 2006, p. 78-81.

65 See *e.g.* Lin & Salvendy 2000; Van Duijne 2005, p. 236 who uses the model to explain findings of her studies and Lehto & Papastavrou 1993, p. 581, who evaluated findings of the warning research from the perspective of levels of performance.

safety information accompanying a pesticide product. There was a group of amateurs and a group of professionals that were instructed to use a houseplant insect killer. The results showed that the professionals produced higher levels of compliance with the safety information than the amateurs. The model can offer an explanation for this research finding. The researchers reasoned that the professional operators work at a skill-based level and can slip from skill-based to ruled-based behaviour when necessary. Amateurs, on the other hand, are less likely to be familiar with the task. They are more likely to be operating at a knowledge-based level and therefore more prone to making errors during task performance, such as failing to comply with the warning.[66]

3.4.5 Conclusion

Models that represent the warning process can help us better understand how warnings work, why they failed to work and how they can be improved. Three types of models and the corresponding implications were described in this paragraph.

Information processing models of the warning process identify several sequential stages of information processing and show that the stages need to be successfully performed by the human for a warning to flow through to the nest stage and to ultimately change behaviour. Another important implication is that effectiveness will not be greater than the least effective stage in the sequence.[67] A leading model of the warning process is the C-HIP model. This model is favoured for its explicitness. It incorporates communication and information processing components that are relevant to the warning process and it extensively considers a variety of factors, such as environmental stimuli and receiver characteristics that can produce a bottleneck at a stage of the model. Hence, this model can be particularly of value to function as a framework to organise the research findings, which has been done by Wogalter, and as an investigative tool to find out in a systematic manner the causes of a warning's failure. Because of these benefits, this model will also be used as a basis in the following paragraphs of this chapter.

The decision making models adopt a cost-benefit approach, in that the decision of an individual is made by considering the costs and benefits of taking the behavioural action. These can be applied to warnings, as warning receivers must make decisions about the risks involved and whether to act safely as expressed in the warning. The model of Edworthy is such an example. In this model, the warning complements the cost-benefit analysis of people by providing information about the level of risk involved. The models from the persuasion literature based upon the value-expectancy theory can also be used to explain the findings of the warning research and to offer implications for the design of

66 Edworthy e.a. 2004, p. 24.
67 Lehto 2006, p. 66, 68.

effective warnings. They emphasise that warnings should not only serve to alert and inform product users, but also to persuade them that they are at risk and that they need to change their attitudes and beliefs and engage in self-protective behaviour. The models address the main factors that are important determinants for health behaviour and thus may be of importance to behavioural compliance with warnings. These are the costs of performing an action, treat-related beliefs, self efficacy and response efficacy, and normative expectations.

Thirdly, Lehto's model of human behaviour that builds upon the model of Rasmussen and that provides a warning design methodology was described. The primary insight from this perspective of levels of performance is that, to reduce human error, warnings should be designed to match the operators' level of performance. A warning is likely to have the greatest influence when product users operate at a knowledge-based level. Regrettably, most behaviour occurs at skill or rule-based level. Unfortunately, since the design of the most effective warning, in terms of format and content, is fundamentally different at each level of performance, trade-offs in the design of warnings need to be made.

3.5 Research Methods of the Warning Process

3.5.1 *Introduction*

Various research methods are available to cognitive psychologists and product ergonomists to study the functioning of the mind independently or combined with the design of a product.[68] Before dealing with the findings of the empirical warning studies, I believe it necessary to consider the methods employed by researchers to obtain these results regarding the effective processing of warnings.

Because of the benefits of the popular C-HIP model, which were highlighted in the previous paragraph, this theoretical model is used to describe the common methods that have been carried out by warning researchers to investigate the processing of visual warning messages at each of the intermediate cognitive stages within warning receivers. Furthermore, the relevant methods used in the warning research to study the behavioural stage are discussed. Each of those methods has its weaknesses and strengths which must be considered by researchers prior to implementation.

It is beyond the scope of this thesis to discuss them in detail here. The following parts of this chapter aim at giving a general impression of the research methods that are generally employed to study the intermediate receiver stages of the warning process (paragraph

68 Solso, Maclin & Maclin 2008, p. 23 ff, Breakwell e.a. 2006; Stanton 1998b.

3.5.3) and the stage of behavioural compliance (paragraph 3.5.4).[69] The following paragraph briefly deals with some methodological considerations that need to be mentioned in this respect; paragraph 3.5.6 closes this paragraph. But first, some general comments are made in paragraph 3.5.2 on studying the human mind.

3.5.2 *Studying the Human Mind: General*

As noted earlier, the most fundamental assumptions made by cognitive psychologists are that mental processes exist and that the functioning of these internal processes can be investigated scientifically. Traditionally, cognitive psychologists conduct experiments to test their ideas about how the cognitive processes of the mind work and to deepen our understanding of the human mind in real life. However, mental processes are unobservable and cannot be studied directly. Therefore, to test their hypotheses, researchers use indirect measures that are believed to reflect the mental processes. By studying how people respond when performing specific tasks, psychologists draw inferences about the functioning of the human mind.[70] From this, it follows that the mental activities that occur during the processing of warnings cannot be studied directly. Therefore, they are measured by an indicator that can be seen. Naturally, overt behaviour in terms of following the warning's directions can be measured directly.[71]

Experimental research can be carried out in the field, that is to say in a real-world setting, but experiments are preferred to be conducted in a laboratory since the conditions of laboratory experiments can be better controlled. Basically, experiments with a group of participants have the purpose to achieve empirical evidence with regard to the causal effect of one or more variables on another *variable*[72] The former are independent variables, since they can be manipulated or controlled by the experimenter in order to measure the effects on the dependent variable. The dependent variable is the one being measured during a study. So, the primary advantage of conducting laboratory experiments is that cause and effect can be explored and determined.[73] The results of these studies can be *significant*, which means that the finding is reliable, trustworthy. It does not say anything about whether the difference is large or small, merely that the finding is detected.[74]

---

69 See for an extensive discussion of the methods for evaluating the effectiveness of warnings, Smith-Jackson & Wogalter 2006; Young & Lovvoll 1999; Wogalter & Dingus 1999 and Kalsher & Williams 2006.

70 Payne & Wenger 1998, p. 18.

71 Smith-Jackson & Wogalter 2006, p. 24.

72 A variable is any characteristic that can vary across people or situations and that can be of different levels or type, see Breakwell e.a. 2006, p. 69.

73 See for more details on the experimental method in psychology, Breakwell e.a. 2006, p. 64-87.

74 Dunn 2009, p. 292.

Even though experimental cognitive psychology has been an influential approach within psychology for many years, there are limitations.[75] A fundamental concern of laboratory experiments called *ecological validity* concerns the issue of whether the findings are applicable to the real world. If a task or situation within a laboratory experiment has high ecological validity, then the cognitive processes employed in performing the task are similar to those in the real world. Under such conditions, the results from laboratory studies to make predictions about the real world can be generalised.[76] Many variables can influence behaviour in the real world and it is hard to manipulate them all as well as to adequately assess the relative contribution of each variable. Thus, researchers need to be cautious regarding their conclusions about the cause-effect relationships, as the laboratory setting remains a simplified simulation of reality. Another disadvantage of experimental research is that the participants are usually students, since students are easy to recruit for experiments in the university laboratory. It has been questioned whether they are a good representation of mankind.[77] If not, it seems undesirable to generalise the findings to the population in general.[78] On the other hand, the stages of information processing may well be considered to be similar between individuals. Furthermore, experimenters need to consider the issue of *demand characteristics*; for example, participants of an experiment take on the role of good participant and are more likely to perform the desired behaviour, because they are aware of the experimenter's expectations.[79]

Contrary to laboratory experiments are field studies. These studies are usually done to describe the phenomena of interest, whereas experiments are used to explain phenomena.[80] The benefit of field studies is that they are more similar to real life situations than laboratory studies and hence their findings have more *external validity*, which means that the findings are robust and stay valid in changed and more realistic conditions.[81] However, it is more difficult to control the conditions and consequently to have solid evidence on the causal effects of the manipulated variables.[82]

Relatively few warning research has taken place in the field.[83] The majority of the warning studies are employed under highly controlled laboratory conditions and are primarily based on measuring the effects of one or more variables on an intermediate information processing stage. The independent variables are the variables that the researcher

75 Eysenck & Keane 2000, p. 514, 530.

76 Payne & Wenger 1998, p. 9; Breakwell e.a. 2006, p. 86.

77 See Wogalter & Dingus 1999, p. 60.

78 See Stewart, Folkes & Martin 2001, p. 358.

79 See Young & Lovvoll 1999, p. 60; Dunn 2009, p. 264; Cox III e.a. 1997, p. 300. Demand characteristics refer to cues in an experiment that tell the participant what behaviour is expected, see Myers 2004, p. G-2.

80 Solso, Maclin & Maclin 2008, p. 23.

81 Eysenck & Keane 2000, p. 531; Breakwell e.a. 2006, p. 86. External validity is related to settings as well as to participant populations.

82 See Young & Lovvoll 1999, p. 61.

83 See *e.g.* Goldhaber & DeTurck 1988b; Vredenburgh e.a. 2005; Frantz & Miller 1993.

wants to investigate and that are often related to the design of a warning, such as the warning's size or the presence or absence of symbols, colour or signal words. These variables are being manipulated in order to assess their impact on the dependent variable like attention to warnings or warning compliance.

3.5.3 *Measuring the Intermediate Stages of the Warning Process*

3.5.3.1 Attention Switch and Attention Maintenance

The upcoming paragraphs discuss the common methods that have been used by researchers to measure the stages of the warning process.

The first two cognitive stages with regard to processing warning information concern attention switch and maintenance. There are different methods to measure attention. For example, the use of eye-tracking equipment to record eye movements. It assesses the direction of the eyes to a warning and how long participants fixate on the warning when exposed to it.[84] Another method is measuring the response time to see how quickly an individual can search for and detect a target.[85] It is assumed that if participants are able to find a stimulus (*e.g.* symbol) in one condition faster than in another condition, that warning design attracts attention more easily. The research method of looking behaviour entails the empirical observation of head movements to assess whether a certain feature of a warning, such as colour, influences the noticeability of a warning. Observers record whether participants looked at a warning as well as the amount of time spent examining the warning.[86]

Furthermore, attention maintenance in particular can be investigated by examining the legibility of warning text and symbols. The purpose of such studies is to assess which warning stimuli remain discernable under various degraded conditions. Participants must attempt to identify the warning when its size is reduced or when they are exposed to it only for a short duration.[87]

The above-mentioned methods measure attention more or less directly.[88] An indirect and less preferred approach is through recall measures.[89] In those studies, it is reasoned that people cannot remember and recall information that they do not notice and comprehend.[90] In addition, the above-mentioned methods are also objective, which means that the measures are reported by someone or something that is external to the participants.

84 See *e.g.* Laughery & Young 1991a.
85 See *e.g.* Bzostek & Wogalter 1999.
86 See *e.g.* Wogalter & Rashid 1998.
87 See *e.g.* Wogalter e.a. 2002; Nilsson & Kaiserman 2005; Wogalter e.a. 2006, p. 162.
88 Young & Lovvoll 1999, p. 28.
89 See *e.g.* Goldhaber & DeTurck 1988b.
90 Young & Lovvoll 1999, p. 31.

Subjective measurement on the other hand, *e.g.* questionnaires and interviews, refers to data collected by participants who give their subjective judgements.[91] An advantage is that the data of subjective measures are generally easy to collect and not very costly, unlike with objective measures. However, the researcher is dependent on the personal assessment of participants for the cognitive process. Their assessment may be subject to influences, such as decreased memory, that can diminish accuracy of the data.[92]

Attention switch and maintenance can be measured on the basis of subjective measures. After having performed a task during an experiment, participants can be asked to complete a questionnaire that includes the question whether or not they noticed the warning.[93] In other studies, participants are shown a number of warning designs and are instructed to rate the likelihood of noticing the warning.[94] These ratings are usually based on a so-called Likert-type scale.[95] Participants need to respond to close-ended questions on the basis of a rating scale with several anchors that form the response options, for example a 9-point rating scale with associated anchors varying from 0 (not at all likely to notice the warning) to 8 (extremely likely to notice the warning). In addition, to assess attention maintenance in a subjective way, participants can be asked to judge whether they would be willing or likely to read a particular warning design or if the warning design is easy to read (in view of *e.g.* the font size or the amount of white space between text).[96] Other studies have measured attention maintenance by asking participants to provide a rank order of warning labels according to their perceived readability.[97]

3.5.3.2 Comprehension and Memory

The second stage of warning processing concerns whether the potential user population understands and remembers the meaning of the warning message. Understanding warnings is linked to the amount of knowledge people possess. Comprehending textual information involves having knowledge of the language, the terms and sentences and understanding the coherence of the message itself.[98] If people lack knowledge about the characteristics of the hazard that is needed for safe use, a warning must fill this knowledge gap. Consequently, it is apparent that warning designers pre-test their prototype warnings in order to find out what knowledge is already available with the target audience of the warning. To determine

91 Smith-Jackson & Wogalter 2006, p. 30.
92 Kalsher & Williams 2006, p. 326.
93 See *e.g.* Duffy, Kalsher & Wogalter 1995 (described in paragraph 3.6.4.8); Frantz & Rhoades 1993 (described in paragraph 3.6.4.7).
94 See *e.g.* Harris & Wiklund 1989.
95 Likert 1932.
96 See *e.g.* Silver & Braun 1993; Kalsher, Wogalter & Racicot 1996.
97 See *e.g.* Wogalter & Vigilante 2003.
98 Young & Lovvoll 1999, p. 33.

what information needs to be present in warning messages, interviews or focus groups can be used to evaluate a potential warning design.[99]

To assess whether people comprehend warning information, researchers have asked people to define the meaning of hazards attached to certain consumer products. The questions can be open-ended, for example asking participants to describe the hazards, but also close-ended, using a multiple choice format to assess people's hazard-related knowledge.[100] Furthermore, much research has measured whether people understand the hazard level that is being communicated by different signal words, colour or symbols as intended by designers. For instance, participants are instructed to rank signal words, such as CAUTION and DANGER, from most to least hazardous or to provide understandability ratings of the terms.[101] Explicit language is an important factor for warning comprehension and a number of studies have used questionnaires to investigate whether messages that contain detailed warning information improve comprehension compared to non-explicit warnings.[102]

There are also various techniques to assess the understandability of warning symbols.[103] International standards, such as those of the International Standard Organisation (ISO) and the American National Standards Institute (ANSI), provide methods for designing and evaluating warning symbols.[104] A method to assess symbol comprehension involves showing participants the symbols and have them define their meaning, usually on the basis of open-ended responses or a multiple choice procedure. This can be done in absence of or with written information.[105] The latter is more preferred.[106] By providing contextual information, such as the inclusion of a photograph of the product alongside the warning symbol, testing the comprehension of the warning symbol is more realistic, which in turn enhances its ecological validity.[107] Another method is the phrase generation procedure which involves presenting the symbols to participants and having them write down as many phrases as come to mind when viewing each symbol.[108] Identifying the concepts or representations an individual has for a symbol which are being activated when the symbol is first viewed provides valuable information about how symbols are initially interpreted.[109]

To increase symbol comprehension and retention, several studies have used training methods. First, a pre-training comprehension test is required to determine the extent to

---

99 Young & Lovvoll 1999, p. 36; Fischhoff e.a. 1998.
100 See *e.g.* Malouff & Schutte 1992; Leonard & Wogalter 2000.
101 See *e.g.* Wogalter & Silver 1995; Drake, Conzola & Wogalter 1998; Silver & Wogalter 1989.
102 See *e.g.* Trommelen 1997; Laughery & Stanush 1989.
103 Young & Lovvoll 1999, p. 36.
104 See in more detail, Deppa 2006.
105 See *e.g.* Wolff & Wogalter 1998; Collins & Lerner 1982; Wogalter, Sojourner & Brelsford 1997.
106 Leonard, Otani & Wogalter 1999, p. 167; Young & Lovvoll 1999, p. 36.
107 See *e.g.* Davies e.a. 1998.
108 Hancock e.a. 2004.
109 Hancock e.a. 2004, p. 184.

which the symbols are understood. Participants are shown the symbols and can be asked to write down their meaning or symbols and a meaning are presented together and participants need to response whether that meaning is correct. After that, training begins. One type of training is to first show participants symbols paired with additional text explaining the hazard and/or its consequences and then to test their symbol comprehension again at several time intervals (a week or a month later) to see whether the effect of training on comprehension remains over time.[110] Another training method recently employed involves showing participants symbols that are paired with explanatory information on real-world accident scenarios followed by comprehension tests at several time intervals.[111] During the comprehension tests, the time that participants needed to decide accurately whether the text presented on a screen matched the meaning of the symbol (accuracy time) and the speed with which the response was made (reaction time) were measured. It was hypothesised that accurate and fast responses reflect the benefit of training, since it is expected that training improves the ability of the warning symbol to automatically trigger the associated information.

3.5.3.3 Attitudes and Beliefs

Subjective measures are the main method to determine people's pre-existing beliefs and attitudes on warnings.[112] People's perception of the level of risk attached to a consumer product is the most commonly researched attitude. Familiarity with a product has also gained much research attention. Many studies in psychology use a Likert-type rating scale as a measure for attitudes and beliefs.[113] For example, if perceived risk is measured, participants are asked to rate consumer products or warning labels on its hazardousness on a scale from 0 tot 8 with the following anchors of 0 (not at all hazardous), 2 (somewhat hazardous), 4 (hazardous), 6 (very hazardous) and 8 (extremely hazardous).[114] Similarly, respondents can also be asked to judge how familiar the product is, how severely they might get injured with the product, or to which degree they believe that the source that provides the warning information is credible.

Even though ratings are popular among researchers, there are concerns regarding this way of yielding data on risk perception. Because such ratings are subjective and hence consist of people's own descriptions of their perceptions and cognitions, they may not be accurate. People are limited in being able to reveal their cognitive processes, for example because in some instances participants may not even be aware of such processes.[115] More-

110 See *e.g.* Wogalter, Sojourner & Brelsford 1997.
111 See *e.g.* Lesch 2008a.
112 Smith-Jackson & Wogalter 2006, p. 29.
113 Dunn 2009, p. 162.
114 See *e.g.* Wogalter e.a. 1991; Wogalter, Jarrard & Simpson 1994.
115 Van Duijne 2005, p. 63.

over, the findings of studies that have used rating scales only provide general information about people's judgement of product risks. They do not provide insight in the way in which products are actually used, why they are used in such a way and what type of risks may occur during use. In this regard, risk perception is explained as a belief or a judgement. It has been argued by researchers that this view is deficient, since it ignores the influence of contextual factors on risk perception in product use. Rather, risk perception should be explained as an interactive process between user and product.[116]

To tackle the above-mentioned shortcomings and to thus improve our understanding of the role of risk perception in accidents involving consumer products, Van Duijne has proposed a research approach that centres on the users' perspective on risk in product usage and on the way in which users understand product characteristics that are relevant to safe usage.[117] The studies that have applied this research approach do not specifically investigate the effectiveness of product warnings. However, because they study people's perception of the hazardousness of consumer products – an important factor for warning effectiveness – it is argued here that the findings have a bearing on the topic of warning effectiveness.

The approach entails a qualitative observational research methodology that addresses risk and risk perception of users who do not have any recent accident experience with their consumer product. The methodology includes the methods of the measurement of anthropometric characteristics of participants, the observation and recording of individual user activities, followed by open-ended interviews with the participants. The data collected from the observations demonstrate users' actions with the product, while the interviews give details on why participants perform an action and how they perceive the risks. The anthropometric data provide information on the physical characteristics of participants and indicate whether there are constraints that can act as boundaries for usage of the product.[118] The research is carried out in a naturally occurring setting with a minimum of intrusion from the researcher.[119] Participants show how they use their own consumer product or a product that is familiar to them without being exposed to real threats.

The methodology is partly derived from the accident research approach introduced by Weegels. This qualitative observational research methodology identifies the awareness and perception of risk of users who have recently been involved in an accident with a consumer product.[120] The method involves a video-recorded reconstruction of accidents involving consumer products in an investigation on the site of the accident. This makes it possible to collect detailed accident data on the product, the use activities of the consumer

116 Van Duijne 2005, p. 50, 52; Van Duijne, Green & Kanis 2001.
117 Van Duijne 2005, p. 57-80.
118 Van Duijne 2005, p. 62.
119 Van Duijne 2005, p. 64 ff.
120 Weegels & Kanis 2000.

and the situation. Subjects are visited as soon as possible after the accident and are asked to demonstrate what happened and to describe it, followed by an open-ended interview.[121]

3.5.3.4 Motivation

Like with the previous cognitive stage, the most common procedure to measure motivation is on the basis of ratings.[122] To determine whether people have the motivation to comply with a warning, researchers have measured participants' judgement with regard to their intent to act cautiously or their willingness to comply with the warnings presented to them.[123] Other terms used in the warning research to describe behavioural intentions include precautionary intent, intended carefulness or likelihood of complying.[124]

3.5.4 *Measuring Behavioural Warning Compliance*

3.5.4.1 Using a Control Condition

The C-HIP model ends with the stage of behavioural compliance with the warning. If people are sufficiently motivated, they will carry out the behaviour prescribed by the warning. In the warning research, this is seen as the ultimate measure of warning effectiveness.

To assess whether a warning is truly effective in steering compliant behaviour, it is important to determine the extent to which people would perform the required behaviour without the warning being present compared with the situation in which a warning is present. By including a control condition in studies, researchers are able to assess differences between the level of safety behaviour without a warning and the level of behavioural compliance when a warning is present. Researchers can then see whether the warning brings *added value* to the situation. For example, suppose that in the absence of a warning 40% of the participants in an experiment don the protective gloves in order to prevent hand injuries when performing a task. This score can be viewed as the baseline score. When in the warning condition, 60% of the participants wear the gloves as directed in the warning, warning compliance has increased with 20%. This latter score can be called the effectiveness score.[125] Likewise, it is possible to measure the effectiveness score of a warning design feature, such as colour, on behavioural compliance by comparing the scores of the warning without colour with the warning with colour.[126]

121 Weegels 1996, p. 35-52; Weegels 1998.
122 Smith-Jackson & Wogalter 2006, p. 29; Young & Lovvoll 1999.
123 See *e.g.* Laughery e.a. 1993a.
124 Kalsher & Williams 2006, p. 325.
125 Edworthy & Adams 1996, p. 9.
126 Wogalter & Dingus 1999, p. 58.

3.5.4.2 Measures of Behavioural Compliance: Direct/Indirect and Subjective/Objective

Data on behavioural compliance with warnings has been collected in a number of ways.[127] Two categorisations with regard to measuring compliance are particularly of relevance here. Behavioural compliance can be measured directly or indirectly using subjective or objective measures.[128]

Contrary to the internal mental processes of the intermediate processing stages of the C-HIP model, behaviour can be measured directly. This entails observing whether people comply or do not comply with the warning, whereby it is essential to define what type of behaviour entails warning compliance. Observation can be done in the lab under controlled conditions with participants, but also in field settings under less controlled conditions. Naturally, participants of laboratory studies are not told about the true purpose of the warning research.[129]

Studying compliance indirectly concerns measuring the variables that influence the substages of the warning process that occur prior to behavioural compliance. These variables may have an indirect impact on warning compliance. Because it is assumed that behavioural compliance is the result of the successful processing of the subsequent substages within the receiver, studying them can be viewed as indirect measures of warning compliance.

Behavioural compliance can also be evaluated through subjective measures. For example, self-reports in which consumers tell whether they performed the behaviour in the past or by measuring behavioural intentions, which means that participants are asked to give their judgement regarding the perceived effectiveness of a warning or to what extent they would be willing to comply with the warning in a particular situation.[130] In this regard, it is argued that one measures whether there is motivation to act. It assumes that behavioural intentions predict behaviour. Research findings in social psychology indicate that there is a causal link between behavioural intentions and behaviour.[131] Thus, by assessing whether receivers of a warning would have the intention to follow the warning's directions, behavioural compliance can be predicted.[132] Nevertheless, saying that you have the intention is not the same as actually complying with a warning. Therefore, it is preferred to examine

127 Kalsher & Williams 2006, p. 324-328; Young & Lovvoll 1999, p. 61-76.

128 Edworthy & Adams 1996, p. 49-73.

129 This is also referred to as the incidental exposure experimental approach, see Wogalter & Dingus 1999, p. 59.

130 See *e.g.* the rating experiments of Wogalter e.a. 1987.

131 *E.g.* Kim & Hunter 1993a; Kim & Hunter 1993b.

132 See *e.g.* the research of Edworthy e.a. 2004 (described in paragraph 3.6.9.2) which measured compliance with safety information subjectively and objectively in two separate studies. The results of the studies were similar and showed that there was a relationship between participants' subjective judgement with regard to the perceived effectiveness of a variable and the degree in which that variable produced actual compliance behaviour.

compliance objectively through observing actual behavioural compliance in a controlled laboratory setting or in a field study.

3.5.4.3 Difficulties Associated with Studying Behavioural Compliance

Notwithstanding that behavioural compliance is the most important stage, relatively few studies have measured actual behavioural compliance. The majority of the warning research consists of studies that have investigated the effects on the cognitive stages prior to behaviour. Of the studies that have evaluated behavioural compliance, most were carried out in the laboratory.

The explanation for this imbalance between warning studies seems to be that studying behavioural compliance can be difficult. The main reason is that it is unethical to expose participants of a study to real hazards. As a result, researchers must create a product use situation that is safe. Moreover, it must be under realistic conditions of product use for the participants. They must believe that they are in a hazardous situation and that following the warning will prevent the hazard. Another potential obstacle is that of control. It may not always be possible to control or mimic the conditions that are needed for the situation, especially in field or quasi-field studies. Furthermore, behavioural compliance studies can be time and labour consuming as well as expensive. In such instances, using measures of behavioural intentions are more suited.[133]

To deal with the above-mentioned problems, warning researchers have used the incidental exposure experimental paradigm for their experiments. A widely used incidental exposure paradigm in the field of warnings is the chemistry demonstration task, introduced by Wogalter in 1987.[134] Within this approach, participants of a study are not informed of its real purpose (that is to say that the study deals with the effectiveness of the warning). The warning is presented to participants in the context of a set of tasks that they are trying to accomplish. Participants' exposure to the warning occurs incidentally and it simulates how people are most often exposed to warnings in real life.[135]

Even though there are difficulties related to the study of warning compliance, the warning research literature strongly advises researchers not to be discouraged from conducting behavioural research in the future. In the end, studying actual behavioural compliance is the best method of warning effectiveness and several techniques are presented to researchers.[136] This does not mean that subjective measures of warning effectiveness or indirect measures of the cognitive processes that underlie behaviour are of no value. In

133 Kalsher & Williams 2006, p. 325.

134 Wogalter e.a. 1987. Participants are led to believe they will participate in a chemistry demonstration procedure in which they come into contact with potentially dangerous chemicals. Section 6 (*e.g.* paragraph 3.6.3 and 3.6.9.8) discusses studies that have used the chemistry demonstration task.

135 Kalsher & Williams 2006, p. 327.

136 Kalsher & Williams 2006, p. 326.

contrast, it is recommended to collect data that is gathered with a variety of measures in order to bolster our insight into the influential factors that affect the warning process.[137]

3.5.5 *Methodological Considerations*

The warning process is complex; various factors can be influential. It is thus not surprising that researchers experience difficulties in conducting research on warnings. Researchers have drawn attention to the methodological concerns in the warning research and research design recommendations have been given to address the issues.[138]

For example, one issue concerns the different testing methods that have been used across studies to measure the effectiveness of the cognitive stages involved, such as questionnaires, preference ratings or observational methods. As a result of the use of different methodologies, research findings are more difficult to compare and may have contributed to mixed results with regard to the effect of a variable on the dependent measure.[139] Furthermore, the validity and reliability of the data acquired through certain measuring techniques are not as good as with others. For example, questionnaires using rating scales have often been employed to examine risk perception whilst an observational approach to study the perception of risk in consumer products can provide more complete data.[140] Nevertheless, each methodology has its advantages and shortcomings.[141] Another issue is that several studies do not include a control condition, in which no warning is given. More studies with a control condition are preferred, because it creates the possibility to measure whether the presence of the warning increases safe behaviour. A second benefit is that studies can then be compared in order to generalise the findings.[142]

3.5.6 *Conclusion*

This paragraph addressed the common methods that have been used by researchers over the past decades to collect empirical data on the processing of product warning messages. The warning studies have been conducted in the field and in the lab employing a variety of measures. Relatively few warning research has taken place in the field. The majority of the warning studies are employed under highly controlled laboratory conditions and are primarily based on studying the effects of one or more variables on an intermediate information processing stage and on the behavioural stage with objective or subjective

137 Young & Lovvoll 1999, p. 76.
138 Rogers, Lamson & Rousseau 2000, p. 132; Edworthy & Adams 1996, p. 68.
139 See Hancock e.a. 2004, p. 184.
140 Van Duijne, Green & Kanis 2001.
141 Smith-Jackson & Wogalter 2006; Edworthy & Adams 1996, p. 49.
142 Cox III e.a. 1997, p. 201; Edworthy & Adams 1996, p. 8.

measures. The independent variables are the variables that the researcher wants to investigate and that are often related to the design of a warning, such as the warning's size or the presence or absence of symbols, colour or signal words. These variables are being manipulated in order to assess their impact on the dependent variable like attention to warnings or warning compliance.

The paragraph of paragraph 3.5.3 gave a general impression of the research methods that are generally employed to study the intermediate receiver stages of the warning process, such as attention, comprehension and memory, attitudes and beliefs, and motivation. It became apparent that each method has its own benefits and weaknesses. The stage of attitudes and beliefs and motivation are primarily assessed on the basis of subjective measures by asking participants directly what they think. Such subjective measurement has often been done on the basis of a so-called Likert-type scale. In the other stages, more objective methods are available such as measuring the time to respond to a warning to measure attention.

Paragraph 3.5.4 discussed how behavioural compliance can be measured. This can be done indirectly, which concerns measuring the variables that influence the above-mentioned substages or directly by observing actual behavioural compliance in a controlled laboratory setting or in a field study. These latter methods are objective, as there is involvement of someone or something that is external to the participants of the study. Using behavioural intent as a subjective measure to study behavioural compliance has been done frequently. Although it is a first step in evaluating the effectiveness of warnings, warnings should ultimately be tested in real-world settings to gain empirical evidence with regard to behavioural compliance. Furthermore, paragraph 3.5.4 paid attention to the difficulties of studying behaviour. For example, measuring actual behavioural compliance in the lab is challenging, because researchers cannot expose participants to real-world hazards, the laboratory setting must be believable to participants and the design of these experiments takes time and effort.

Finally, paragraph 3.5.5 described a number of methodological issues associated with designing a good warning experiment. Researchers have stipulated that, in order to move forward, future research should address these.

3.6 THE EFFECTIVENESS OF PRODUCT WARNINGS

3.6.1 Introduction

In the following paragraphs (3.6.2-3.6.9), the stages of the C-HIP model are described together with the corresponding variables that have been identified in the literature as potentially influential.[143] Furthermore, I discuss important design implications and guidelines suggested by the warning research for each information processing stage of the C-HIP model including the behavioural effectiveness of warnings.[144] Bear in mind that because the warning process is an interactive process, a certain degree of overlap is inevitable and as a result, a number of warning issues and/or factors are discussed at several locations.

In addition to the discussion of the factors identified by studies and the design implications, a number of principles that facilitate warning effectiveness are also discussed in this paragraph, such as the design principle of employing a warning system in paragraph 3.6.10 and the principle of testing warnings in paragraph 3.6.11. Even though there have been many studies that have investigated the effectiveness of warnings, there is still a need for more research impetus in this field. The directions in which future research should develop are mentioned in paragraph 3.6.12. The final paragraph 3.6.13 provides a summary of the previous paragraphs.

This chapter mainly focuses on visual product warnings.[145] For explanatory and illustrative purposes, I discuss various warning studies in greater detail along the way. A final remark is that not all research findings related to the topic of warning design and warning effectiveness are mentioned here. Nevertheless, an attempt is made to provide a realistic overview of the empirical evidence and literature on this topic.

3.6.2 *Source: Transmitter of the Warning*

Characteristics of the source can influence the effectiveness of a warning message. The source component of the C-HIP model is the initial transmitter of a warning, for instance the producer or the government.[146] Important warning issues such as whether there is a need for a warning and how the warning should be designed and communicated to receivers are dealt with by the source.[147]

Whilst the effects of design variables, such as colour, are investigated intensively, the impact of the characteristics of the warning source on warning effectiveness has been scarcely researched.[148] An explanation for this might be that consumer product warnings are an impersonal means of communication as they usually contain just textual information. It is plausible that transmitting information about the characteristics of the source by means of an expert in a broadcast advertisement is more powerful than the use of source

143 The C-HIP model was introduced in paragraph 3.4.2.

144 See *e.g.* Lehto 1992; Wogalter, Conzola & Smith-Jackson 2002; Wogalter 2006a provides a table with warning design guidelines.

145 For a review of the auditory warning research, see *e.g.* Edworthy & Adams 1996.

146 See for an extensive discussion of this stage Cox III 1999 and Cox III & Wogalter 2006.

147 Cox III & Wogalter 2006, p. 112-114.

148 See *e.g.* Lirtzman & Shuv-Ami 1986; Wogalter, Kalsher & Rashid 1999 (described in paragraph 3.6.2).

information in a written warning. Another reason why little research has examined source characteristics as a factor in warning effectiveness is that the available warnings on consumer products rarely identify the source of the message explicitly.[149]

Fortunately, there is a body of research on persuasive communications that provides a good starting point for understanding the role of the source characteristics in communicating warnings effectively.[150] Although there are inconsistencies in the empirical communications research, it is clear that source characteristics can contribute to the effectiveness of a communication. These characteristics include the credibility and likeability of the source.[151] Credibility can be defined to include expertness and trustworthiness.[152] Studies have reported that the greater the perceived credibility of a source, the greater the persuasiveness of the message.[153] Likeability refers to someone's physical or personal attractiveness. Such people are usually more likely to communicate a message persuasively than less attractive ones. An illustrative example is that in the beginning, tobacco manufacturers used athletes and doctors in their advertisements to spread the message that the use of cigarettes was safe.[154]

On a more general note, preliminary research on risk perception in product use paints a comparable picture with respect to the influence of the characteristics of the producer. The results suggest that product users' belief in quality brand names of products can influence their judgement of how safe they think the product is.[155] Nevertheless, if people believe that products are safe because they are of a well-known high-quality brand, they may exercise less caution than necessary.

The findings of the warning research suggest that when people perceive the source of a warning as reputable and expert, the warning appears more credible and its presence enhances the willingness to comply. The primary benefit of including information that emanates from a trustworthy independent expert is that it may help to change erroneous beliefs and attitudes that are being held by warning receivers.[156] Note that in such event the information processing stage of attitudes and beliefs affects the earlier stage of message source. Adding information about the source characteristics to warnings can also create certain 'side-effects'. It may negatively affect the legibility of warnings. Given that the available space on many product labels is constrained, warning design trade-offs have to

---

149 Cox III & Wogalter 2006, p. 118.
150 See *e.g.* on this topic Lipstein & McGuire 1978.
151 Cox III & Wogalter 2006, p. 115; see also Williams & Noyes 2007, p. 15.
152 Hovland, Janis & Kelley 1953.
153 See *e.g.* McGinnies & Ward 1974.
154 Bohme & Egilman 2006, p. 637.
155 Van Duijne 2005, p. 134, 227.
156 Cox III & Wogalter 2006, p. 119.

be made by the warning designer between providing source information on the label and the inclusion of other important information.[157]

Because of the paucity of warning research on this topic, one must be careful with generalising the findings. Future research is needed to determine if and how the inclusion of source-related terms enhances the effectiveness of warnings and ultimately safe product use. It may be that people do not perceive the warning to be more credible when terms referring to the national or European authorities are being included in the warning message for the reason that they have a negative attitude towards the government. The use of source information would then not be beneficial to warning effectiveness.

> The two experiments conducted by Wogalter, Kalsher and Rashid addressed the effect of the signal word WARNING and the effects of adding various words that referred to the source of the warning on people's judgement of the warning's credibility and their intention to comply.
> The participants viewed the warning messages of alcohol, cigarettes and iron-containing vitamin supplements. There were six conditions in the first experiment:
> (1) no words added;
> (2) WARNING;
> (3) GOVERNMENT WARNING;
> (4) US GOVERNMENT WARNING;
> (5) FEDERAL GOVERNMENT WARNING;
> (6) US FEDERAL GOVERNMENT WARNING.
> Participants were asked to imagine that the signal words were added to the beginning of the warning messages and to assess the credibility of the warning and the likelihood that they would comply with the warning on the basis of a rating scale from 0 (not at all credible/likely) to 8 (extremely credible/likely).
> The results showed that the warning message of the iron-containing vitamin supplements was rated significantly more credible than the other two warnings. The presence of the signal word WARNING produced higher credibility ratings compared to its absence. Furthermore, adding information about the message source increased participants' judgement of credibility. Adding US GOVERNMENT to the term WARNING scored significantly higher on credibility. The three highest combinations were US GOVERNMENT WARNING, FEDERAL GOVERNMENT WARNING, and US FEDERAL GOVERNMENT WARNING. The two highest (FEDERAL GOVERNMENT WARNING and US FEDERAL GOVERNMENT WARNING) were significantly greater than GOVERNMENT

157 Wogalter, Kalsher & Rashid 1999, p. 192.

WARNING. As regards the compliance likelihood ratings, similar results were obtained.

The procedure of the second experiment was identical to the first experiment, only it investigated the effects of three categories of sources on credibility and compliance likelihood. These sources were:

(1) governmental agencies (*e.g.* US FOOD AND DRUG ADMINISTRATION WARNING, US CONSUMER PRODUCT SAFETY COMMISSION WARNING);

(2) scientific groups (*e.g.* AMERICAN MEDICAL ASSOCIATION WARNING, AMERICAN PEDIATRIC ASSOCIATION WARNING);

(3) general statements without a source (*e.g.* MEDICAL HEALTH WARNING, IMPORTANT HEALTH WARNING).

As in the first experiment, the presence of the word WARNING produced higher credibility and compliance likelihood ratings than when it was absent. There were no significant differences between the categories of sources. The highest credibility ratings were for US SURGEON GENERAL'S WARNING, AMERICAN MEDICAL ASSOCIATION WARNING, US FOOD AND DRUG ADMINISTRATION WARNING, and AMERICAN PEDIATRIC ASSOCIATION WARNING. The source information AMERICAN MEDICAL ASSOCIATION WARNING produced the highest compliance likelihood ratings, followed by US FOOD AND DRUG ADMINISTRATION WARNING, US SURGEON GENERAL'S WARNING, and AMERICAN PEDIATRIC ASSOCIATION WARNING.

The findings of both studies support the view that adding source information can influence the effectiveness of a warning. It was shown that the presence of specific, reputable expert source information compared to its absence increased participants' judgements of the credibility of the warning and their intent to follow it.[158]

3.6.3 Channel: The Way of Transmitting the Warning

The characteristics of the channel can benefit the effectiveness of warnings. The channel represents the way in which warnings are transmitted from source to receivers.[159] The source selects how to communicate the message.

The warning channel has two dimensions. The first refers to the *medium* by which the information is sent. Warnings can be presented through product labels, mass media cam-

158 Wogalter, Kalsher & Rashid 1999.

159 See for an extensive discussion of this information processing stage Mazis & Morris 1999 and Cohen e.a. 2006.

paigns, advertisements on television and radio, or electronic media such as DVDs, CD-ROMs and internet. The second dimension of the channel concerns the *sensory modality* of receivers used to capture the warning. The most common way of communicating warnings is through the visual (printed text warnings) and auditory channel (alarm tones, voice warnings). Other sensory channels that can be used to deliver warning information but are rarely considered are the tactile (braille warnings), the olfactory (odour cues) and the gustatory channel (taste cue). The two dimensions are tied to each other: a warning on a videotape can communicate information both via the auditory (voice) and visual (text and symbols) senses.

The warning channel fulfils an important role in the C-HIP model. If warnings are not delivered to the target audience, it is clear that warning effectiveness will not be reached and that consumers are at risk.[160] A problem that can occur during delivery of warnings is for example that receivers are unable to process the warning because of their sensory deficiency or that the warnings are not available because the separate product manual got lost.

The type of sensory channel can affect the processing of a warning. Clearly, an auditory warning is likely to be more effective in conveying a short message than a long one. Similarly, long, more complex warning messages lend themselves to be visually communicated.

Studies have investigated whether one modality is better than another with regard to processing warnings.[161] Their results are somewhat contradictory. Nevertheless, they generally suggest that a spoken voice warning is better in producing compliance behaviour than a print warning.[162] The advantage of voice warnings is that they are attention-getting, even in circumstances where the visual attention of people is occupied and focused on other things. A potential disadvantage is that voice warnings take time to be transmitted and are thus not feasible for presenting long messages.[163] Moreover, the research demonstrates that in many contexts it is more effective to use two modalities to transmit a warning message than one, particularly both a print and a voice warning combined. By presenting the warning in more than one sensory channel, a larger population can be reached, including persons who are visually or hearing impaired. A second reason is that two channels attract more attention than one. Whether a modality can enhance the effectiveness of a warning also depends on the context in which the warning is presented or the product task. For example, the effect of an auditory warning on a construction site may be reduced

160 See also in this regard the stage of delivery in the C-HIP model.

161 See *e.g.* Wogalter & Young 1991 (described in paragraph 3.6.9.6); Wogalter, Kalsher & Racicot 1993; Wogalter e.a. 1991b (described in paragraph 3.6.3); Barlow & Wogalter 1993; Racicot & Wogalter 1992; Racicot & Wogalter 1995 (described in paragraph 3.6.9.12).

162 See *e.g.* Wogalter e.a. 1991b (described in paragraph 3.6.3); Wogalter & Young 1991 (described in paragraph 3.6.9.6).

163 Wogalter e.a. 1991, p. 722.

by the noise on the site or because personnel is wearing hearing protection. A print warning sign may be more appropriate to deliver the information.[164]

> Wogalter, Rashid, Clarke and Kalsher examined the effects of a multi-modal warning sign on compliance behaviour in a controlled laboratory situation.
> A total of 198 undergraduate and high school students followed a set of printed instructions to perform a chemistry task that involved measuring and mixing disguised non-hazardous chemicals. It was measured whether visual clutter around the warning sign, the presence of pictorials, the presence of a voice warning, and the presence of a flashing strobe light influenced compliance behaviour with the warning, which consisted of putting on and wearing protective glasses and a mask before mixing the chemicals. Their effects were measured combined and individually.
> The print warning sign contained black words on a yellow background, the symbol of a triangle exclamation mark and the signal word CAUTION on the top of the sign and below the text 'Skin and Lung Irritant. Improper mixing may result in a compound that can burn skin and lungs. Wear rubber gloves and mask'. The two pictorials illustrated the wearing of gloves and a mask. The strobe light was attached to the sign and flashed for eight seconds. On the right of the warning was a speaker for the voice warning. The voice warning contained a digitised male voice vocalizing the message of the warning sign. The sign was placed near a laboratory table and could be seen while standing at the table. The laboratory material and equipment were presented for the participants on the table. The clutter variable was manipulated by scattering the laboratory table with extra tools and equipment.
> It was hypothesised that the presence of pictorials, a voice warning, and a flashing strobe light would increase the salience of the warning sign in visual clutter and thereby reduce any camouflaging effect clutter might have. After the task, participants were asked to complete a questionnaire that addressed whether they saw the protective equipment, whether they saw or heard warnings, and if they could remember the content of the warning. The questionnaire also asked about the perceived hazardousness of the chemicals and their carefulness in performing the task.
> The results showed that complying with the warning was significantly greater when the warning was presented in an uncluttered environment compared to a cluttered environment. The results also showed that the presence of a voice warning produced a strong and reliable increase in compliance compared to

164 Cohen e.a. 2006, p. 132.

conditions without a voice warning. Although there was a tendency for greater compliance when the pictorials and the flashing strobe light were present, no statistically reliable effect was found. In addition, compliance was significantly related to remembering the warning, perception of hazard, and reported carefulness. The results of the questionnaire showed that if the students complied with the warning that was present, they also reported seeing the protective equipment, seeing or hearing the warning and believing the products were more hazardous and more careful. Nevertheless, the number decreased along the measurements: there were fewer participants who complied with the warning than participants who were aware of the protective equipment and the warning.[165]

3.6.4 *The Receiver Stage of Attention Switch*

3.6.4.1 General

The next paragraphs focus on the mental activities that occur within receivers of warnings. The information processing models imply that the first requirement for processing warnings effectively is that they attract the attention of receivers.[166] Attention is an important consideration in warning research since warnings must be noticeable if they are to have any potential impact on receivers' behaviour. Attention is defined as the amount of cognitive effort and/or capacity that a person directs to a particular stimulus.[167] In other words, it refers to selecting certain stimuli from among many and focussing cognitive resources on those selected.[168]

Most modern theories of attention agree that people have limited pools of cognitive resources that are used for attending and for working (conscious) memory. People are unable to process all sensory cues simultaneously, which explains why people selectively attend to one or several cues and ignore others.[169] Nevertheless, people can perform tasks at the same time.[170] In general, people attend to the most salient visual or audible stimuli in the environment. To switch the attention of the receiver to a warning, the warning must have characteristics that make it stand out from the background and that draw attention away from other stimuli and thoughts. This is important since most environments are noisy and cluttered and often people's attention is divided among various stimuli. A visual

165 Wogalter e.a. 1991; Wogalter, Kalsher & Racicot 1993.

166 See for an extensive discussion of this information processing stage: Wogalter & Leonard 1999 and Wogalter & Vigilante 2006.

167 Kahneman 1973.

168 Kellogg 2007, p. 63.

169 Solso, Maclin & Maclin 2008, p. 90.

170 Payne & Wenger 1998, p. 153.

warning must be salient, conspicuous or prominent relative to its background, particularly since many consumers are not always actively seeking warning information.[171]

Over the past 30 years, empirical research has shown a variety of factors that can influence the salience of a visual warning and subsequently the noticeability of a warning.[172] The more important ones are discussed here.

3.6.4.2 Size

Evidently, the size of the overall warning message and the font size of the wording are of importance to capture attention.[173] Consideration must be given to the size of the warning message relative to other non-warning information and to the environment in which the warning occurs. Visual clutter on the label and in the environment can diminish noticeability. A warning will be less salient if other non-warning information on the label is larger. Furthermore, the salient features of the warning are particularly of importance when the environment in which the product is used is cluttered by various stimuli.[174]

3.6.4.3 Brightness and Colour Contrast

Another relevant design feature that can produce a more prominent warning is high brightness and colour contrast. Brightness contrast refers to the noticeability of an object against its background. An example of high brightness contrast is black print on a white background or vice versa.

Colour is one of the most common features that can make warning information stand out from its background.[175] To be effective in attracting attention, a coloured warning message must be distinguishable from its background, *e.g.* the remainder of the printed label, and from surrounding colours in the environment. Certain combinations, such as black print against a white background, have a high contrast and are likely to get noticed. Likewise, combinations of black on a saturated yellow and white on a saturated red produce a salient effect. In contrast, yellow on white is not a successful combination.[176] Furthermore, the use of fluorescent colours has gained interest in recent years. They have the ability to give extra emphasis compared to non-fluorescent colours. However, additional research is needed to examine the impact of the use of certain fluorescent colours in product warnings.[177]

171 Wogalter & Vigilante 2006, p. 245.
172 Wogalter & Vigilante 2006; Rogers, Lamson & Rousseau 2000.
173 See *e.g.* Barlow & Wogalter 1993; Young & Wogalter 1990; Adams & Edworthy 1995; Davies e.a. 1998.
174 See *e.g.* Wogalter e.a. 1991b and Wogalter, Kalsher & Racicot 1993 (described in paragraph 3.6.3); Wogalter & Laughery 2006, p. 896.
175 Kline e.a. 1993; Laughery e.a. 1993b; Braun & Silver 1995; Griffith & Leonard 1997.
176 Wogalter & Vigilante 2006, p. 246.
177 Wogalter e.a. 1999a.

3.6.4.4 Borders

Another method to draw attention is the use of a border around warning information, for example a red or black border around the text. One would suggest that using a border would be better than nothing at all. A study has shown that the use of warning signs with thick red and thick yellow borders were noticed and examined better than signs with thin or no borders.[178] Nevertheless, research findings have shown that this does not always hold. The effect of adding a border appears to be weak and it can even produce a detrimental effect on the noticeability of warnings.[179]

3.6.4.5 Signal Words

Signal words also tend to attract attention.[180] Three signal words are commonly used in warnings in the United States. These are 'WARNING', 'DANGER', and 'CAUTION'. In addition to attracting attention, signal words can also be used to connote different levels of hazard. This aspect is dealt with in the paragraph regarding the comprehensibility of a warning.

3.6.4.6 Warning Symbols

Warning symbols are attention-getting. Warning symbols, also called pictorials, graphics, pictographs, pictograms, icons, and so on, are words used to describe printed nonverbal presentations intended to convey specific information.[181]

There are all sorts of warning symbols. Although warning symbols are primarily intended to communicate information, they also have the ability to call attention to themselves, seeing that they generally have more prominent features than textual information. Several studies have shown that adding pictorials helps to increase the likelihood that a warning will be noticed, especially when the pictorial is used with text.[182] On the other hand, there are also studies that report a low rate of noticeability of warning signs.[183]

> In a study of Laughery, Young, Vaubel and Brelsford, the noticeability of warnings on alcohol containers was examined on the basis of three experiments. As of November 1989, a public health warning is required for all alcohol beverage containers marketed in the United States. While the wording and content of the mandated warning are prescribed by law, its design is not specified, except for the letter height and the requirement that the warning must be

178 Wogalter & Rashid 1998; Rashid & Wogalter 1997.
179 Laughery e.a. 1993b; Laughery & Young 1991b.
180 See *e.g.* Laughery e.a. 1993b; Bzostek & Wogalter 1999.
181 Wogalter e.a. 2006, p, 159; Johnson 2006, p. 464.
182 See *e.g.* Laughery e.a. 1993b; Kalsher e.a. 1994; Davies e.a. 1998; Bzostek & Wogalter 1999.
183 See *e.g.* Goldhaber & DeTurck 1988a.

conspicuous and prominent on the label. As a result, a wide variety of warning formats on alcohol containers were available.

In the first experiment, the researchers examined the manner in which design features of actual warning labels from the marketplace affected the noticeability of the warning. The method employed included the use of a response time measure. Subjects were shown a variety of actual alcoholic beverage containers and were instructed to respond whether or not they saw a warning appear on the label. The time necessary to determine the presence or absence of a warning was measured.

The results indicated that the location of the warning was an important feature. Warnings printed on the front were found more quickly and thus more salient than when the warning was placed left or right of the label. Furthermore, warnings printed horizontally were more noticeable than those printed vertically. Overall, the experiment demonstrated that the actual alcoholic beverage labels were poorly designed.

The second and third experiments manipulated the design features (pictorial, colour, signal icon and border) to assess their impact on attention. The features were examined individually, but also combined to determine whether their presence attracted attention compared to their absence.

The second experiment manipulated the warning in four ways. Firstly, a pictorial was present to the left of the warning or it was absent. Second, the warning was printed in red or in black. When red was used, only the entire warning was printed in red. Thirdly, an icon was present or absent above the warning. Fourthly, a border was placed around the warning or was absent. In total, there were 48 different warning combinations. The subjects were shown 96 labels, with and without a warning. It was assumed that the labels containing a more salient warning would be found more quickly.

The results of the second experiment demonstrated that warnings with a pictorial had significantly faster responses than warnings without a pictorial. In addition, red warnings were located more quickly than warnings printed in black. Equally, warnings with an icon were found more quickly than warnings without an icon. However, the effect of a border surrounding a warning resulted in a slight decrease in response latencies in comparison with warnings without a border. The researchers also compared the baseline warning (without the salience features) with a warning that had one or more salience features. Respondents' time to find the baseline warning was the longest of all the combinations. The use of all four salience features produced the greatest improvement over the baseline warning. The second-best label contained the coloured warning with a pictorial and an icon.

The third experiment measured attention by using eye-scan equipment to record visual scanning patterns of subjects searching for warning information on simulated labels. Such equipment can measure the location and the decision time for seeing the presence or absence of a warning. Location time is the time from the start of the label display until the subject's eyes arrive at the warning. Decision time is the time from the eyes' arrival at the warning until the response is given by the participant. Researchers presumed that shorter response times would indicate the degree in which a feature would enhance the noticeability of the warning.
The responses to a warning were faster when the warning included combinations of the features than when there was only one feature present. The warning with colour, a pictorial and an icon produced the lowest total response time, the warning without any salience features the highest. With regard to the location as well as the decision time, the pictorial was the only individual design variable that had a significant effect on the noticeability of the warning, compared to the other features. The lowest location time was produced by the warning with a pictorial, colour and icon. The lowest decision time was produced by the four features together.
Overall, the results of experiments 2 and 3 demonstrated that adding salience features such as colour, pictorials and icons to a warning can substantially improve its noticeability.[184]

3.6.4.7 Location

Where the warning information is located affects the probability that consumers will notice a warning. Placement is concerned with how the product is used and the environmental context in which the warning occurs. Obviously, warnings are not likely to grab people's attention if they are placed on a location where product users are unlikely to look.

A general guideline from the research literature is to place the warning in close proximity to the danger in place and in time unless this creates a potentially dangerous situation because warning receivers see or hear the hazard too late to avoid it.[185] Proximity refers to the physical distance between the warning and the product or user, but it can also relate to the time that the warning appears whilst the user performs a task with a product, for example, during the different steps of installing a product.

184 Laughery e.a. 1993b.

185 Wogalter & Laughery 2006, p. 896; Rogers, Lamson & Rousseau 2000, p. 113.

First of all, it follows from this that warnings should be physically placed in close proximity to the product and the hazardous characteristic. Hence, placing warnings directly on the product package is preferred to its presence in a separate instruction manual. This corresponds with the finding that users expect warnings on hazardous products to be located in close proximity of the product. Failure to find a warning in an instance may lead people to assume the product is less hazardous than it actually is which could lead to unsafe handling of the product.[186] Nevertheless, space limitations can require an additional manual. Because manuals may not be read or available when needed, it is recommended in the literature to place the most important warning information on the product and to direct users to a secondary location, like a manual, for more detailed information.[187]

Secondly, it is important that warnings are located close to the hazard in time, which means that a warning is given when users are exposed to the hazard during the use of the product. Placement in time is likely to be more effective than physical placement, because it considers users' cognitive and behavioural activities during product use. Note that placing a warning too far in advance will cause product users to forget the warning message.[188]

The optimal location for a warning varies across products. Whether the location is aesthetically pleasing for consumers can for example be of relevance too. Consideration of how users behave with the product is of value for finding an effective warning location.[189] Given that more than one location can be suitable candidates, it is useful to carry out a task analysis to determine where a warning should be located.[190]

The task-analytical approach of Frantz and Rhoades is helpful in this respect. It involves analysing the user's interaction with the product by decomposing the product task into subtasks in which exposure to a hazard exists. The task-analytical approach includes the cognitive and behavioural activities of users during product use to examine and identify which warning locations are temporally and spatially close to the critical task element and would prompt users to notice and comply with the warning. The cognitive aspect of the task analysis involves examining the subtasks and predicting the extent to which users will be seeking external warning information, as opposed to applying their prior knowledge to the use of the product. To achieve this, scripts that users are likely to invoke during interaction with the product are considered. *Scripts* are schemas stored in long-term memory. They are mental representations of the usual sequence of events and consequences of events, such as the steps involved in getting ready for work.[191] Script theory suggests

186 Wogalter e.a. 1991, p. 75.
187 Wogalter & Vigilante 2006, p. 250.
188 Wogalter & Vigilante 2006, p. 250.
189 Wogalter, Barlow & Murphy 1995, p.1088.
190 Frantz & Rhoades 1993 (described in paragraph 3.6.4.7).
191 Eysenck & Keane 2000, p. 381.

that after experience, people tend to use behaviours based on that experience in future events. With repeated experience, these sets of behavioural sequences become well-learned. Frequently, as a result of the development of a well-learned script, behaviour occurs automatically without much conscious thought. Interrupting a well-learned script by introducing a new element into the situation of product use that emphasises a warning's location increases the likelihood that product users notice and ultimately comply with the warning.[192] The advantage of this analysis is that warnings that are well-integrated into the product task being performed can trigger the retrieval of relevant information from long-term memory.

The task-analytical approach can also be beneficial to determine the location of important warning information relative to other information on the product label itself. Many consumer product labels typically have several sections with pieces of information.[193] The findings of the warning research do not show a clear picture with respect to where important hazard-related information that is embedded in other safety or product information, like on the label or in an owner's manual, should be placed.[194] It seems that effective placement is related to how people attend to and process information during product use. A general design guideline following from the research is that warnings should be integrated into the flow of task-related information that users are likely to process.[195] This guideline results from the finding that people tend to read only portions of certain sections on a product label rather than process each and every section in its entirety. Placing warning information (such as safety precautions) in the directions for use section led to higher levels of noticing, reading and compliance than when it was placed in a separate section, because participants were looking for directions on how to use the product. Likewise, studies have revealed that once users realised that they were reading a warning, they tended to skip to another section on the product label. The subjects stopped reading the warning section because they were not interested in learning about the dangers of the product, but wanted to know what the product could be used for. Apparently, the product users' goal was simply to seek information on how to use the product, not on how to use the product safely.[196] Hence, this indicates that people selectively attend to information that is compatible with their goals.[197] Field studies that investigated risk perception in consumer product

192 Duffy, Kalsher & Wogalter 1995, p. 160.

193 These sections that usually appear on the back panel of a product can concern information about the use instructions, precautionary statements containing warning information on the danger and how to avoid the danger and a section with other mandatory product information.

194 Wogalter & Vigilante 2006, p. 257; Wogalter, Barlow & Murphy 1995, p. 1082. Some studies indicate that it is preferable to put warnings before the instructions for use in a manual. Placing important warning information in the section of the directions for use on a product label also appears to be a good location, see *e.g.* Wogalter e.a. 1987; Wogalter, Barlow & Murphy 1995; Edworthy e.a. 2004.

195 Lehto 1992, p. 129; Frantz 1993, p. 152; Frantz 1994, p. 544.

196 Strawbridge 1986; Friedmann 1988, p. 514; Frantz 1994, p. 544.

197 Lehto 2006, p. 69; Stewart, Folkes & Martin 2001, p. 336, 343.

use paint a similar picture as regards users' goals of product use. The findings imply that the beneficial results that can be achieved with consumer products are the main reason for consumers to interact with them. Safety and protection against the risks can be incompatible with the users' goal and lead to unsafe actions.[198]

> Frantz and Rhoades used a task analysis in their research to investigate and measure the effects of various candidate warning locations and task interference on attention and warning compliance.
> Within the context of realistic usage conditions, 60 subjects had to unpack and arrange office furniture, including a two-drawer metal file cabinet. Their behaviour was recorded on video and they were not informed of the true purpose of the experiment. The warning message, which was an actual warning, was in black print and stated:
>
> 'WARNING. TO AVOID TIPPING YOUR FILE: (1) ALWAYS LOAD THE BOTTOM DRAWER FIRST, AND WHEN FULL, FILE IN THE NEXT DRAWER ABOVE. (2) ALWAYS LOAD THE TOP DRAWER AFTER ALL OTHER DRAWERS ARE FULL. (3) NEVER OPEN MORE THAN ONE DRAWER AT A TIME.'
>
> There were four warning conditions. Condition 1 involved the presence of a warning located on the side of the shipping carton of the file cabinet along with other information. The other three warning conditions were determined on the basis of a task analysis by predicting the likely interaction of the user while unpacking and loading the file cabinet. The task performance was divided into the following subtasks:
>
> (1) remove file cabinet from shipping container;
> (2) place cabinet at desired location in the room;
> (3) remove packing tape from drawers;
> (4) open one or both drawers;
> (5) gather files or materials in drawer(s);
> (6) place files or other materials in drawer(s);
> (7) close drawer(s).
>
> Thus, in the first condition, the warning is placed at step 1. In the second warning condition, the warning label was affixed to the bottom of the top

198 Van Duijne 2005, p. 228.

drawer. In this location, the warning was visible as soon as the user opened the top drawer. The user would probably encounter the warning after he started to load the drawer. In the third condition, the warning was put on the front of the file cabinet and on the bottom of the top drawer. The first could be removed after initial use, while the latter was located permanently on the bottom. According to the researchers, the performance of the task consisted of two scripts, *i.e.* unpacking the file cabinet and loading the file cabinet. The danger warned against was associated with loading the cabinet and this script probably starts at step 4. The experimenters hypothesised that physically interrupting this script at its beginning might prompt the subjects to attend to the warning more than to a warning presented later in the script. In condition 4, the warning was designed to physically obstruct the placement of files in the top drawer. The warning was printed on white cardboard and was placed in the top drawer. The cardboard formed a bridge across the width of the drawer. In addition, a warning was also positioned on the bottom of the top drawer. The cardboard bridge obscured the warning on the bottom of the top drawer. The researchers assumed that the warning bridge would be more effective than the warning placed on the shipping carton or on the bottom of the top drawer, because in the fourth condition the warning formed a physical obstruction that interrupted the user at the point of hazard. Furthermore, they hypothesised that the fourth warning condition would be more effective than the third, since the bridge appeared to more substantially interfere with the task.

The results showed that the effects of warning location on noticing, reading and complying with the warning were significant. Across the four warning conditions, the percentage of subjects who noticed the warning ranged from 0% to 93%, and the percentage who complied with the warning ranged from 0% to 53%. More subjects noticed and read the warning in the third (front and bottom) and fourth (warning bridge and bottom) conditions than in condition 2 (bottom). Also, more subjects noticed and read the warning in the second condition than the warning located on the carton. With regard to warning compliance, there was a significant difference between the warning in condition 2 and in condition 4. Furthermore, the findings confirmed that the two warnings that interfered with task performance were significantly more effective than those of the first and second condition.

The research illustrates that the effects between the warning locations can differ considerably. Moreover, they point out the importance of a task analysis to determine the best location.[199]

199 Frantz & Rhoades 1993.

3.6.4.8 Interaction with a Warning

As demonstrated in the above-mentioned experiment, noticeability can also be enhanced through interaction with the warning.[200] An interactive warning requires product users to interact with the warning label in order to use the product.

Locating warnings such that psychical manipulation of the warning is required to use the product safely can improve noticeability compared to placing a traditional warning onto the product. As an interactive label introduces a new step in the task performance, it disrupts script driven behaviour and increases the probability that a warning will be seen, read and complied with. In addition, an interactive warning provides a tactile cue since the warning must be touched or moved in order to use the product. A disadvantage of interactive warnings is that they can be too intrusive and as a result annoying to product users because they interrupt task performance. Thus, careful consideration of the level of interaction is advisable.[201]

> The study of Duffy, Kalsher and Wogalter examined the effects of interactive warnings, task load and colour on warning noticeability, recall of the warning content (as an indicant of reading), and warning compliance. Task load can be described as the number of tasks an individual is carrying out at any given time. The hypothesis under psychology is that an increased level of task load negatively influences performance. Such an effect can also be applicable in situations of product use where users carry out multiple tasks at once. This could distort a successful processing of warning information. Attention and recall were measured with a post-task questionnaire; compliance was assessed by observing the behaviour of the participants.
> Two interactive warning labels (with and without a colour component) on a familiar product were compared with the condition of a standard tag label and no label in the context of a realistic product-use task. The task involved the set-up of video equipment. Participants were asked to plug the electrical cord of the television and videocassette recorder into outlets using a set of extension cords. During this task, they were incidentally exposed to one of three warnings attached to the extension cords. Each extension cord had a removable outlet cover which was permanently attached near its female end. In the standard warning condition, the warning was attached to the extension cord and located 5 cm above the female receptacle. In the interactive condition, the warning was affixed to the outlet cover of the female receptacle. The warning panel on the

200 See *e.g.* Gill, Barbera & Precht 1987; Hunn & Dingus 1992; Frantz & Rhoades 1993 (described in paragraph 3.6.4.7); Duffy, Kalsher & Wogalter 1995 (described in paragraph 3.6.4.8); Visschers e.a. 2004.

201 Duffy, Kalsher & Wogalter 1995, p. 165; Visschers e.a. 2004, p. 1496.

cord contained a border surrounding the warning message with a signal icon, the signal word 'WARNING' and underneath 'Electric shock and fire. Do not plug more than two items into this cord' inside. In the colour condition, the signal word was in black print on an orange background. Researchers expected the interactive colour warning label under low task load conditions to be most effective.

Participants were initially told they would be evaluating instructional media. They were put into a room with the equipment left unplugged. While the experimenter was preparing the equipment, he remarked that he had left the videotape in another room. The experimenter then explained to the participant the tasks that they would be asked to do. Participants in the low task load condition were told that they would be watching a job training videotape and then would complete two questionnaires. Participants in the increased task load group were told that they would listen to a portion of an audiotape lecture on industrial control rooms (the extra task), would then watch a job training videotape and would later complete two questionnaires. In the higher task load condition, the experimenter started the audiotape and told the participant that he would return shortly with the videotape. The low task load condition lacked the audiotape and all procedures associated with it. Before exiting the room, the experimenter asked each participant if he or she would mind 'helping out' by plugging in the television, video cassette recorder, and videotape rewinder using the extension cords. In the higher task load condition, participants had to carry out the task while they listened to the audiotape. The experimenter left the room and returned after four minutes with the tape. Participants were shown the videotape in order to avoid drawing explicit attention to the tasks they had just performed and to help disguise the true purpose of the study. Finally, participants were taken into another room to complete the questionnaires.

The results of the study showed significant effects of label type on noticeability, recall and compliance. Participants said that they noticed, recalled and complied with both interactive warnings (with and without colour) more often than the standard warning and when there was no warning attached to the cord. With regard to the compliance rates, 6.7% complied when there was a tag, 43.3% complied with the coloured interactive warning, 53.3% of participants complied with the interactive warning without colour, and when no warning was present no one complied. No significant differences were measured between the interactive warning with and without colour, although there was a positive non-significant effect for colour on noticeability and recall. Additionally, the study failed to find a significant difference between the normal warning condi-

tion and the absence of a warning. The higher task load showed no significant influence on any of the measures.
The results indicate that an interactive label facilitates the capturing of attention and thus increases the potential for further processing of the warning message.[202]

3.6.4.9 Hazard Perception

Person variables that have consistently been proved to influence the likelihood that a warning will be noticed are hazard perception and familiarity with a product.[203] These receiver characteristics have an influence on people's expectations of a product and on safe use and emanate from the processing stage of attitudes and beliefs.[204]

Hazard perception refers to the threat posed by a particular product, object or activity.[205] Many warning researchers use this term instead of risk perception, since findings have indicated that people's subjective perception of a risk is mainly influenced by the severity of the negative consequences rather than the probability that these consequences happen. Research results reveal that the more hazardous consumers perceive a product to be, the more likely they will look for, notice and read a warning.[206] This illustrates that processing at an earlier stage can be affected by a later stage of the C-HIP model.

As the noticeability of a warning can be affected by lowered hazard perception, increasing the level of hazard perception with the use of warning design features that have shown to affect hazard perception may be a solution.[207]

3.6.4.10 Familiarity Belief

In the warning literature, the *familiarity belief* is defined as product users' individual knowledge of and/or prior experience with the product, object or activity in question.[208] Hence, product familiarity is closely related to prior experience and frequency of use. Nonetheless, product familiarity is not always akin to product experience. People may have some familiarity with products or classes of products that they seldom or never use themselves.[209]

Research indicates that people who are less familiar with a product may be more likely to actively seek warning information, notice it and subsequently comply with the warning

---

202 Duffy, Kalsher & Wogalter 1995.
203 Rogers, Lamson & Rousseau 2000, p. 105.
204 Paragraphs 3.6.7.2 and 3.6.7.6.
205 DeJoy 1999a, p. 192.
206 See *e.g.* Godfrey e.a. 1983; LaRue & Cohen 1987; Otsubo 1988 (described in paragraph 3.6.9.8); Wogalter e.a. 1991.
207 Rousseau & Wogalter 2006, p. 150. This will be discussed in more detail in the section on attitudes and beliefs, paragraph 3.6.7.
208 DeJoy 1999a, p. 202.
209 DeJoy 1999a, p. 203; Riley 2006, p. 292.

because they do not know what to expect of the product's risks. Likewise, people who are more familiar with a product are less likely to look for, see and read the warning on the product's label than are consumers who are less familiar with the product.[210] However, findings suggest that when people perceive a familiar product as dangerous, their hazard perception and not their familiarity with the product is likely to determine whether they will seek for information.[211]

There is, however, another set of findings, originating from American surveys, which have shown that people who frequently used a certain product, like cigarettes or alcohol, were more aware of the warning label and were more able to recall the content of the warning than people who did not or seldom used these products.[212] One explanation for this is exposure: people who frequently drink alcohol or smoke cigarettes come into contact with the warning, are repeatedly exposed to it and thus are more likely to be aware it.[213] Even though the survey results are seemingly conflicting with the previous findings that reflect the concept of 'familiarity breeds contempt', they are quite easy to reconcile. It is not surprising that those subjects of a survey that frequently use the product report being more aware of its warning compared to non-users. However, if those people were brought into the laboratory to perform a task with that product, non-users would be more likely to read and follow the warning messages.

Hence, in general, familiar or experienced users can be expected to be more knowledgeable about the product's hazards and more confident in using the product. This knowledge and confidence also means that they will probably not be motivated to seek additional information about the product. In contrast, inexperienced users are more likely to look for and notice warning information.[214]

3.6.4.11 Information Seeking Mode

As was touched upon just now, related to familiarity and hazard perception is the information seeking mode of people. If people are actively looking for warnings, they are more likely to attend to them. This particularly holds in the event that users are novice and unfamiliar with the product and/or when the product's characteristics are perceived as dangerous.[215] It suggests that attention will often be goal driven. If people want to use a product safely (because the product is new to them and/or they perceive it as dangerous),

210 See *e.g.* Goldhaber & DeTurck 1988a; Otsubo 1988 (described in paragraph 3.6.9.8); Godfrey e.a. 1983; Godfrey & Laughery 1984; Wogalter e.a. 1991.
211 Wogalter, Desaulniers & Brelsford 1986.
212 See *e.g.* Kaskutas & Greenfield 1992; Mazis, Morris & Swasy 1991; Graves 1993; Kaskutas & Greenfield 1992.
213 DeJoy 1999a, p. 204.
214 DeJoy 1999a, p. 205.
215 Wogalter & Vigilante 2006, p. 259; Rogers, Lamson & Rousseau 2000, p. 111.

they will seek out and read warning information, because warnings can provide assistance to consumers in achieving safe product use.[216]

The levels of performance model supports this mode of processing.[217] One of the conclusions based on this model is that warnings are likely to have the greatest influence when the behaviour is at a knowledge-based level, because people performing at this level are likely to be seeking out information and may not have developed conflicting behaviour patterns. Users who perform at a knowledge-based level are inexperienced and are learning or they are experienced, but in a novel complex situation. This emphasises that the users who operate at this level need a warning that explicitly informs the user what actions to follow to use the product safely.[218] In contrast, the model also concludes that warnings are unlikely to be noticed when people behave at a skill-based level which level is at the bottom of the hierarchy model. People who operate at a skill-based level are highly experienced: their behaviour corresponds to the following of a well-learned schema or script, in which there is little need for conscious control.[219]

3.6.4.12 Environmental Conditions

Unsurprisingly, environmental conditions such as smoke, fog, too much sun light or low illumination can also negatively influence the noticeability of textual information and symbols. Their visibility can decrease over time as a result of long-term exposure to environmental conditions, such as sunlight.[220]

3.6.4.13 Channel

Lastly, a fairly consistent finding of the warning research is that the use of more than one sensory channel to present the warning message enhances the likelihood that the warning information will be attended to.[221] The use of, for instance, an auditory warning in conjunction with a visual one provides an extra alerting cue. Unlike visual warnings, auditory warnings are omnidirectional, which means that receivers do not have to be looking at a particular location to be alerted.[222] Another benefit of having an additional channel is that the warning design takes people who have vision or hearing impairments into account.[223]

216 Lehto 2006, p. 69; Frantz 1994, p. 544; Stewart, Folkes & Martin 2001, p. 336, 343.
217 This theoretical model was discussed in 2.4.4. See Rasmussen 1986; Lehto 1991.
218 Lehto & Papastavrou 1993, p. 587, 590; See DeJoy 1999a, p. 199; Van Duijne 2005, p. 235.
219 Lehto & Papastavrou 1993, p. 584, 590.
220 Wogalter & Vigilante 2006, p. 247; Vredenburgh & Helmick-Rich 2006, p. 378.
221 Lehto 2006, p. 68.
222 Wogalter & Leonard 1999, p. 143.
223 Cohen e.a. 2006, p. 128.

3.6.5 *The Receiver Stage of Attention Maintenance*

3.6.5.1 General

Once attention is attracted, it needs to be maintained.[224] Because attracting people's attention to the warning and keeping it there can be affected by different variables, the stage of attention is split into two substages.

Attention maintenance refers to the process of holding attention onto the warning long enough so that users can encode the message. Encoding a warning means that the external information on the warning must be translated into some internal representation through reading verbal information, recognising pictorial information, or decoding pictorials and symbols.[225] The encoding process puts information into memory.[226] This stage fails if a person noticed the warning, but switched their attention away to something else. Whereas brief warnings have the potential to be encoded quickly, long messages must possess characteristics that make the encoding process likely.

A significant factor that facilitates the maintenance of attention is legibility. Legibility concerns whether the letters of words or symbols are discernable/distinguishable so that they can be identified and recognised. Product users will not spend much time and effort on a warning that is difficult to read and takes a large amount of time to process. Hence, warnings must possess qualities that make the reading process enjoyable. Although legibility and readability both refer to the ease with which people read, legibility concerns the functioning of design variables such as font, print size and format, while readability mainly is concerned with comprehension of a larger group of words.[227]

Admittedly, various design characteristics that attract attention also help make the reading process less effortful and more likely. A coloured label is perceived more enjoyable to read than a black and white warning label.[228] In addition, research indicates that attention maintenance is also enhanced by the use of warning symbols. People have also rated labels with graphic pictorials as more likely to read and easier to read compared to labels without pictorials.[229] What's more is that symbols can facilitate encoding especially in situations where there are time constraints, given that well-designed pictorials quickly communicate

224 See for an extensive discussion of this information processing stage Wogalter & Leonard 1999 and Wogalter & Vigilante 2006.

225 Rogers, Lamson & Rousseau 2000, p. 116.

226 There are three logically separable parts of any memory task. The first is encoding (also called acquisition or learning) and refers to the process of putting information into memory. The second part of the task is called retention, the process of holding on to information once it has been stored in memory. The final part is retrieval, the actions required to get information out of memory, see Payne & Wenger 1998, p. 178.

227 Frascara 2006, p. 396; Wogalter & Vigilante 2006, p. 253.

228 Braun, Mine & Clayton Silver 1995; Kline e.a. 1993; see also Nilsson & Kaiserman 2005.

229 Kalsher, Wogalter & Racicot 1996.

information.[230] Environmental conditions can also affect the legibility of textual information and symbols of warnings.[231] Furthermore, person variables, like familiarity and hazard perception, influence whether a warning is encoded and read.[232] Results indicate that consumers are more willing to read a warning on more hazardous products than on products that are not perceived as dangerous and consumers who are more familiar with a product are less likely to read the warning.[233] The perceived level of hazard by persons is the primary determinant of the likelihood that warnings will be read.[234] Finally, evidence also indicates that the placement of warnings, *e.g.* when embedded in other information, can also affect whether users will take the effort to read the information after having noticed it.[235]

Several factors in particular have been demonstrated to improve the legibility of a warning. The paragraphs below pay special attention to the variables that have empirically shown to have an influence on the receiver stage of attention maintenance.

3.6.5.2 Letter Case

Warnings containing words that are all printed in upper case are more difficult to read than warnings that have a mix of upper case and lower case letters. This can be explained by the visual angle of the letters. Lower case letters produce smaller visual angles than upper case letters. If the visual angle is small, the letters of the words may not be distinguishable from each other.[236]

3.6.5.3 Font Style

A design variable that may have an effect on the reading process is font style. Although the number of studies on choice of font in warning labels are small and the results ambiguous, it is recommended to use sans serif fonts such as Helvetica, Arial and Universe over fonts with serifs such as Times New Roman, New Century Schoolbook and Goudy.[237]

3.6.5.4 Increasing the Surface Area: Alternative Label Designs

Many consumer products have small containers. This is particularly true for pharmaceuticals and food packages. In view of the limited amount of space available on these labels,

---

230 Laughery 2006, p. 471.

231 Wogalter & Vigilante 2006, p. 247.

232 Rogers, Lamson & Rousseau 2000, p. 116.

233 See *e.g.* Otsubo 1988 (described in paragraph 3.6.9.8); Friedmann 1988 (described in paragraph 3.6.9.7); Leonard e.a. 1989; Wogalter, Desaulniers & Brelsford 1986.

234 Wogalter, Desaulniers & Brelsford 1986; Wogalter e.a. 1991.

235 Laughery 2006, p. 470; Wogalter & Vigilante 2006, p. 257. See *e.g.* Frantz 1994; Strawbridge 1986; Friedmann 1988; Laughery 2006, p. 470.

236 Wogalter & Vigilante 2006, p. 253.

237 Wogalter & Vigilante 2006, p. 254; Silver & Braun 1993.

producers often reduce the print size of warning information so that all the relevant information fits on the label. This in turn negatively affects the legibility of the label. People with poor vision, for instance the elderly, especially cope with reading difficulties.

There are methods to remedy the problem of limited space. One way would be to decrease the amount of information on the label by putting only the essential warning information on the label and the remaining elsewhere in a separate insert. The problem with this method is that consumers may lose or disregard the insert. Another approach is to increase the available surface area of small labels and as a result a larger more legible print size can be used for the warning information. This can be accomplished by alternative labels such as a tag label, wings on the container or a label that folds out.

Alternative labels have shown to be a desirable solution to enhance the process of attention switch and maintenance.[238] Alternative label designs are especially more preferred than traditional labels by older adults, because such labels allow the use of larger text.

> The purpose of the study of Kalsher, Wogalter and Racicot was to examine the user preference effects of alternative design labels of prescription drugs, such as a tag label and a fold-out.
> A tag and a fold out label were compared to standard pharmaceutical label design. Participants were asked to rate the labels on ease of reading the label, likelihood of noticing the warnings, likelihood of reading the warnings, preference for a label and likelihood of recommending a label to a friend. Secondly, researchers investigated the presence versus absence of pictorials. Participants were asked to rate the effectiveness of the pictorials in remembering or understanding the warnings.
> Six prescription drug bottles were constructed and shown to participants. All of the labels contained the same fictitious printed information, but were different in size and pictorials. Besides other information, the label also contained instructions for use ('TAKE 1 TABLET AT EACH MEAL AND 1 AT BEDTIME' and 'TAKE WITH WATER') and warnings ('MAY CAUSE DROWSINESS' and 'DO NOT TAKE WITH ALCOHOL'). The pictorials communicated information about the instructions and warnings. The print size of the traditional label was 8 point Times Roman. In the tag label condition, the instructions and warnings were put on a tag attached to the side of the bottle. The font size was 25% larger than that of the standard label. In the fold-out condition, the label was unfolded outward from the side of the bottle and then down. The surface area was 40% larger.

238 See *e.g.* Barlow & Wogalter 1991; Wogalter & Young 1994 (described in paragraph 3.6.9.3); Kalsher, Wogalter & Racicot 1996; Wogalter e.a. 1999a.

Across all dimensions, the tag label was measured the highest and the standard label the lowest. These effects were significant except for the question of label preference. The results showed that the tag label was rated significantly higher than the fold-out and the standard label, except for the ratings of likelihood of reading and recommending. The results with regard to the presence of pictorials were also convincing. For every measure, the presence of pictorials was rated significantly higher.[239]

3.6.5.5 Layout

Another factor that can enhance encoding of the warning message is its layout, that is, the spatial structure of the warning. The warning should have an aesthetically pleasing design to look at or else people may decide that reading the warning will take too much effort.[240] Most of the textual warning information is presented as continuous flowing text, in paragraph form.

A feature that affects a good format is the amount of white space in the printed warning. Sufficient white space (*e.g.* between the letters, between the lines or between different information components) is more likely to make the reading easier than a warning that has a large amount of compressed text. Studies indicate that people have a preference for and are more likely to actually read warning information presented in an outline format than information presented in paragraph layout.[241] Horizontal spacing, which is concerned with the amount of white space between the letters and the words, is also relevant. Justified text, in the sense that the text is aligned equally along both the left and right margins, can slow down the reading process. Left justification of the text, *i.e.* ragged right, is preferred, because the spacing between the letters and words is consistent.[242]

3.6.5.6 Potential Problem of Information Overload

A potential obstacle in this stage of attention is an overload of information, as a result of which consumers fail to encode the message. *Information overload* refers to the situation when the amount of information in a given situation is more than a person is able or willing to process.[243]

Many consumer products pose more than a single hazard. Consequently, a lengthy and complex warning message expressing multiple hazards attached to the product has the potential to reduce the legibility of the text, as it is likely to impose a high work load

239 Kalsher, Wogalter & Racicot 1996.
240 Wogalter & Vigilante 2006, p. 255.
241 See *e.g.* Desaulniers 1987; Morrow e.a. 1998; Wogalter & Shaver 2001; Wogalter, Shaver & Chan 2002; Wogalter & Vigilante 2003.
242 Wogalter & Vigilante 2006, p. 256.
243 Wogalter & Vigilante 2006, p. 258.

on individuals, which in consequence may require too much effort of people to attend to and process it successfully.[244]

The capacity of the short-term memory of humans is limited. The *short-term memory* which forms part of the working memory can only hold information in memory briefly for a few minutes and is then discarded.[245] Basically, this means that if people read a sequence of letters and are then asked to recall the letters, only a limited number can be retained in memory and recalled. The general rule is that short-term memory can deal with seven plus or minus two units of information at a time.[246] This means that seven separate numbers or letters can be processed through short-term memory and recalled in the correct order. Research has also shown that when these letters or digits are chunked into meaningful units of information, it is possible to remember seven *chunks* of information.[247]

Although research has not yet clearly answered the question regarding when overloading occurs with warnings, the implications for warning design following from the short-term memory limitations is that the content of a warning message should be as brief as possible and easy to read with simple, short sentences so that all the items in the warning can be successfully put into memory and recalled later when needed. Having irrelevant information in a warning message can thus affect someone's ability to recall the important information.[248]

Because many consumer products pose more than a single hazard, providing a relatively short warning message that expresses multiple hazards can be problematic.[249] Take for example the large amounts of information accompanying prescription drugs.[250] Warning researchers recommend the principle of *prioritisation, i.e.* ordering the hazards and their warnings, to help counter the problem of overloading. Placing high priority warnings onto the product label and less relevant information in an accompanying manual or a package insert increases the likelihood that important warning information will be seen, read and eventually followed.[251]

On the basis of their study, Vigilante and Wogalter have provided a number of considerations with respect to how to prioritise the product hazards that need to be warned of. These pertain to (1) injury severity; (2) injury likelihood; (3) not known by the target audience; (4) important to know; and (5) practicality. Thus, the more likely the injury, the

244 Lehto 2006, p. 68; Bettman, Payne & Staelin 1986, p. 9.

245 See Solso, Maclin & Maclin 2008, p. 158; Payne & Wenger 1998, p. 179, 199.

246 Miller 1956.

247 Solso, Maclin & Maclin 2008, p. 160; Kellogg 2007, p. 106. For example, the information 123456789 is easier to remember than 629479876, because the former set of digits composes a chunk, that is, a meaningful pattern or unit of information. Thus, as a result of chunking more information can be recalled.

248 Frantz e.a. 1999, p. 91; Hancock, Fisk & Rogers 2005.

249 Wogalter & Vigilante 2006, p. 258; Magat, Viscusi & Huber 1988.

250 Schommer, Doucette & Worley 2001.

251 Vigilante & Wogalter 1997; Wogalter & Laughery 2006, p. 906.

higher the priority for that warning. Likewise, the more severe the adverse consequences, the higher the priority for that warning. If there is some familiarity with the hazard, it has less priority than warnings of hazards that are unknown. Furthermore, warnings that are perceived by users as containing important information should be given higher priority than warnings that are perceived as less important. For example, warnings that contain information concerning the initial and safe operation of the product are usually perceived as more important than warnings with respect to hazards related to the maintenance and storage of the product. Lastly, it may not be practical to place all the warnings on the product's label (*e.g.* space constraints can impair legibility). In such a case the label can be enlarged or another component of the warning system can be used to convey the warning information that is of less priority.[252]

3.6.5.7 Potential Problem of Habituation

Another potential problem related to the stages of attention switch and maintenance is habituation.[253] The psychological process of *habituation*, that is to say, repeated exposure to a warning, is a long-term effect that can negatively affect attracting and maintaining attention of a warning over time and it can consequently decrease the likelihood of following that warning. Furthermore, it has been suggested that this process can even extend to other warnings that look similar.[254] The warning literature has used this process to explain why the familiarity belief can have a decreasing effect on attention and behavioural compliance.

Habituation can occur even with well-designed warnings. On the positive side, habituation indicates that there is information available in memory. Unfortunately, this knowledge may not be complete or accurate. Habituation shows that a salient, conspicuous warning design is essential. Reducing this process may thus be facilitated by taking account of the size of the warning and the use of multiple design features in a warning that increase attention. Another method to reduce and possibly prevent habituation is by changing the appearance and the content of a warning every so often. The latter approach will be particularly helpful for non-durable consumer products that are purchased on a fairly consistent basis. Habituation may also be counteracted by using an interactive warning.[255] The reason for this is that these warnings provide an additional, tactile cue that can disrupt the script-driven, automated behaviour thereby increasing the chance that the warning will be noticed and processed at a meaningful level.[256]

---

252 Wogalter & Laughery 2006, p. 906; Wogalter 2006a, p. 5; Wogalter, Conzola & Vigilante 2006, p. 490.

253 Habituation is a condition in which attention to a stimulus declines as a result of repeated presentation of the stimulus, see Payne & Wenger 1998, p. G-6.

254 See *e.g.* Thorley, Hellier & Edworthy 2001; Amer & Maris 2007.

255 Wogalter & Vigilante 2006, p. 259; Wogalter & Leonard 1999, p. 140; Leonard, Otani & Wogalter 1999, p. 181; Amer & Maris 2007.

256 Wogalter & Vigilante 2006, p. 258; Frantz & Rhoades 1993; Wogalter, Barlow & Murphy 1995.

Related to this issue is the standardisation of warning formats and designs. On the one hand, the availability of (non-binding) legal standards on this topic generally improves the noticeability of warnings, since it provides producers with general criteria for warning design that should be considered. Given that due to standardisation the appearance of a warning will be relatively consistent, people will eventually learn what a warning looks like and will be able to more easily pick out the warning from a cluttered environment. On the other hand, a potential problem of standardisation is that it promotes similarity across all types of warnings which in turn is likely to exacerbate habituation to a specific warning and to similar-looking warnings. If all warnings look and sound the same, it is quite possible that over time they will lose their attention-getting value.[257]

3.6.6 *The Receiver Stage of Comprehension and Memory*

3.6.6.1 General

Understanding the meaning of the content of the warning is the next stage in information processing.[258] Evidently, product users that have attended to a warning must understand the information they are reading, remember it and retrieve it at the appropriate time, otherwise that warning message will not be processed further. Giving sufficient and adequate information to enable consumers to make an informed decision on the product's hazardousness and how to use it safely is a principal purpose of warnings as was discussed in paragraph 3.3.

Warning comprehension and warning retention are dependent on knowledge already stored in memory.[259] Knowledge is important for understanding warning text, because it provides the necessary background and context within which the information being read can be interpreted.[260] Because many product packages have limited space available, producers cannot convey every safety aspect of the product's hazard in the warning message. As a result, they rely on consumers to draw the correct inferences from the warning text to understand its full meaning. To make such inferences, people use their prior knowledge and general knowledge about the world stored in memory.[261] Comprehending and

257 Wogalter & Vigilante 2006, p. 259; Rogers, Lamson & Rousseau 2000, p. 120; DeJoy 1999a, p. 205.

258 See for an extensive discussion of this information processing stage Leonard, Otani & Wogalter 1999 and Hancock e.a. 2006.

259 Warning comprehension refers to the understanding of the meaning of the warning intended by the designer. Warning retention refers to the successful storage and subsequent recall of warning information over a time interval. Memory is a multidimensional construct comprising processing of information in real-time, general knowledge, and personal experiences that affect how warnings are understood, see Hancock e.a. 2006, p. 268.

260 Hancock e.a. 2006, p. 268.

261 A form of long-term memory consisting of general knowledge about the world, language and so on is referred to as semantic memory, see Hancock e.a. 2006, p. 268.

remembering a warning message is also influenced by people's expectations of products and their hazards, *e.g.* based on past experience, including schemas and scripts.[262]

Consumers have to remember the warning message and retrieve it at the appropriate time. There are limitations attached to human memory as noted above and these can affect the understandability of warnings. When people have to rely heavily on their memory, warning comprehension is attacked. Research suggests reducing memory demands by designing well-organised warnings that have simple, short sentences and as little amount of information as possible.[263]

Comprehension of warning information can be influenced by factors related to characteristics of the warning design and the warning receiver.[264] These are discussed in the six following paragraphs.

3.6.6.2 Abilities of Receivers

Understanding information correctly is dependent on the cognitive capabilities of the recipients, such as their language skills, reading abilities and level of knowledge.[265] Research indicates that consumers are more willing to continue to read a warning if they understand the warning text.[266]

For recipients to comprehend the content of a warning message as is intended by the warning designer, it is imperative that they have knowledge of the language. Given the growing trade of consumer products worldwide and today's multicultural society, the problem of having to deal with receivers that do not speak and read the language of the warning is real. Supplying warning information in only one language can cause processing difficulties for those receivers of whom it is not their primary language. Presenting more than one language in a message can solve the problem. Nevertheless, this might create the subsequent dilemma of having too little space available to succeed in designing a legible warning. Altering the design of the label, for example a fold-out, offers a solution according to the warning research literature.[267]

Even if people have knowledge of the language of the warning, their skills and competence in understanding it may be poor. Warning designers must consider that the levels of reading comprehension vary across the target population of a warning. Some readers have good reading skills, others have low literacy levels and a small percentage of the

---

262 A schema is an organized packet of information about the world, events, or people, stored in long-term memory. A script is a type of schema for representing typical events (*e.g.* going to a restaurant) proposed by Schank and Abelson, see Eysenck & Keane 2000, p. 536; Hancock e.a. 2006, p. 270. See also paragraph 3.6.4.7.

263 Hancock, Fisk & Rogers 2005.

264 Hancock e.a. 2006; Rogers, Lamson & Rousseau 2000.

265 Wogalter & Laughery 2006, p. 898, Laughery & Brelsford 1991.

266 Leonard e.a. 1989.

267 Leonard, Otani & Wogalter 1999, p. 152.

population is even illiterate. It follows that warnings should be readable; their content must be simple, direct and easy to understand.[268] The reading level of a warning must match the reading level of the target audience. It has been suggested in the literature to base the text of warnings on a low reading level, such as a reading level between fourth and sixth grades if the warning aims at reaching the general public.[269] Others advise a reading level at the eleventh or twelfth grade.[270] Although these suggestions are not based on empirical evidence, it makes sense to use simple words and short sentences in order to create a warning that can be understood and retained by the general public.[271] Research has shown that lay people have difficulty with understanding certain words that are commonly used in warnings.[272] For instance, a warning that says 'Keep combustible material away' might not be helpful to product users who do not understand the precise meaning of the word 'combustible'.[273]

It is critical to find out what hazard-related knowledge the target audience of a warning possesses. Incorrect assumptions about the knowledge level of receivers to whom the warning is directed can produce dangerous situations when important information is left out or when words are misunderstood.[274] Researchers warn designers of warnings not to assume that the average consumer knows as much about the hazardous characteristics and consequences of a product as they do.[275] People's knowledge is often incomplete, especially with regard to non-apparent hazards that are hidden in technological consumer products.[276] If there is such a knowledge gap, new memory must be formed so that people's existing knowledge becomes consistent with the knowledge about the hazard. This process of assimilating new information is easier if the gap is small, *i.e.* if extensive related memory already exists, as the formation of new memory generally requires mental effort. If the knowledge gap is large, it is less likely that the warning will be effective.

Because of the amount of mental effort involved in information processing and because people may be distracted by other things, a warning should be constructed so that the information being transmitted is easy to grasp and does not overload the system. Thus, to facilitate the comprehension and retention of warnings, warnings should be well matched with what people already know.[277] For instance, if warnings are designed for a specific

---

268 Leonard, Otani & Wogalter 1999, p. 156.

269 Wogalter & Laughery 2006, p. 898. Children in the sixth grade of school in the United States are usually 11-12 years old.

270 Pyrczak & Roth 1976.

271 Rogers, Lamson & Rousseau 2000, p. 123.

272 See *e.g.* Leonard, Creel & Karnes 1991; Main, Frantz & Rhoades 1993.

273 Rogers, Lamson & Rousseau 2000, p. 121.

274 Leonard, Otani & Wogalter 1999, p. 152.

275 Laughery 1993.

276 Leonard, Otani & Wogalter 1999, p. 150.

277 Leonard, Otani & Wogalter 1999, p. 151. See also paragraph 3.7 on this subject.

target audience, *e.g.* experts, then it is acceptable to use technical vocabulary since it is to be expected that this subpopulation possesses knowledge of these terms.[278]

To assure that the assumptions about the knowledge level of product users are valid, producers should investigate what the potential target population already knows about the product's hazards and the circumstances that can lead to injury. This can be done by testing the adequacy of the warning design.[279] Testing a prototype warning with the potential user population to evaluate the extent to which it is understood can reveal what information can be left out and what needs to be altered in order to improve the warning's comprehensibility.[280]

3.6.6.3 Signal Words

Signal words are used to communicate the level of hazard associated with using a product. In addition to signal words, warnings usually contain several other information components such as a warning symbol, colour, written information about the hazard, the consequences, and instructions to denote the product's hazardousness. For successful processing to occur, the warning components must provide information that allows warning receivers to understand and thereby respond to the level of hazard present in a particular situation. The concept of matching the strength of the hazard implied by warning components, such as a signal word, to the specific level of hazard of the situation being warned about is termed *hazard matching*.[281]

For hazard matching to succeed, warning recipients must interpret the relationship between the signal words and the hazard levels consistently. The primary advantage of hazard matching is that such a warning not only informs that a threat is present, but also indicates the severity of the hazard involved. A second advantage is that it helps prevent habituation caused by the inappropriate use of only the most hazardous warnings.[282] In the auditory warning research, this process is called *urgency mapping*. The urgency of the warning sound indicates the urgency of the situation signalled.[283] Findings have shown that warnings that are faster or higher in pitch are perceived as being more hazardous.[284]

In the United States, the standard Z535.4 of the ANSI has adopted a hazard-matching approach.[285] This standard is voluntary, as well as other standards such as those of the ISO.

278 Knowledge that is specific to a particular topic or field is called *domain-specific knowledge*, see Hancock e.a. 2006, p. 269.
279 Wogalter 2006a, p. 9. See also paragraph 3.6.11 on testing.
280 Wogalter & Laughery 2006, p. 907.
281 Hellier & Edworthy 2006, p. 407; Hellier e.a. 2000.
282 Hellier e.a. 2000, p. 579.
283 Edworthy 1994.
284 Edworthy, Loxley & Dennis 1991; Hellier, Edworthy & Dennis 1993.
285 ANSI 2002a.

The ISO does not recommend the use of signal words.[286] Thus, producers are not obliged to comply with these considerations on warning label design. The standard recommends that warning labels and signs contain a panel that includes a signal word (DANGER, WARNING or CAUTION) that is coupled with a specific colour background (red, orange, and yellow, respectively) and an alert symbol (a triangle surrounding an exclamation mark). Hence, the signal words are intended to provide information about the probability of damage as well as the severity of injury.[287]

> DANGER is used for hazards in which serious injury or death will occur if the warning is not followed.
> WARNING is used when serious injury might occur.
> CAUTION connotes hazards where minor injury or damage to property might occur.

Research findings have demonstrated that signal words and colours produce different ratings of perceived hazard. While the ANSI standard advises the use of signal words and colours to represent a hierarchy of decreasing potential for damage, it is of importance that this practice is borne out by empirical evidence. Research findings have confirmed that the signal word DANGER connotes significantly higher ratings of perceived hazard than the words CAUTION and WARNING. However, the distinction in perceived hazard between the latter two words is less clear, absent, or even contrary to the recommendations of the voluntary guidelines.[288]

The effects of other signal words on perceptions of hazard have been investigated in research as well. For example, it seems that the alternative term DEADLY produces higher ratings of perceived hazard than all other signal words.[289] Wogalter and Silver have investigated the effects of a wide range of signal words more than once.[290] In these studies, participants were asked to rate the signal words on six dimensions such as perceived likelihood and severity of injury and carefulness on the basis of a Likert-type scale technique. As a result of the intercorrelations among the ratings of the dimensions, the authors suggested that the dimensions measured the same thing, *i.e.* the arousal strength of the signal words. The term *arousal strength* can be seen as a general indicator of the overall importance of a signal word perceived by warning receivers. The definition incorporates both the likelihood and the severity of the potential danger.

---

286 ISO 2004.

287 Hellier & Edworthy 2006, p. 409.

288 See *e.g.* Braun & Silver 1995; Chapanis 1994; Wogalter & Silver 1990; Wogalter, Jarrard & Simpson 1994; Young 1998; Silver & Wogalter 1989; Hellier e.a. 2000.

289 See *e.g.* Wogalter & Silver 1990; Silver & Wogalter 1989; Wogalter e.a. 1995; Drake, Conzola & Wogalter 1998.

290 Wogalter & Silver 1990; Silver & Wogalter 1991; Wogalter & Silver 1995.

Researchers suggest expanding the list of signal words to reduce the potential problem of overusing those limited signal words because overuse could result in habituation and affect attention negatively.[291] They have developed and proposed a list of 20 signal words that represent a wide range of arousal strengths. The signal words are in descending order of arousal strength: DEADLY, FATAL, POISON, DANGER, HAZARD, VITAL, SEVERE, SERIOUS, URGENT, BEWARE, WARNING, HARMFUL, CAUTION, ALARM, ALERT, CAREFUL, PREVENT, NEEDED, NOTICE and NOTE. Studies have been followed up to examine the consistency in arousal strength of the alternative list. The results showed small differences in perceived hazard, but the rank order of the signal words was consistent across different US populations such as college students, young children, the elderly and non-native English speakers.[292] The arousal strength and understandability of signal words has also been tested in the UK population. The research has shown that previous findings with regard to the US population can be generalised to the UK population.[293]

Overall, the research evidence on signal words is fairly robust and generally supports the use of signal words to facilitate hazard matching as people interpret signal words as implying differential levels of hazard. Words such as DEADLY and DANGER are consistently interpreted as implying high levels of hazard, whereas terms like NOTICE and IMPORTANT are consistently interpreted as implying less hazard. Because of a lack of empirical consistency as regards the differential hazard connotation of WARNING and CAUTION and because of the limited number of signal words to be used for many hazards, several researchers have suggested a revision of the voluntary standards. It is suggested to replace CAUTION by a term much lower in arousal strength than WARNING, such as ALERT or CAREFUL to increase the probability that people will clearly perceive the difference between the hazard levels of the signal words.[294]

> An experiment of Wogalter, Jarrard and Simpson investigated the influence of signal words and a signal icon on perceived hazard of warning labels of real consumer products.
>
> Under the guise of a marketing research study, 90 participants examined the labels of household products and answered a number of questions including the important one of how hazardous they perceived the product to be on a scale of 0 to 8. The experimenters used 16 labels from actual household products (*e.g.* aspirin, contact lens cleaner, paint thinner): nine carried the experimental conditions, and seven labels were labels of products that never carried a signal word or a warning message. There were five label conditions with the signal

291 Wogalter & Silver 1990; Hellier e.a. 2000.
292 Silver & Wogalter 1991; Wogalter & Silver 1995.
293 Hellier e.a. 2000.
294 Hellier & Edworthy 2006, p. 410, 411.

words NOTE, CAUTION, WARNING, DANGER, or LETHAL as a header of the warning message. In two other conditions, a symbol (exclamation mark surrounded by a triangle) was paired with the signal words DANGER or LETHAL to investigate the influence of the signal icon by comparing them with the condition without the symbol. The signal word LETHAL was used instead of DEADLY, because the use of DEADLY would not be realistic on these consumer products. In the final two control conditions, the label lacked a signal word but with the associated warning message, the other label missed both.
Results showed that the presence of a signal word increased hazard perceptions compared with its absence. The no signal word condition produced significantly lower hazard ratings than the conditions with a signal word. Significant differences were noted between extreme terms such as NOTE and DANGER, but not between CAUTION and WARNING, which are usually suggested in warning design guidelines. CAUTION and WARNING produced significantly lower hazard ratings than LETHAL, which was rated the highest. Lastly, the signal icon showed no significant effect on hazard perception.[295]

3.6.6.4 Colour

In addition to facilitating attention, colour can also be used to improve comprehension. Warning labels in colour are perceived more hazardous than labels presented in black and white.[296] Some colours have stronger effects on people's hazard judgement than others.

Standards recommend warning labels on a red, orange, and yellow background paired with the signal words DANGER, WARNING, and CAUTION, respectively to indicate decreasing hazard levels. Studies have been carried out to assess whether these propositions are congruent with how people respond to them. Research findings provide robust evidence that red is associated with the highest level of hazard, followed by orange and yellow. However, the distinction in hazard connotation of orange and yellow is less clear and less consistent with the recommendations in guidelines.[297] Previous research also suggests that black may communicate a high hazard level.[298] As regards the combinations of colour and signal words, red is the only consistent indicator of hazard. Red paired with DANGER is perceived to indicate the greatest amount of hazard. The association of orange and yellow with the prescribed signal words is less reliable.[299]

295 Wogalter, Jarrard & Simpson 1992.
296 Braun, Mine & Clayton Silver 1995; Kline e.a. 1993.
297 *E.g.* Chapanis 1994; Braun & Silver 1995; Griffith & Leonard 1997; Leonard 1999.
298 *E.g.*Braun e.a. 1994; Lesch e.a. 2009.
299 See *e.g.* Griffith & Leonard 1997; Leonard 1999; Chapanis 1994.

Braun and Silver carried out two experiments to examine the interaction of signal words and colours on perceived hazard and warning compliance.
In the first experiment, 30 undergraduates were asked to scale the perceived hazard of signal words printed in specific hazard colours. Results indicated that the degree of hazard varied as a function of the signal word and the colour in which it was presented. Printing words in red connoted the highest level of perceived hazard followed by orange, black, green and blue. The signal word DEADLY in green conveyed less hazard than when it was printed in red ink.
Experiment 2 examined the effect of colour on actual behavioural compliance with printed warnings. 65 participants interacted with a pool-water test kit and a two-part adhesive. The warning on each product was factorial for colour (*i.e.* red, green and black). Behavioural compliance was assessed by indicating if subjects donned protective gloves as directed by the warning. The results indicated that warnings printed in red produced higher levels of compliance than green and black combined.[300]

3.6.6.5 Warning Symbols

An important purpose of symbols is to communicate comprehensible warning information.[301] With the internationalisation of markets, the use of symbols has flourished in recent years because well-designed symbols can reach people who are not able to read text labels due to poor vision, poor reading skills, or language barriers. Another benefit is that they can convey large amounts of information to be picked up at just a glance, which is especially useful in situations where time is limited.[302] Furthermore, it seems that pictorials are easier to recall than text.[303]

Research has however shown that symbols are not always well understood or are misinterpreted.[304] When symbols convey meanings other than intended, consumers might neglect to take the appropriate measures to avoid an accident. Many pictorials are poorly understood because they have never been tested for comprehension prior to their widespread introduction. The American standard on safety symbols recommends using symbols only if testing has shown that they are comprehended by 85% of a sample of 50 participants, and with no more than 5% critical confusion errors.[305]

As with text, symbols can communicate several information components, such as information about the potential hazard (*e.g.* fire), the potential consequences (*e.g.* electro-

300 Braun & Silver 1995.
301 Wogalter e.a. 2006.
302 Wogalter, Sojourner & Brelsford 1997.
303 Young & Wogalter 1988.
304 See *e.g.* Collins & Lerner 1982; Davies e.a. 1998; Easterby & Hakiel 1981; Lesch 2008b; Rother 2008; Van Duijne e.a. 2008 (described in paragraph 3.6.7.6).
305 ANSI 2002b.

cution) or what to do or not to do to avoid the hazard (*e.g.* wear protective glasses).[306] Concepts represented by symbols must be recognised and understood from general experience and knowledge. Some pictorials are by their very nature easier to understand than others and some concepts may never be sufficiently interpreted. Concrete, specific symbols are generally easier to comprehend than are abstract, general symbols, because the meaning of the former is easier to depict than the latter. For example, the abstract symbol indicating a biohazard is less direct than a symbol conveying a fire hazard. The meanings of such symbols can be learned to be understood.[307]

Studies have provided evidence that comprehension and retention of warning symbols can be improved through training.[308] The findings suggest that training is favoured, especially for warning symbols of which the meaning is not readily apparent. During training, symbols are paired with associated text describing the meaning of the symbol and/or the nature of the hazard or for example real-world accident scenario information. After training, comprehension is tested at several time intervals to determine potential training effects over time. It is expected that the more detailed the associated text, the greater the recall. Previous warning studies failed to show this additional effect relative to a simple description.[309] However, a recent study has confirmed the prediction.[310] Moreover, research does show that presenting the same warning information both verbally and symbolically benefits memory.[311] This is in line with Paivio's dual coding theory, which generally says that there are two distinct coding systems (verbal and nonverbal) for the representation and processing of information. Memory for visual words is distinct from memory for visual objects such as graphic symbols. Text is more likely to be memorised using the verbal system, whilst the nonverbal system is used for the memory of visual objects including symbols. Thus, accompanying symbols with associated written information is likely to enhance memory, because when one system is lost or unavailable, memory can rely on the other one.[312]

3.6.6.6 Explicit Information: Information Categories

Another design feature that has been researched extensively and that influences the understandability of textual warnings is the *explicitness* of information. Explicitness of information in warnings can be defined as the specificity or detail with which information is described.[313]

306 See *e.g.* Trommelen & Zwaga 1998.
307 Wogalter e.a. 2006, p. 167.
308 See *e.g.* Lesch 2003; Lesch 2008b; Wogalter, Sojourner & Brelsford 1997; Young & Wogalter 1990.
309 Wogalter, Sojourner & Brelsford 1997; Lesch 2003.
310 Lesch 2008b.
311 Paivio 1971.
312 Lesch 2003, p. 496; Eysenck & Keane 2000, p. 262.
313 Laughery e.a. 1993a, p. 598.

The concept of explicitness of information exists at two levels.[314] Firstly, it raises the question of which information categories must be communicated so that consumers can make informed decisions concerning how to use the product safely. Secondly, explicit information refers to the way in which the information itself is disclosed. This paragraph deals with the first form of explicitness, the next paragraph deals with the latter.

For consumers to be adequately informed, they must comprehend the hazard, know how to avoid it, and know the potential consequences of unsafe behaviour. Consequently, warnings in general must contain information about the hazard, instructions on how to avoid it, and information on the potential negative consequences associated with the hazards.[315] Some warning messages, however, may not require all of the information statements. A general principle is that unnecessary information should not be included in warnings, because *brevity* is also an important criterion for warning effectiveness: warnings should be as brief as possible.[316] Given that, researchers have argued that an information category (*e.g.* the nature of the hazard) can be omitted if it is obvious or if the other available information in the warning is sufficient to cue knowledge that is already available to the target audience.[317]

Whether an information category is obvious to warning receivers or already stored in existing knowledge should therefore be an important question for the warning designer.[318] The concept of obviousness refers to circumstances where the appearance and/or the function of the product or environment already communicate the warning information. The extent of mechanical hazards with regard to pieces of moving machinery is easier to comprehend, whilst the extent of chemical hazards cannot readily be inferred from their appearance.

Furthermore, consumers may already have existing knowledge of safety information. People can acquire knowledge through experience with products, training, or other sources of information. If the target audience already has some of the information, it may not be necessary to provide full information. On the other hand, if there is a large knowledge gap, it is important to communicate the missing information. Hence, the type and amount of information put in a warning largely depends on the knowledge level of the target population. Warning information is especially useful in situations where users are novice and want to learn how the product should be used, as implied by the levels of performance model.[319]

---

314 Laughery & Smith 2006, p. 420 ff.

315 See *e.g.* Wogalter e.a. 1987, Vredenburgh e.a. 2005 (described in paragraph 3.6.6.6).

316 Wogalter & Laughery 2006, p. 905; Laughery & Smith 2006, p. 421.

317 Laughery & Smith 2006, p. 420; Wogalter, Conzola & Smith-Jackson 2002, p. 222; Wogalter & Laughery 2006, p. 899.

318 See Leonard & Wogalter 2000.

319 Lehto & Papastavrou 1993, p. 587; Lehto 1991, p. 606; Rasmussen 1986.

Even though people may have knowledge of a hazard, they may not be aware of it at the time they are at risk. Note that there is a distinction between awareness and knowledge. Every year, trained personnel who work with hazardous industrial machines suffer injuries in spite of the fact that they know about the dangers attached to the equipment. They 'forgot' it, precisely because they were not conscious or aware of the risky situation at the critical time. Researchers have explained this distinction between awareness and knowledge by making the analogy with the short- and long-term memory distinction in cognitive psychology. Short-term memory or working memory can be associated with consciousness whereas *long-term memory* is people's knowledge of the world. Specific circumstances can call for a *reminder warning*, such as (1) intense mental work load and involvement in the product task; (2) when foreseeable distractions are expected to occur during product task performance; or (3) where the non-obvious hazard is encountered infrequently and forgetting plays a role.[320]

> The study of Vredenburgh e.a. tested health care workers' prior knowledge of latex allergies and evaluated three warning labels designed to alert users to the hidden risk of latex exposure while wearing latex gloves. The goal of the research was to develop and evaluate a warning for latex gloves by comparing the relative effectiveness of three different label configurations placed on boxes of latex gloves.
>
> A total of 85 health care workers (such as nurses, firemen and medical students) were given latex glove boxes to examine. The first label contained the statement that the gloves contained natural rubber latex. The second label also included this statement as well as the information that it may cause allergic reactions and the signal word Caution. The third warning label was consistent with the ANSI guidelines and provided information about the nature of the hazard, the potential consequences and their severity, and the steps people should take to avoid the hazard. The word CAUTION was used on a yellow background with the warning symbol of a triangle surrounding an exclamation mark. A latex box was shown to the participants and then they were asked to complete a questionnaire concerning the gloves, and the packaging. Participants' prior knowledge of the hazards associated with latex was also assessed before they were exposed to the box.
>
> 65% of the participants reported that they had heard of a natural rubber latex allergy, 9% of them had had an allergic reaction to rubber latex gloves. However,

320 Wogalter & Laughery 2006, p. 901; Laughery & Hammond 1999, p. 7; Laughery & Smith 2006, p. 421; Leonard & Wogalter 2000.

when they were asked to describe the allergy and its symptoms, most associated it with a hand rash, which is the less dangerous type of latex allergy.

In the second part of the study, participants were asked to rate (on a seven-point Likert-type scale) each of these labels as to their ability to capture people's attention, their ability to alert and inform people of the specific hazards associated with exposure to natural latex gloves, and their ability to change behaviour. The findings showed that for each dimension, the ANSI warning label was rated most effective. Participants then completed the third section of the questionnaire, which included an evaluation of their perceptions of dangers regarding latex gloves both prior to and following their participation in this study. After participation, the gloves were perceived significantly more dangerous: the warnings thus influenced the perceived hazard perception of participants. The results indicate that a well-designed warning can be an important method of informing people and influencing them to exercise appropriate precautionary behaviour to avoid the risks.[321]

3.6.6.7 Explicit Information: The Way in which Information Is Expressed

The second aspect of explicit information refers to the way in which the information is disclosed.[322] An important design guideline from the warning research literature is that textual and graphic information about the hazard, the potential negative outcomes and the instructions should be explicit, as explicitness results in more effective product warnings. 'To prevent electrical shock, turn power off before servicing' is better than 'Turn power off before servicing'.[323] The phrase 'Using this drug while drinking or taking other drugs may result in severe effects' is less explicit than the sentence 'Exceeding the recommended dosage may result in brain damage, prolonged coma or death'.[324] Also, the precautionary measure 'Avoid contact with skin and eyes' is vaguer than the corresponding explicit version 'Wear rubber gloves and protective glasses'.[325]

The benefit of explicit warnings is that they give consumers a better understanding of the product's safety because of the increase in detailed information and, accordingly, enables them to exercise a sufficient amount of care. This is evidenced by research. Research findings report that the information in explicit warnings is generally better understood and remembered than is non-explicit information.[326] Additionally, with explicit warnings

321 Vredenburgh e.a. 2005.

322 Laughery & Smith 2006, p. 421ff.

323 Braun & Shaver 1999.

324 Laughery e.a. 1993a.

325 Frantz 1994.

326 See *e.g.* Trommelen 1997; Martin 2000; Malouff & Schutte 1992; Laughery e.a. 1993a.

people perceive products to be more hazardous and the injuries as more severe.[327] Moreover, the results indicate that more explicit warnings lead to an increase in intent to act cautiously than do non-explicit warnings, especially where the severity of the potential hazard is great.[328] Research also suggests that explicit information about the hazard will produce little effect when the hazard is obvious.[329] In such an event, detailed hazard information can be viewed as superfluous.[330] Another important implication of the studies is that especially the consequence-related information should be made explicit when the potential consequences are severe, since research has demonstrated that people's perception of the severity of the consequences is a good predictor of how they perceive the safety of a product and whether they have the intention to act cautiously.[331]

There might also be disadvantages associated with the use of explicit information in warnings. First, increasing the explicitness of information also increases the length of the message. Secondly, producers may reason that explicit warnings negatively influence consumers' decision to purchase the product. This hypothesis has been tested in several studies.[332] The majority do not show a clear relationship between explicit information and purchase decisions. It appears that in general, manufacturers' concern that consumers' intention to purchase the product will be affected by the presence of explicit warnings on the consumer product compared to non-explicit warnings is unwarranted.[333]

> The field demonstration of Dingus, Wreggit and Hathaway of 1993 investigated the effects of the variables of cost of compliance and the content of a warning with regard to the use of racquetball eye guards.
> A total of 420 racquetball players participated in the experiment. The dependent measure was whether the players wore eye protection. In the low cost condition, the eye protection devices were provided in a box just outside the racquetball courts in a salient location. In the medium cost condition, eyewear was available at a checkout station located 60 feet away from the entrances to the courts. In order to access the courts, participants needed to pass by the checkout stand where a sign indicating the availability of the eyewear was prominently displayed. There were three conditions. In the first control condition, there was no warning sign. The second condition involved a warning sign according to

327 See *e.g.* Trommelen 1997; Heaps & Henley 1999; Laughery e.a. 1991; Laughery e.a. 1993a; Laughery & Stanush 1989.

328 See *e.g.* Laughery e.a. 1993a; Laughery e.a. 1991; Heaps & Henley 1999; Trommelen 1997; Laughery & Stanush 1989.

329 *E.g.* Shaver & Braun 2000.

330 Laughery & Smith 2006, p. 423.

331 Wogalter e.a. 1999b (described in paragraph 3.6.9.8); Laughery e.a. 1993a, p. 599.

332 See *e.g.* Laughery e.a. 1993a; Vaubel & Brelsford 1991; Vaubel 1990.

333 Laughery & Smith 2006, p. 427.

the ANSI standard. It contained a pictorial that showed eyewear protection, a triangle with an exclamation mark, the signal word WARNING and the text underneath: WEAR EYE PROTECTION. SERIOUS EYE INJURY SUCH AS BLINDNESS, RETINAL DETACHEMENT, SECONDARY GLAUCOMA, AND EYE GLOBE RUPTURE MAY OCCUR WHEN NOT WEARING EYE PROTECTION. The other warning condition involved the ANSI warning plus a sign that contained specific consequence information stating racquetball facts:

1. Wearing appropriate eyewear while playing racquetball can prevent 99% of all eye injuries.
2. There are 70.000 eye injuries per year, 40% of which occur while playing racquet sports.
3. Odds for obtaining an eye injury while playing racquetball is one in four.
4. Racquet ball can penetrate open eye guards.
5. Eye glasses and contact lenses offer no protection from racquetball eye injuries. They can shatter and be lodged in the eye.

The results of the experiment show the interaction between cost of compliance and warning information. When the cost of compliance was at a medium level, the compliance percentages were near 0%, regardless of the information provided. However, in the low cost condition, the compliances rates increased as the explicitness of the information increased. The compliances percentages were 11% for the no warning condition, 30% for the ANSI warning condition and 38% for the condition with the ANSI warning plus consequence information.[334]

3.6.7 *The Receiver Stage of Attitudes and Beliefs*

3.6.7.1 General

The information processing stage following comprehension and retention is the cognitive stage of attitudes and beliefs.[335] Discrepant attitudes and beliefs held by warning receivers hinder subsequent processing of warning information, even if the warning is understood.

In the warning literature, *beliefs* are defined as someone's convictions about phenomena or objects that are accepted as true, although they may in fact not be.[336] Beliefs are often viewed as the building blocks of attitudes. *Attitudes* are similar to beliefs, but have greater

334 Dingus, Wreggit & Hathaway 1993.
335 See for an extensive discussion of this information processing stage DeJoy 1999a; Riley 2006 and Vredenburgh & Zackowitz 2006.
336 DeJoy 1999a, p. 189; Riley 2006, p. 289.

emotional involvement.[337] Because beliefs and attitudes are closely related, they are dealt with together in this stage.

Beliefs and attitudes can have powerful effects on warning compliance as they are important ingredients in the formation of expectations, and expectations are the anticipatory outcomes of behaviour. Consequently, the expectations that persons bring to the situation influence their decision making about whether and how to behave with respect to a warning. Hence, the factors that affect people's attitudes and beliefs play an essential role in the effectiveness of warnings.[338]

On the basis of the C-HIP model, a warning will be processed successfully if the beliefs and attitudes of the warning recipient concur with the information in the warning message. If this is not the case, the warning's design needs to alter the existing attitudes and beliefs that the person brings to the product use situation in order to be effective. Unfortunately, people's attitudes and beliefs are often firmly held and difficult to modify. Warnings must thus be sufficiently persuasive to override those discrepant beliefs and to motivate the recipient to comply.[339]

In the warning research, value-expectancy theory has been used and a value-expectancy model applied to warnings has been built to better understand the contribution of attitudes and beliefs to the warning process. This theory from the persuasion literature was already mentioned in paragraph 3.4.3.2. Value-expectancy theory holds that people estimate the seriousness of the potential risk, evaluate the costs and benefits of actions and then choose a course of action that will maximise the expected outcome. It assigns central importance to the individual's subjective evaluation of the risky situation. This approach assumes that people engage in a conscious and a fairly rational decision making process involving the weighing of costs and benefits before taking action or not. The findings of the warning research are generally consistent with the predictions of this theory. When product users receive a warning, they must decide whether or not to comply with the warning. A fairly robust finding of the warning research is that warning effectiveness is affected by people's subjective perception of hazard associated with the product and that the effectiveness of warnings increases as hazard perception increases. Furthermore, studies have also shown that warning effectiveness decreases as the costs of compliance (*e.g.* time and effort) increase, which indicates that people make trade-offs with regard to their decision to comply with a warning.[340]

---

337 In the glossary of Myers 2004, an attitude is defined as a favourable or unfavourable evaluative reaction toward something or someone, exhibited in one's beliefs, feelings, or intended behaviour.

338 DeJoy 1999a, p. 190; Vredenburgh & Zackowitz 2006, p. 345.

339 DeJoy 1999a, p. 189; Vredenburgh & Zackowitz 2006, p. 350. Persuasion refers to the process by which a message induces change in beliefs, attitudes, or behaviours, see Myers 2004, p. G-6.

340 DeJoy 1999a; DeJoy 1991; cf. Laughery 2006, p. 473.

The factors that are influential at this major stage can be categorised into threat-related expectations, outcome-related expectations and receiver characteristics. Threat-related expectations concern the individual's beliefs about the potential adverse consequences associated with a hazard, including their assessment of the likelihood and the severity of the possible outcomes. As will follow from the discussion below, people's perceived hazardousness of a product is an important determinant for behavioural compliance with warnings. Outcome-related expectations emphasise the individual's beliefs about the effectiveness of recommended precautions and the costs associated with complying with the warning. Receiver characteristics emphasise that individuals differ in the ways they interact with warnings. Individual differences such as familiarity and experience with a product, gender, age and so forth can influence warning compliance.[341]

3.6.7.2 Risk/Hazard Perception: Injury Severity Controls Perception of Product Hazards

A most important factor in this stage is hazard perception. The warning literature generally uses the term hazard perception interchangeably with risk perception, perceived threat or perceived hazardousness to connote a person's subjective judgement of the hazard posed by a product.

In general, products containing warnings are perceived to be more hazardous than products without warnings.[342] Moreover, there is strong evidence that the behavioural effectiveness of warnings is influenced by individuals' expectations about the level of hazard and the consequences associated with a product. The more hazardous a product is thought to be, the greater the likelihood of behaviour compliant with the warning.[343] Conversely, people who do not perceive a product as being unsafe are less likely to look for, notice, willing to read and comply with a warning.[344] Hence, attitudes and beliefs, particularly hazard perception, also influence information processing at earlier stages of the C-HIP model. It is thus not surprising to see that warning researchers have claimed that, in relation to the other processing stages, this stage is more powerful, since attitudes and beliefs broadly determine how individuals will react to warnings.[345]

Subjective risk/hazard perception can be distinguished from objective risk perception: that is, the ratio between an accident or injury rate and a measure of exposure.[346] In the risk perception literature, it is agreed that risk is primarily determined by a combination of these dimensions of injury severity and injury likelihood. The perceived likelihood

341 DeJoy 1999a, p. 191. For more on individual differences see paragraph 3.7.

342 *E.g.* Wogalter & Barlow 1990; Wogalter e.a. 1991; in contrast: Ursic 1984.

343 See *e.g.* Godfrey e.a. 1983; Otsubo 1988; Friedmann 1988; Wogalter e.a. 1991; Frantz 1994.

344 See *e.g.* Otsubo 1988; Wogalter e.a. 1991.

345 DeJoy 1999a, p. 190; Young & Lovvoll 1999, p. 45; Riley 2006, p. 297.

346 Weegels & Kanis 2000, p. 365.

appears to be more important according to the risk perception literature.[347] As regards many consumer products, findings of the warning research suggest that lay persons do not interpret risk in this way. How people perceive the risks associated with using a consumer product depends largely on whether they perceive the severity of the potential injuries as great.[348] Observational research on product users' perception on risk paints a similar picture: participants of the empirical studies referred to the severity of the injury when describing the risks.[349]

Some findings thus suggest that likelihood of injury plays an essential role, whereas other results indicate that hazard perceptions are principally determined by the severity of injury. A possible explanation for these conflicting results is that the risk perception research has examined people's judgement with regard to items that people do not encounter on a daily basis. Participants not only rated several consumer products such as lawn mowers and pesticides but also technologies and activities such as hunting, railroads, X-rays and nuclear power. The warning studies have merely examined participants' evaluations of common household consumer products that are representative like battery alarm clocks, sewing machines, oven cleaners, baby powder, milk and so forth. It seems that if an item is extremely hazardous, very likely to result in very severe injury, unfamiliar, not controllable, and/or catastrophic, injury likelihood plays a substantial role in judgements about risks and consequently, the intent to act cautiously. However, the risks of consumer product use situations are usually not characterised in this way which offers an explanation for the deviant findings of the warning research. Nevertheless, warning researchers have suggested that when injury severity reaches a certain level, for example fatal injury, the only remaining uncertainty is the probability of injury. It is then likely that the perceived probability of the injury is the dominant factor in judgements of perceived threat associated with a consumer product.[350] Another potential reason for consumers not to consider accident likelihood is that the frequency of accidents and injuries with many consumer products is extremely low. As a result, it is difficult for people to make meaningful distinctions between one injury per 10.000 product uses or per 100.000 uses, as both chances appear to be zero. Thus, the severity of the consequences becomes the overriding criterion for hazard judgements.[351]

---

347 See *e.g.* Slovic, Fischhoff & Lichtenstein 1979; Slovic, Fischhoff & Lichtenstein 1980.

348 See *e.g.* Young, Wogalter & Brelsford 1992; Wogalter e.a. 1999b (described in paragraph 3.6.9.8); Wogalter & Barlow 1990; Wogalter e.a. 1991; Young, Brelsford & Wogalter 1990; Wogalter, Brems & Martin 1993.

349 Van Duijne 2005, p. 225, 230.

350 Wogalter e.a. 1999b, p. 157; Young, Wogalter & Brelsford 1992; Vredenburgh & Zackowitz 2006, p. 346.

351 Wogalter e.a. 1991, p. 81.

3.6.7.3 Cognitive Heuristics and Biases: General

The broader risk perception literature has devoted considerable time to understanding how people judge uncertain events, for example the occurrence of environmental or health risks. These findings may play a role in understanding how consumers judge risks associated with product use.[352] People tend to rely on mental shortcuts, called *heuristics*, when making complex judgements under conditions of uncertainty.[353] Heuristics make it possible to simplify and deal with a decision, and thereby put less of a burden on the working memory.[354] They can thus be useful, but also problematic when they lead to systematic erroneous judgements: namely, biases. These heuristics and biases can influence an individual's belief about how hazardous a consumer product is.[355]

Although the heuristics of availability, overconfidence, and suppression have not been explored systematically by warning researchers, it seems that these in particular can influence consumers' judgement of the hazards of using the product.[356]

The availability heuristic means assessing the frequency of an event by the ease with which occurrences can be brought to mind.[357] Previous research on risk perception implies that people tend to overestimate the likelihood of infrequent causes of death, such as tornados and accidents in nuclear power plants, and to underestimate the likelihood of frequent causes, such as a stroke and diabetes. This bias in probability judgements can be explained by the availability heuristic. Whilst more frequent events are often easier to imagine, it is reasoned that because less frequent causes of death often receive wide media coverage and their injuries are more dramatic and vivid, they can be more available.[358] The finding in the warning research that participants underestimated the risk of common consumer products with higher objective injury frequencies and overestimated the risk of consumer products with lower injury frequency concurs with this. Furthermore, the results also showed that peoples' estimation accuracy of the injury frequencies associated with common consumer products did not improve when additional time was given or when participants constructed and analysed possible accident scenarios with the products (the ways one might get injured) before making the estimates. Accordingly, it seems that people do not consider various accident scenarios in determining risk. Information about the severity of a product accident scenario that initially comes to mind provides the basis for hazard

352 Van Duijne 2005 argues that even though heuristics can explain why certain risks receive more attention than others, they do not provide insight into which type of product-user activities are perceived to be safe or risky. She offers a research methodology to better understand risk perception in consumer product use.

353 Kahneman, Slovic & Tversky 1982.

354 Helander 2006, p. 76. A heuristic can be defined as a rule-of-thumb strategy that enables quick and efficient judgements, see Myers 2004, p. G-4.

355 DeJoy 1999a, p. 197; see Van Duijne 2005, p. 43, 232.

356 DeJoy 1999a, p. 197-200; Riley 2006, p. 293.

357 Tversky & Kahneman 1973.

358 Lichtenstein e.a. 1978; Slovic, Fischhoff & Lichtenstein 1977.

judgements as participants' quick estimates of risk were the same as those made after extra time and information were given. Hence, if the most readily imaginable scenario is not representative of the severity of the potential negative consequences associated with the product, biases in people's subjective judgement of product hazards can occur. Providing explicit information about the severity of the injury might reduce biased judgements.[359]

Another cognitive heuristic of interest in relation to product hazards is overconfidence. People tend to have excessive confidence in their own assessments and performance.[360] This can be recognised in consumers' feeling of personal control, as a result of which they fail to consider the potential product risks they face.[361] Related to this idea of personal immunity, is people's unrealistic optimism in judging their own risk of injury ('It won't happen to me'). It is the belief that the risks of a given behaviour are lower for themselves than for others which provides an explanation why individuals consistently overestimate the likelihood of positive life events and underestimate the likelihood of negative ones.[362] This overly positive perception that negative events are less likely to happen to them and rather to others, may as a result decrease people's motivation to engage in self-protective behaviour. For example, smokers are more likely to believe that their risk of experiencing any illnesses or death from smoking is lower than for other smokers, even though they acknowledge that smoking is unhealthy.[363] Similarly, many adolescent and also adult smokers believe that they personally can smoke for a few years and then quit when they want to. This confirms that they are biased in their optimism, because they believe that the addictiveness of smoking that applies to most people does not apply to them. It is suggested that warnings could be used to counteract adolescents' optimistic bias with regard to smoking addiction by stressing the risk of addictiveness and the difficulty of quitting later on.[364]

The suppression bias relates to the tendency of people to selectively discount or ignore information that conflicts with pre-existing interpretations of a situation.[365] As a result, it may entail that consumers who have preconceived notions about a product's risk, will

359 Wogalter, Brems & Martin 1993; Wogalter e.a. 1991, p. 81.

360 Fischhoff, Slovic & Lichtenstein 1977.

361 Van Duijne 2005, p. 232; McKenna 1993.

362 Weinstein 1989.

363 Arnett 2000; Weinstein 1998.

364 Arnett 2000, p. 631. It seems that the EU Directive on tobacco warnings supports this type of reasoning, see Directive 2001/37/EC of the European Parliament and of the Council of 5 June 2001 on the approximation of the laws, regulations and administrative provisions of the Member States concerning the manufacture, presentation and sale of tobacco products (*OJ* 2001, L 194/26). Article 5 obliges that besides the general warning of 'Smoking kills/smoking can kill' or 'Smoking seriously harms you and others around you', an additional warning must be included. One of the additional warnings is the statement 'Smoking is highly addictive. Don't start'.

365 DeJoy 1999a, p. 199.

ignore warning information if it is not consistent with their thinking.[366] This heuristic may contribute to the effect of familiarity: warning effectiveness decreases as familiarity with a product increases.[367]

3.6.7.4 Message Framing

The framing effect is also a cognitive bias that has been found to affect the beliefs and attitudes people have about a hazard.[368] In framing studies, identical choice situations are differently described and the resulting preferences of people are studied.[369] For example, a risky decision problem can be framed either as a gain or as a loss. Tversky and Kahneman have argued that, on the basis of prospect theory, people are risk averse, that is, they avoid risky behaviour when benefits are made salient but prefer taking risks when they consider the losses. Consequently, gain-framed messages are generally more effective in trying to persuade people to avoid risky behaviour.[370]

Drawing on findings from the health prevention literature, there is modest evidence indicating that in order to promote health behaviour (such as wearing seat belts and using sunscreen) the potential positive consequences of a certain action or behaviour (gain perspective) should be presented in the message rather than the potential negative consequences of not performing an action (loss perspective).[371] For instance, a gain-framed message that emphasised the benefits of using sunscreen resulted in higher intentions to request sunscreen and to use it than did the loss-framed message that highlighted the potential loss, even though the two frames described objectively equivalent situations. The gain-framed message was: 'Use sunscreen to help your skin stay healthy'. The loss-framed message stated: 'Without sunscreen you increase your risk of developing skin cancer'.[372]

In contrast, with regard to detection behaviour, such as breast self-examination, findings suggest that negatively framed messages are more effective than positively framed messages, as this type of behaviour may be perceived as risky at the time of engaging in that behaviour: they may discover that the risk has materialised.[373] Nonetheless, there is also research that casts doubts on the degree in which framing can induce positive attitudes, intentions and

366 See Van Duijne 2005, p. 150, 233.
367 DeJoy 1999a, p. 199.
368 Tversky & Kahneman 1981; Williams & Noyes 2007, p. 13.
369 Eysenck & Keane 2005, p. 559, defines framing as the influence of irrelevant aspects of a situation on decision making.
370 Kahneman & Tversky 1979; Tversky & Kahneman 1981.
371 Rothman & Salovey 1997; see *e.g.* Kiene e.a. 2005.
372 Detweiler e.a. 1999.
373 Detweiler e.a. 1999, p. 190. See *e.g.* Meyerowitz & Chaiken 1987; Banks e.a. 1995.

healthy behaviour.[374] This indicates that the impact of framing on risk perception and decision making is influenced by other factors.[375]

It has been noted by warning researchers that very little research has addressed the framing effect in relation to warnings.[376] In the context of warnings, gain-framed messages may entail emphasising the benefits of performing a certain action ('Wear safety glasses to protect your vision'). Since, in most product use situations, people are being asked to perform preventive behaviour, it is conceivable that gain-framed warning messages would be more beneficial in persuading users to comply with the warning.[377]

3.6.7.5 Perceived Control over Product Risks

The control perception is also a factor that can affect how individuals personally perceive risks and how they react to warning information.

Perceived control can be described in terms of internal and external locus of control.[378] People who have an internal locus of control believe that they can control outcomes by their own efforts. People with an external locus of control, on the other hand, perceive outcomes as in the hands of powerful others, luck or chance. People with an external locus of control would tend to believe that an accident was beyond their personal control and therefore not their fault, whereas people with an internal locus of control would generally believe that an accident was under personal control and thus one's own fault.[379] It is assumed that people who believe that they encounter a hazard voluntarily and who have high internal locus of control are more likely to attend to and follow a warning. Those with an external locus of control are more prone to accidents.[380] However, it appears that if the task is perceived as very dangerous, people with a more external locus of control may be more likely to comply.[381]

Perceived control can also be described in terms of an illusion of control.[382] People's subjective perception of the product's risks can be inaccurate as a result of their belief that

374 See *e.g.* Van Assema e.a. 2001 and Brug, Ruiter & Van Assema 2003 who conducted three studies and found no framing effects on participants' attitudes and intentions to perform the preventive nutrition behaviour.
375 Rothman & Salovey 1997; Williams & Noyes 2007, p. 14.
376 DeJoy 1999b, p. 238; Silver & Braun 1999, p. 249.
377 DeJoy 1999b, p. 238.
378 Rotter 1975; see also Myers 2004, p. 58, G-5.
379 Smith-Jackson 2006a, p. 342; Vredenburgh & Helmick-Rich 2006, p. 377.
380 Young & Lovvoll 1999, p. 42.
381 Vredenburgh & Helmick-Rich 2006, p. 377.
382 Illusion of control can be defined as perception of uncontrollable events as subject to one's control or as more controllable than they are, see Myers 2004, p. G-4.

the risk can be personally controlled by them.[383] As a result, less attention may be paid to warning messages and safe handling of the product may be omitted.[384]

Findings of this perception of control have been observed in the field studies of Van Duijne that pertain to risk perception in consumer product use. Even though these studies are not primarily focused on the effectiveness of product warnings, I believe that they are of relevance here, since this factor can have an influence on risk perception and in all probability on producing compliant behaviour too. The results that concerned the use of a power file, a blender, a chip pan, a gas lamp and electrically powered gardening tools indicate that product users who perceive the risks to be under control, seem more willing to carry out risky use actions than users who feel less certain about the controllability of the perceived risks. This is especially true for skilled product users. Hence, Van Duijne argues that next to perceived severity of injury, perceived controllability is essential to comprehending how product users perceive and describe risk.[385]

> A field study of Van Duijne e.a. addressed consumers' risk perception in the use of gardening tools. The researchers visited ten participants and asked them to demonstrate their everyday use of their electrically powered gardening tools. They were asked to explain their use activities and their understanding of the safety aspects of the tools in order for the researcher to get insight into how users perceive the risks that arise from the usage of their electrically powered gardening tools. Participants were not involved in an accident with their tool recently and most of them used their product frequently. All observed actions and interviews were recorded. The products addressed in the study were three lawn mowers, four grass trimmers and five hedge trimmers.
> The results showed that when participants were explaining the risk of their use of the tools, they referred to product characteristics of the tool that they considered hazardous. The owners of the hedge trimmers explained that the blades of a hedge trimmer can make serious cuts. One participant feared that he might cut his legs. The owners of the grass trimmers perceived the fast spinning line as dangerous, because the line can cause injury if it touches one's foot. The tools were perceived to be hazardous, with the hedge trimmer as most hazardous. Participants also mentioned the positively evaluated protective features

383 This belief can be linked to unrealistic optimism bias (see paragraph 3.6.7.3), as in both situations people tend to underestimate their personal probability of encountering negative events. They can also be distinguished from each other, in that unrealistic optimism refers to a generalised expectancy for positive outcomes independent of the source of the outcomes, whereas the illusion of control locates the source of the expected outcome in terms of personal control, see McKenna 1993, p. 42.

384 See *e.g.* Laux & Brelsford 1989; Friedmann 1988 (described in paragraph 3.6.9.7); Van Duijne e.a. 2008 (described in paragraph 3.6.7.5); Van Duijne 2005.

385 Van Duijne 2005, p. 230.

of the tools, such as the double-hand-switch activation mechanism of the hedge trimmers that requires both hands to activate the product in order to prevent the hands touching the blades during use. The lawn mowers and two grass trimmers contained a switch that requires both a button to be pressed and a lever to be pulled. This prevents unintentional activation of the product by one button.

The risky use action of failing to unplug electricity and accidentally activating the product when touching the blades, for example during cleaning, was not perceived by all participants. Four participants did not consider it possible that they might unintentionally activate the product while having their hands between the blades. Six participants were aware of this risk and unplugged their tool before removing the grass and leaves from the blades of the grass mower or from the hood of the trimmer.

The risk of cutting the power cord was perceived by all participants. They explained that they took precautionary measures in order to control the risk, such as being very cautious or by hanging the cord over their shoulders. However, the observations showed that some of these precautions would not always be sufficient enough to avoid the risk. Four participants had previously cut the cord of their gardening tool without being injured. Nevertheless, some of them did not show safer behaviour.

All participants wore protective footwear, because they wanted to prevent foot injury. They also said they wore gloves during the use of their tool. Not for safety reasons, but to keep their hands clean. Participants seemed unaware of the risk of eye injury as a result of debris being propelled from the tool. However, injury reports show that users of electrically powered gardening tools have suffered this type of injury.

The power tools contained warning symbols and texts on them. The texts were displayed only in foreign languages. Although all participants indicated that they understood the languages, only five of them said that they paid attention to the warning texts and symbols. They said that they had read and filed the instruction manual that was available in their own language. One participant said that she always reread the manual when using the hedge trimmer again every half year. Not everyone understood the symbols correctly or knew what they meant. Three people told the researcher that they had never even looked at the warnings. They believed, as one of them said, that these displayed 'just the standard safety messages'.

The results suggest that participants used product characteristics and the severity of the potential injuries associated with those characteristics to explain the risks. Some risky actions were overlooked by participants. Furthermore, participants perceived that they can control the risks associated with using the tools by paying attention to the particular hazardous product characteristics and by taking actions that were perceived to prevent the risk. There is also evidence that participants were willing to perform risky activities in order to get the desired result. Even though they perceived the tools as dangerous, they also considered them very useful to get the job done. Furthermore, it seems that participants accepted certain risks, because of the perceived benefit of less effort. They took short cuts, such as neglecting to study the warning information, failing to unplug electricity or leaving the power cord tangled.[386]

3.6.7.6 Familiarity Belief

An individual difference of product users that may account for the hazard judgement and that has been regularly studied in the warning research is the belief of familiarity.[387] Familiarity is, next to hazard perception, a highly influential factor in this stage, because it can affect earlier information processing stages, especially attention. Familiarity produces the belief that everything that needs to be known about a product or situation is already known. Even though familiarity beliefs are derived from prior knowledge and/or experience by repeated exposure, it does not mean that they are accurate.

Familiarity with one product can influence expectations about other, similar products. If newer versions of familiar products are more dangerous in the sense that they pose new, hidden risks, familiarity might lead to hazardous product use. Familiarity is closely related to experience. These two concepts are not necessarily identical. People may have familiarity with products or classes of products that they seldom use.[388]

Many studies have investigated the relationship between familiarity and hazard perception. The results indicate that people who are more familiar with a certain product generally perceive it to be less hazardous.[389] Consequently, they are less likely to notice, read, and comply with the warning label and vice versa.[390] There are also results that point to the other direction. It appears that hazard perception is more important than familiarity.[391] In most situations, hazard perception will determine whether people are seeking warning information and are motivated to act as instructed.[392] For example, with regard to familiarity

386 Van Duijne 2005, p. 127-151; Van Duijne e.a. 2008 (described in paragraph 3.6.7.5).

387 Vredenburgh & Zackowitz 2006, p. 347-350.

388 DeJoy 1999a, p. 203.

389 See *e.g.* Godfrey e.a. 1983; Wogalter e.a. 1991.

390 See *e.g.* Wogalter e.a. 1991; Godfrey e.a. 1983; Godfrey & Laughery 1984; Wogalter, Barlow & Murphy 1995 (described in paragraph 3.6.9.10).

391 Wogalter e.a. 1991, p. 206.

392 DeJoy 1999a, p. 206; see also Stewart & Martin 1994, p. 6.

with certain dangerous activities, people who are more familiar with high-risk activities generally have higher expectations of risk.[393]

Three explanations have been proposed in the literature for the familiarity effect on warning effectiveness.[394] The first concerns prior benign experiences. Benign experiences occur when people perform a specific behaviour without being harmed. The more familiar a person is with a product through various benign experiences, the more confident he will be in using the product and the less likely that person will have expectations of danger, be in an active information-seeking mode and comply with a warning.

As with the levels of performance model, the warning research has used script (or schema) theory to interpret the finding why familiarity generally has a negative effect on attention and compliance.[395] Scripts are schemas stored in long-term memory. They deal with knowledge about the usual sequence of and consequences of events.[396] Based on this theory, it is hypothesised that the behaviour of more experienced product users is likely to be script-driven and is thus well-learned; they perform the product's task without much conscious thought. Consequently, it is less likely that warnings are being noticed and complied with. Less experienced users are more likely to rely on external information that is provided, because their script of performing the task is not well developed yet.[397] The literature offers the example of driving a car. Most experienced drivers follow a series of events to start a car. These steps which are necessary to get on the road often occur without much conscious thought. One of these steps can be seat belt use. If putting on seat belts is part of the script of driving a car, this step will be automatically performed. On the other hand, if this is not the case, the script-driven behaviour must be disrupted in order to improve safe driving.[398]

The process of habituation also offers an explanation.[399] People who become habituated with the warnings are overexposed to them through repeated exposure, causing their salience to be reduced. As a result, the warning is less likely to attract and to hold attention, and subsequently to be followed.[400]

3.6.7.7 Prior Injury Experience with the Product

Whilst repeated exposure to benign experiences tends to decrease warning compliance, prior personal experience of an injury or having personal knowledge of someone else being

393 Vredenburgh & Zackowitz 2006, p. 347.
394 DeJoy 1999a, p. 205; Vredenburgh & Zackowitz 2006, p. 348, 349.
395 Schank & Abelson 1977.
396 Eysenck & Keane 2000, p. 381. See also paragraph 3.6.4.7.
397 Wogalter, Barlow & Murphy 1995; Duffy, Kalsher & Wogalter 1995, p. 160.
398 Vredenburgh & Zackowitz 2006, p. 348.
399 See also paragraph 3.6.5.7.
400 Wogalter & Vigilante 2006, p. 259; Vredenburgh & Zackowitz 2006, p. 348. See *e.g.* Thorley, Hellier & Edworthy 2001; Amer & Maris 2007.

injured may lead to safer behaviour.[401] The availability heuristic can account for this effect. Because such a negative experience creates a salient memory, it is more easily remembered.[402] Research indicates that previous injury experience with a product increases one's hazard perception.[403] Participants with negative personal experience gave significantly higher injury estimates for products compared to those without any injury experience.[404]

3.6.7.8 Risk-Taking Attitude

Another factor that may influence behavioural compliance with warnings is risk-taking or sensation-seeking attitude.[405] It seems that risk-taking is associated with lowered risk perception. Risk-taking has been defined as the need for varied novel and complex sensations and experiences, and the willingness to take physical and social risks for the sake of such experiences.[406]

The presence of a warning message may be attractive for risk seekers.[407] It follows that compliance rates are likely to be lower for warning recipients who are willing to take risks than for people who do not seek out new experiences or thrills.[408] For high risk seekers, warnings could even produce the opposite behavioural effect of what is intended by the warning. This is referred to as *psychological reactance*.[409] There is modest evidence that warnings are capable of producing boomerang effects and only for certain categories of products.[410] For instance, one study regarding alcohol warning labels found that participants who were exposed to the warning label in an add rated the benefits of drinking more favourably and male drinkers reported higher intentions to drink than did the control group that was not exposed to the warning.[411] Note that the boomerang effect consisted of drinking intentions and not actual drinking behaviour. To see whether this counterproductive effect could be replicated, succeeding research was conducted and no effect was observed.[412]

---

401 Wogalter 2006b, p. 57.
402 Vredenburgh & Zackowitz 2006, p. 349.
403 See *e.g.* Chy-Dejoras 1992; Karnes Edward, Leonard & Rachwal 1986; Van Duijne 2005, p. 170.
404 Wogalter, Brems & Martin 1993; in contrast: Lehto & Foley 1991.
405 Vredenburgh & Helmick-Rich 2006, p. 377.
406 Zuckerman 1979. Note that risky behaviour is not always the result of consciously having considered the risks. People may also perform risky behaviour as a result of routine and automatic behaviour. In those instances people generally do not evaluate the risks, see Wagenaar 1992, p. 279.
407 Stewart & Martin 1994, p. 12.
408 See *e.g.* Purswell, Schlegel & Kejriwal 1986; Weaver, Helmick & Burke 2003; Weaver e.a. 2003.
409 Stewart & Martin 1994, p. 11. Reactance can be defined as (1) a motive to protect or restore one's sense of freedom. Reactance arises when someone threatens our freedom of action. (2) The desire to assert one's sense of freedom, see Myers 2004, p. G-6.
410 Riley 2006, p. 294.
411 Snyder & Blood 1992.
412 MacKinnon & Lapin 1998.

Risk seeking and the phenomenon of psychological reactance have not been investigated systematically in the context of warnings. Questions regarding its interaction with hazard perception and its relative contribution to warning effectiveness remain open.[413] Framing important information as a loss may have some effect to target the attitudes of this subgroup.[414]

3.6.7.9 Personal Relevancy of the Warning to Warning Receivers

When individuals believe warnings to be relevant to them or the task that they are performing, they are more likely to pay attention to and follow a warning.[415] The belief of relevancy means that people believe something is applicable to them. If consumers think that the warning is not personally relevant to them, they may ignore it and the warning will have little behavioural effect. Increasing the relevancy of a warning message to a person by personalising it, for example by using his name on a warning display, has shown to enhance compliance behaviour.[416] However, it must be recognised that personalising warnings on consumer products is difficult to apply to real-world situations. This finding would be of more interest to work place safety. Relevancy may also be achieved by seeing others complying with the warning.[417]

3.6.7.10 Costs of Compliance: Time, Effort and Convenience to Carry Out the Behaviour

Expectations that relate to the outcome are important in the stage of attitudes and beliefs. For example, people's belief with regard to the costs of complying with a warning clearly has an impact on a person's intent to comply and actual compliance. The *cost(s) of compliance* refer to the amount of time, effort, convenience, and/or money required to follow the warning. A strong and consistent finding in the warning research is that warning effectiveness decreases as the costs of compliance increase.[418] This factor is especially a significant factor for motivation and is discussed there in more detail.

---

413 DeJoy 1999a, p. 209; Rogers, Lamson & Rousseau 2000, p. 128.

414 Riley 2006, p. 294.

415 Wogalter 2006b, p. 58; Vredenburgh & Zackowitz 2006, p. 350.

416 See *e.g.* Wogalter e.a. 1994; Wogalter, Kalsher & Racicot 1992.

417 Racicot & Wogalter 1995. See in this respect, paragraphs 3.6.8.4 and 3.6.9.12 on the effect of social influence.

418 See *e.g.* Wogalter e.a. 1987; Dingus, Wreggit & Hathaway 1993 (described in paragraph 3.6.6.7); Hunn & Dingus 1992 (described in paragraph 3.6.9.11); Dingus, Hathaway & Hunn 1991.

3.6.7.11 Response Efficacy: Belief Regarding the Effectiveness of Recommended Precautions

Value-expectancy models imply that the expectations people have of the effectiveness of precautions directed in a warning can also impact warning related behaviour. This type of belief is also called response efficacy.[419]

Although little attention has been paid to this factor in the warning research, findings imply that information about the actions that will reduce or eliminate the hazard is important for warning compliance.[420] If the precautionary measures recommended by the warning are being perceived as insufficient or unlikely to counteract the danger, it is plausible that the warning directed behaviour will not be produced in spite of the perceived threat. In this respect, it should be noted that it is also of relevance to what extent people believe the recommendations to be adequate in relation to the level of hazard. If product users perceive a mismatch between the recommended precautions and the level of threat posed by the product, they might not be motivated to engage in the appropriate behaviour. Hence, this belief emphasises the importance of adequately communicating the actions that are effective in avoiding the hazard. [421]

3.6.7.12 Self Efficacy: Belief Regarding the Ability to Perform Specific Behaviour

Another type of expectation that has been seldom studied in the warning research is the belief of self-efficacy.[422] Self efficacy refers to the extent in which people believe that they have the ability to perform specific behaviour. Thus, people who have low self efficacy expectations should be less likely to engage in the behaviour recommended by a warning. This may be especially true for situations of product use where the instructions are not easy to perform. Hence, it is of relevance that warnings give receivers the impression that they are able to perform the behaviour. Nevertheless, the actions that need to be undertaken to use a consumer product safely are often simple.[423]

419 Cameron & DeJoy 2006, p. 305. This factor was already discussed in the context of the value-expectancy models of paragraph 3.4.3.2.

420 See *e.g.* Wogalter e.a. 1987; Frantz 1994.

421 DeJoy 1999a, p. 201; DeJoy 1999a, p. 228.

422 Bandura 1986; see also Myers 2004, p. 57 and Cameron & DeJoy 2006, p. 305. Self-efficacy can be defined as a sense that one is competent and effective, distinguished from self-esteem, one's sense of self-worth, see Myers 2004, p. G-7. Value-expectancy models from the persuasion literature indicate the potential relevancy of this factor to the effectiveness of warnings, see paragraph 3.4.3.2.

423 DeJoy 1999a, p. 202; Riley 2006, p. 295; Vredenburgh & Helmick-Rich 2006, p. 377; DeTurck 2002, p. 361.

3.6.8 *The Receiver Stage of Motivation*

3.6.8.1 General

Motivation is the last stage before behaviour is achieved.[424] Once a warning is noticed and understood and is consistent with receivers' beliefs and attitudes, it is then essential for safe behaviour that the warning motivates receivers to comply with it. The ability of a warning to motivate is based on either informing people about unknown hazards or by reminding them about known hazards.

It is difficult to define motivation and there is disagreement on the right definition.[425] In the warning literature, motivation has been viewed as a momentum driving one to take a specific action. It is a set of processes that links beliefs with action or inaction, comprising a range of emotional and decision factors.[426]

Several factors have shown to be essential in this stage of information processing, particularly the factors of the previous paragraph, such as people's expectations about the threat and their beliefs about the effectiveness of the precautions expressed in the warning. Beliefs and attitudes can be motivations in themselves. In the warning literature, it is argued that hazard perception provides the initial motivation for self-protective behaviour, since expectations of the threat and the consequences are likely to arouse fear in individuals which in turn increases the likelihood that they comply with the warning. In this regard, warnings can be viewed as communications that arouse fear. Fear is the emotional reaction that occurs when a serious threat is perceived by someone. The more fear aroused in people, the greater the probability of altering attitudes and producing self-protective behaviour.

Whilst research on the topic of fear arousal has provided evidence of the relationship between fear arousal and behaviour change, the results are not straightforward. Messages arousing too much fear can be counter-productive. Moreover, even though perceived threat is a significant factor influencing motivation to comply with a warning, it is not sufficient for warning effectiveness. Other factors like the costs of complying, social modelling and stress play a role too.[427]

3.6.8.2 Costs of Compliance: Time, Effort and Convenience to Carry Out the Behaviour

A key determinant of motivation is the perceived costs of compliance: namely, the amount of time, effort, convenience, and/or money required to produce behavioural compliance. For instance, product users can consider wearing protective clothing or gloves as too costly, because they take too long to find or are less comfortable to wear.

---

424 See for an extensive discussion of this information processing stage DeJoy 1999b and Riley 2006.
425 Smith-Jackson & Wogalter 2006, p. 29.
426 Riley 2006, p. 289.
427 DeJoy 1999b, p. 222, 238; see also DeTurck 2002, p. 347.

Value-expectancy theory and models derived from this general approach assume that individuals' decision to produce precautionary behaviour or not is based on a weighing of the perceived costs and benefits of taking action.[428] Consequently, the theory suggests that the decision to follow a warning depends on considering the benefits and the costs of complying with the prescribed precautions. If people perceive the costs of complying with the warning to be greater than the benefits, they will be less likely to comply than if the benefits are perceived to exceed the costs.

Following this reasoning, warning effectiveness is expected to be low when the costs are high and compliant behaviour should increase if the benefits of warning compliance can be heightened or the costs of compliance reduced.[429] Warning studies have confirmed this. Findings demonstrate that compliance is much lower when the costs are high than when the costs of compliance are low. Moreover, studies show quite clearly that even a minimal amount of extra time, effort or discomfort can reduce warning compliance considerably.[430] More generally, the observational studies pertaining to risk perception in consumer product use support the finding that product users consider the costs and benefits of their actions. If the perceived costs of taking precautionary actions are high, because it takes much effort or because the protective equipment feels uncomfortable, and if the perceived benefits in terms of risk reduction are unclear, product users are less likely to take precautions to guard against the accident scenarios that they know about.[431]

> In the laboratory experiment of Wogalter, McKenna and Allison, 23 college students were asked to perform the chemistry task using a set of instructions that contained a warning directing them to wear a safety mask and protective gloves. The cost of compliance was manipulated by locating the mask and gloves in either an accessible location (low cost) or a less accessible location (high cost) that involved participants having to walk 8 meters to another room. Results showed that the subjects were significantly more likely to comply in conditions of low cost (73%) than of high cost (17%).[432]

3.6.8.3 Costs of Non-Compliance: Severity of Consequences

Predictably, beliefs about the *costs of non-compliance* affect motivation too. These costs relate to people's expectations about the potential negative consequences (*e.g.* personal injury or property damage) associated with failing to follow the warning.

---

428 DeJoy 1991; Weinstein 1993.

429 DeJoy 1999b, p. 227.

430 See *e.g.* Wogalter e.a. 1987; Dingus, Wreggit & Hathaway 1993 (described in paragraph 3.6.6.7); Hunn & Dingus 1992 (described in paragraph 3.6.9.11); Dingus, Hathaway & Hunn 1991.

431 Van Duijne 2005, p. 238. See *e.g.* Van Duijne e.a. 2008 (described in paragraph 3.6.7.5).

432 Wogalter, McKenna & Allison 1988 and Wogalter, Allison & McKenna 1989.

As discussed earlier, a robust finding in the warning research is that the more hazardous a product is perceived to be, the greater the likelihood of precautionary intent and behaviour.[433] As hazard perception is closely tied to people's judgement of injury severity, these costs can exert a powerful influence on motivation. This is especially true when the possible injury is serious. Hence, it is logical to presume that increasing the perceived costs of non-compliance should enhance behavioural compliance as this will likely lead to an increase in perceived threat. As shown previously, giving explicit consequence information may be of help. Furthermore, studies have shown that certain warning design features, such as signal words or an interactive label, have the ability to influence hazard perception.[434]

3.6.8.4 Social Influence of Others

Social influence – or role model effects – is another factor that can persuade people to change their behaviour. A consistent finding is that the behaviour of role models influences the safety behaviour of others, both positively and negatively.

The warning research has shown the effect of social influence. The findings indicate that in the event that product users see another individual not using the product as required by the warning, and no accidents occur, it is likely that they will behave accordingly. Similarly, observing others complying with a warning's safety recommendations can have a positive influence on the motivation of potential victims to follow the warning.[435]

This effect of social influence can be incorporated into the warning design by depicting a role model using the product on label graphics or in instructional videos.[436] Although less relevant in relation to consumer safety, the role model effect also implies that training courses can be a useful way to influence behaviour, as showing people what to do seems to be more effective than merely telling people what to do.[437]

> DeTurck, Chih and Hsu conducted three laboratory experiments to determine the effect of a role model's compliance behaviour with warning messages on observers' safety behaviour across a variety of conditions.
> In the first study, role models (confederates) used a cleaning product requiring them to wear safety gloves. Under the guise of a marketing study 100 subjects were paired with a confederate to examine and test the effectiveness of a new cleaning product called 'QuikClean'. The subjects were informed that they

433 See *e.g.* Wogalter, Brems & Martin 1993; Wogalter e.a. 1991.
434 DeJoy 1999b, p. 229.
435 See *e.g.* Wogalter e.a. 1987; DeTurck, Rachlin & Young 1994; DeTurck, Chih & Hsu 1999 (described in paragraph 3.6.8.4); Wogalter, Allison & McKenna 1989; Racicot & Wogalter 1995; Racicot & Wogalter 1992; Chy-Dejoras 1992; Edworthy & Dale 2000.
436 Riley 2006, p. 296.
437 Wogalter, Allison & McKenna 1989, p. 140.

would be working with a confederate that was described to them as friendly or unfriendly.

The warning label on the back of the container read:'WARNING: Wear rubber gloves when using this cleaner. Cleaner is extremely hazardous to skin and eyes. May cause severe chemical burn if left untreated; rinse skin thoroughly. If cleaner contacts eyes, rinse eyes thoroughly for 15 minutes. May cause permanent loss of vision if left untreated. If swallowed, drink plenty of water and notify a physician immediately. May cause permanent damage to internal organs.'

After examining the label, the pair was asked to test the product. The confederate always tried the product first on a dirty stain on a table and then the subject. In addition to the cleaner, a pair of rubber gloves, a roll of paper towels and a basin of water were present by the stains. The conditions of safety behaviours were:

1) Gloves condition: In this condition, the confederate put on the safety gloves prior to cleaning the stain and did not experience any negative consequences from the product.
2) No gloves/no burn condition: The confederate tested the product without wearing the safety gloves and did not experience any burns.
3) No gloves/mild burn condition: The confederate did not wear the safety gloves when testing the product and accidentally got a small amount of cleaner on his or her hand while using the cleaning product. The confederate mildly exclaimed, 'Oh, I can feel that burns' and quickly wiped the cleaner off using a paper towel. The reaction of the confederate was designed to indicate that he or she experienced only mild discomfort.
4) No gloves/painful burn condition: The confederate did not wear the gloves and experienced a severe chemical burn yelling 'Ow, it really burns'. The reaction of the confederate was designed to indicate that he or she experienced a substantial amount of pain. The confederate immediately washed the skin where the burn occurred.
5) Control condition: Participants examined and tested the product without being paired with a confederate.

It was hypothesised by the researchers that the observers are more likely to behave consistently with a role model's safety behaviour when the role model receives a painful rather than a mild chemical burn. Furthermore, the role model effect should be particularly pronounced when the role model is likable. As predicted, participants who observed the role model experience a painful chemical burn were more likely to wear the safety gloves than participants who observed the role model experience only a mild chemical burn. However, this effect was not more pronounced when the role model was likable. In the control condition, 50% of the participants wore the safety gloves when testing the product, in the gloves condition 75%, in the no gloves/no burn condition 15%, in the no gloves/mild burn condition 21%, and in the no gloves/painful burn condition 75%. As regards likability, subjects were slightly more likely to comply with the likable role model (51%) than the unfriendly role model (42%). The finding that role models' friendliness did not affect participants' compliance suggests that the imitation of others' safety behaviour also pervades to less personal relationships.

In the second study, the procedure was identical, only this study also varied the severity of the hazard communicated in the warning message of the household cleaning product. Subjects examined and tested either a cleaning product with a warning expressing a low hazard level or a high hazard level. The goal of this study was to determine whether the degree of hazard communicated in the message mediated the role modelling effect observed in the first experiment.

The low hazard level warning read: 'CAUTION: Wear rubber gloves when using product. Cleaner is mildly irritating to skin and eyes. May cause slight irritation if left untreated; rinse skin thoroughly. If cleaner contacts eyes, rinse eyes thoroughly for fifteen minutes. May cause temporary blurring if left untreated. If swallowed, drink plenty of water and notify a physician. May cause slight internal discomfort.'

The warning label in the high hazard condition read: 'DANGER: Wear rubber gloves when using cleaner. Cleaner is extremely hazardous to skin and eyes. May cause severe chemical burn if left untreated; rinse skin thoroughly. If cleaner contacts eyes, rinse eyes thoroughly for fifteen minutes. May cause permanent loss of vision if left untreated. If swallowed, drink plenty of water and notify a physician immediately. May cause permanent damage to internal organs.'

The results showed that the compliance rates were similar to the results of the first study; there was a significant effect obtained for the role models' safety behaviour on participants' compliance. However, the level of hazard did not influence subjects' compliance behaviour. Thus, the results of both studies indicate that participants used the painful burn as a cue with regard to the product's hazard and consequentially decided not to imitate the unsafe behaviour of the role model. The pain may have reminded them of the warning's content.[438]

3.6.8.5 Time Stress and Mental Work Load

Two other situational factors that have the potential to frustrate motivation and behavioural compliance are time stress and mental work load.[439] Under conditions of high stress or when the task a person is performing demands a great proportion of cognitive resources, information may not be properly processed. High stress levels are common when the time to perform a product task is limited. High mental work load is present when the task performance is difficult, for example because different activities occur concurrently.[440] Under these conditions, product users may rely on simplified information processing strategies such as heuristics to decide how to handle the product.[441] As a result, task performance may decrease. It is then less likely that consumers will be motivated to behave according to the warning compared with a situation of lower stress or work load.[442]

3.6.9 The Receiver Stage of Behavioural Compliance

3.6.9.1 General

If each of the cognitive stages is successfully processed, the result is behaviour that is compliant with the warning. As noted in paragraph 3.5.4, relatively few studies have measured actual compliant behaviour, even though many consider it the ultimate measure of warning effectiveness. Behavioural compliance can be evaluated through subjective measures of behavioural intentions (or precautionary intent), which means that, for instance, on the basis of a questionnaire, participants are asked to what extent they would be willing to comply with the presented warning in a particular situation. Even though prior research indicates that behavioural intentions predict behaviour, having the intention is not the same as actually complying with a warning. An objective method to examine compliance is through observing actual behavioural compliance in a field study or in a controlled laboratory setting. But, unfortunately, measuring behaviour in the lab and in the field can be difficult, which can explain the limited amount of studies that have addressed the behavioural effects of warnings.[443]

---

438 DeTurck, Chih & Hsu 1999.

439 Wogalter 2006b, p. 58.

440 Cognitive work load can be defined as a measure of the extent to which the information processing system is involved in performing a task, see Payne & Wenger 1998, p. G-3.

441 See *e.g.* Helander 2006, p. 80; Helander 2005, p. 1-8.

442 Wogalter e.a. 1998 (described in paragraph 3.6.9.13); Wogalter & Usher 1999 (described in paragraph 3.6.9.14); Duffy, Kalsher & Wogalter 1995 (described in paragraph 3.6.4.8).

443 For more on research methods see paragraph 3.5.4 of this chapter.

The existing empirical studies that did investigate the behavioural effectiveness of warnings have shown that product warnings can bring an added value to the situation to which they refer; the appropriate safe behaviour was higher when a warning was present than when it was not.[444] However, the results also show that the levels of behavioural compliance with the warning vary among the studies. In addition, there are studies that have found no or only little effect of the warning on compliant behaviour. Some would even argue that warnings are of no use as a prevention tool since a number of research findings demonstrated that warnings do not have a measurable impact on people's behaviour and product safety.[445] Nevertheless, *meta-analyses*[446] of the warning research have confirmed that warnings are effective in modifying behaviour.[447]

The previous overview of information processing stages and the factors affecting the stages prior to behaviour shows that the warning process is complex. Naturally, factors that influence the earlier information processing stages such as attention and motivation are indirectly of importance in determining whether the warning will have a positive effect on behaviour. In general, the variables that influence the initial stages are expected to be positively correlated with the likelihood of compliance.[448] Below the factors are presented that have shown to have a direct effect on behaviour and of which the warning literature and research think that they are influential.[449] It is worthy to note that this list is not exhaustive. Other potentially influential factors that seem less important are not worked out in detail here.[450] However, this does not mean that they cannot be powerful. It only suggests that others have received more attention in the literature and in research, *e.g.* because their measurement is less troublesome, or because others may intuitively be viewed as more important.

Each factor described below is illustrated by a study that measured the relative effect of the specific variable on *actual* behaviour. As noted earlier, researchers have also employed subjective measures. This usually concerns the measurement of attitudes and beliefs. The advantage of having such results is that they can function as a 'back-up' to see whether

444 *E.g.* Wogalter e.a. 1994; Wogalter e.a. 1999b (described in paragraph 3.6.9.8).

445 McCarthy e.a. 1984.

446 A meta-analysis combines the results of several studies to examine the impact of a measure.

447 *E.g.* Cox III e.a. 1997; Argo & Main 2004.

448 Laughery 2006, p. 473. See *e.g.* the field demonstration experiment of the water fountain warning in which adding vivid-enhancing features to the warning sign caused a significant increase in behavioural compliance. The vivid-enhancing characteristics consisted of the use of colour, a pictorial and the increase in size of the warning sign, see Wogalter e.a. 1987, p. 609.

449 For a detailed overview of the warning compliance literature, see Silver & Braun 1999; Kalsher & Williams 2006 and the earlier work of Lehto & Miller 1986. See also Lehto & Papastavrou 1993; Rogers, Lamson & Rousseau 2000; Lesch 2006 and Laughery 2006.

450 Other variables that have been mentioned in the literature are perceived appropriateness (whether people perceive the warnings as appropriate), affordance perception (what the immediate situation allows a person to do), layout of the warning information, information processing objectives, risk-taking style, age and gender.

these findings paint a similar picture like results of behavioural experiments using objective measures. In some cases, subjective measures are the only suitable means. There are plenty of studies available that have measured the impact of a range of variables on people's reported willingness to comply or the perceived effectiveness of the warning. These potentially influential factors include the effect of adding information components,[451] adding explicit information,[452] the effect of perceived hazard,[453] perceived control of the hazard,[454] adding information about the warning's source,[455] adding warning symbols[456] and prior injury experience.[457] The effect of certain variables on behavioural compliance has only been measured subjectively, for instance by adding information about the characteristics of the warning's source to a warning message. Others have been measured both objectively and subjectively.

3.6.9.2 Location

Location or placement of warnings can have a substantial influence on warning compliance.[458] The effect of location is dependent on the context in which the warning appears. A visually cluttered environment decreases warning compliance compared with a less cluttered surrounding, as there are other stimuli in the environment that reduce detection of the warning.[459] Furthermore, a study showed that a warning was more effective when it was positioned in a set of task instructions than when a similar, but much larger warning appeared on a posted sign that was nearby, though outside the field of view.[460]

Clutter on the consumer product's package, label or in the manual, can also distract someone's attention from the warning message and thereby reduce the likelihood that people will comply with the warning. With regard to the location of warning information relative to other information, the results are rather mixed. For example, a study showed that placing important warning information at the beginning, before the procedural instructions for use was significantly more effective than when it was placed after the two page set of instructions.[461] Other studies showed that compliance behaviour was highest

451 See *e.g.* Wogalter e.a. 1987; Wogalter, Kalsher & Rashid 1999, Kim, Cowley & Wogalter 2007; Vredenburgh e.a. 2005 (described in paragraph 3.6.6.6).

452 See *e.g.* Heaps & Henley 1999; Laughery e.a. 1993a; Laughery e.a. 1991.

453 See *e.g.* Vredenburgh e.a. 2005 (described in paragraph 3.6.6.6); Wogalter e.a. 1991.

454 Friedmann 1988 (described in paragraph 3.6.9.7); Rogers, Lamson & Rousseau 2000, p. 128.

455 Wogalter, Kalsher & Rashid 1999.

456 See *e.g.* Davies e.a. 1998; Vardavas e.a. 2009.

457 See *e.g.* Wogalter, Brems & Martin 1993.

458 See *e.g.* Wogalter & Young 1994 (described in paragraph 3.6.9.3); Frantz & Rhoades 1993 (described in paragraph 3.6.4.7); Wogalter e.a. 1998; Wogalter, Kalsher & Racicot 1993; Wogalter e.a. 1991b (described in paragraph 3.6.3); Visschers e.a. 2004; Wogalter, Barlow & Murphy 1995 (described in paragraph 3.6.9.10).

459 Wogalter e.a. 1991b (described in paragraph 3.6.3).

460 Wogalter e.a. 1998; Wogalter, Kalsher & Racicot 1992.

461 Wogalter e.a. 1987; see also Strawbridge 1986.

when important safety information was presented in the directions for use section on the product's label relative to other locations on the label and/or in an additional leaflet.[462] Because of differing results, a task-analytical approach to determine which warning locations are suitable is encouraged.

> The effects of the location of safety information, of the language used to designate hazards and of the status of participants (professional or amateur) on behavioural compliance with regard to the use of a pesticide product were investigated in the research of Edworthy e.a. 2004.
> Pesticide products are potentially hazardous and they usually are accompanied with considerable written information, either somewhere on the label and/or in an additional leaflet. The safety information can be expressed in various linguistic ways, like probabilistic phrases such as 'may harm' and 'may be dangerous' and the use of the personal pronoun ('you'). Furthermore, two groups of participants were tested in the research: an amateur group and a group of 'professionals' who used pesticides regularly during the course of their work. The third study of the research described here, measured the potential effects of these three variables (location, linguistic variation, status) on actual compliance with the safety information. The first two studies were used to find the best and worse case conditions for location and linguistic variation.
> Participants (65 amateurs and 24 professionals) were observed and videotaped during their demonstration of using a (placebo) house plant insect killer to measure the extent in which they complied with the safety information. There were two conditions for location: placement of the safety information in the directions for use section on the product's label (best case) and the additional leaflet (worst case). The label directed the user to the leaflet, an A5 paper size sheet enclosed inside the box in which the product was supplied. There were also two conditions for linguistic variation. In one condition the safety information was presented as a personal instruction statement (best case) or in the other condition as a probabilistic hazard (worst case). For example, regarding washing, the information on the label was presented as either 'You must wash out all containers coming into contact with this product thoroughly before and after use' (personal instruction statement) or 'Contamination of containers may be hazardous' (probabilistic hazard). As regards protective equipment, the phrases were 'You must wear gloves and a respirator mask' and 'May be harmful to people without gloves and a respirator mask'.

462 See *e.g.* Friedmann 1988 (described in paragraph 3.6.9.7); Frantz 1993; Frantz 1994; Edworthy e.a. 2004; Strawbridge 1986.

> The findings showed that compliance was highest in the best case linguistic condition and also for the best case location condition. Generally, the personal instruction statements produced higher levels of compliance than did the probabilistic hazard statements for both professionals and amateurs. However, this effect was clearer for professionals than for amateurs overall. For amateurs, placing the information in the directions for use produced higher levels of compliance than did the placement in the leaflet. For professionals, both locations appeared acceptable ways as they both elicited reasonable levels of compliance. An interesting result is the additive effect of linguistic variation and location on behavioural compliance: when the safety information was presented in the directions for use section and presented as a personal instruction, the highest level of compliance was produced relative to the effectiveness scores of the other manipulations.

3.6.9.3 Limited Surface Area: Alternative Label Designs

A factor that is strongly related to the placement of a warning message and to size is the use of an alternative warning label. Because of limited space on small product containers, manufacturers are often forced to decrease the amount and/or the size of the information presented. To possibly remedy this problem additional information can be provided in a separate product insert or manual, or an alternative warning label such as fold-outs, wings and tags can be designed. Besides that people generally favour alternative labels to standard labels in view of the greater surface area that is available and as a result their increased noticeability, they have also been shown to improve warning compliance.[463]

> In the experiment of Wogalter and Young, two alternative warning label designs, a tag and wings, were compared with a conventional label (control condition) printed on a very small container of a glue product to measure their effect on behavioural compliance.
> The warning in the tag condition was provided as a tag attached to the mouth of the bottle. The wings label extended on both sides of the bottle. Warning compliance was reached when participants wore the protective latex gloves as was directed by the warning. A total of 44 participants received instructions and were asked to put together a part of a model airplane using the glue. They were randomly distributed across the three label conditions: the control and tags condition each had fifteen participants and the wings condition had 14 participants. The participants were not informed of the study's real purpose. After completing the task, participants filled in a questionnaire that asked

463 Wogalter & Young 1994, p. 53.

> whether they noticed, read and recalled the instructions and the warning statement: 'WARNING: Glue can burn and kill skin on contact. Wear supplied gloves when using glue. KEEP OUT OF REACH OF CHILDREN'. Participants were also asked to rank their preference for the three warning label designs. Results showed that the tag design produced significantly greater compliance than the other 2 designs: 2 of the 15 participants complied with the control condition, 5 complied with the wings condition and 12 of 15 participants wore the gloves when the tag was attached to the glue bottle. Measures of noticing, reading and recall of the warning mirrored the compliance results. While participants generally preferred the control label, they most preferred the tag warning. Overall, the results suggest that alternative designs like the tag can enhance warning communication and compliance in cases where surface area is limited.[464]

3.6.9.4 Interaction with a Warning

Another related factor to warning location is interactivity. The results are not entirely consistent with regard to whether an interactive warning facilitates warning compliance. There are studies that have shown a positive effect on noticeability, but not on enhancing compliance.[465] Nonetheless, several studies have yielded the finding that an interactive warning facilitates safer behaviour, meaning that a product warning was more effective when users had to psychically interact with it during product use.[466] The fact that interactivity interrupts script driven behaviour probably accounts for its effect on warning effectiveness. Paragraph 6.9.11 discusses the study of Hunn & Dingus 1992 that examines the impact of interactivity and cost on compliance.

3.6.9.5 Colour

Using colour can produce higher levels of compliance, but not all research on colour has found an effect on behavioural compliance. It seems that red is more effective than other colours, as people associate red with danger.[467]

> For example, 65 undergraduate students in a laboratory experiment were instructed to use two consumer products, a pool-water test kit and a two-part adhesive, under the guise of a marketing study. The warnings were presented

---

464 Wogalter & Young 1994.

465 Gill, Barbera & Precht 1987.

466 Frantz & Rhoades 1993 (described in paragraph 3.6.4.7); Wogalter, Barlow & Murphy 1995 (described in paragraph 3.6.9.10); Duffy, Kalsher & Wogalter 1995 (described in paragraph 3.6.4.8); Visschers e.a. 2004.

467 See *e.g.* Braun & Silver 1995; Rudin-Brown e.a. 2004; Wogalter e.a. 1987. No effect of colour was found in Shaver & Braun 2000.

> on the front and back of the products and stated: 'DANGER: HAZARDOUS CHEMICAL WILL BURN SKIN. Wear rubber gloves when using'. The warning was either printed in red, green or black. Behavioural compliance was assessed by observing whether the subjects wore the gloves as directed by the warning. The rubber gloves were not included in the package but were within reach on the testing table.
> Although the differences were not significant, the results indicated that warnings printed in red resulted in more compliance than green and black. For example, the compliance scores with regard to the two-part glue were 74% for the red warning, 57% for the green warning, and 57% for the black warning. The compliance scores for the pool test kit were 68%, 38% and 52%, respectively. This can be explained by the finding that participants rated the likelihood of being injured higher for products with warnings printed in red than for products with warnings in green or black.[468]

3.6.9.6 Channel

The channel that conveys the warning information has been shown to affect warning effectiveness.[469] Which sensory modality (visual, auditory, olfactory, tactile, gustatory) is most effective to reach potential victims depends on the context in which the warning is presented. Studies illustrate that one modality can be more effective than the other. For example, an auditory warning is not appropriate in a noisy environment or to transmit long messages. However, such a warning can be particularly useful to alert and attract attention in a visually cluttered environment.[470] Furthermore, using a single channel to deliver a warning message can be problematic, especially for people with sensory deficiencies. Research has demonstrated that providing warnings through more than one channel can enhance behavioural effectiveness and therefore recommends using more than one modality to deliver the message.[471]

> The study of Wogalter and Young examined the effect of warning modality on compliance behaviour in two laboratory experiments and a field experiment. The field experiment simulated a slippery-floor hazard in a shopping mall. Researchers defined an area in which the hazard of falling was present. There were cues such as orange traffic cones, a bucket and a mop that indicated the presence of the slippery-floor danger. Behavioural compliance was measured

468 Braun & Silver 1995.
469 Cohen e.a. 2006, p. 128; Kalsher & Williams 2006, p. 316.
470 Wogalter & Vigilante 2006, p. 250 ff.
471 See *e.g.* Conzola & Wogalter 1999; Wogalter, Kalsher & Racicot 1993.

by the number of people that did not enter the dangerous area during the experimental session.

A total of 531 people entered the shopping mall and encountered one of four warning conditions. In the first control condition, there was no warning present, only the cues. In the second condition, a printed warning sign was present. The sign was attached to the cone. In the third condition, a voice warning was provided on a tape that played continuously. The fourth condition contained the print warning and the voice warning.

The field experiment confirmed the findings of the two laboratory experiments showing that compliance was significantly higher when the warning was presented in both modalities compared to the print-only condition. When no warning was present, 20% complied. The compliance scores for the second condition was 42% and for the third condition 64%. Compliance behaviour peaked when a print warning and a voice warning were provided (76%).[472] Note that the effectiveness score in the latter most successful condition is up to 56%.[473]

3.6.9.7 Warning Symbols

Warning symbols (also called pictorials) play an essential role in warnings being noticed and comprehended, but their ability to change behaviour has been less clear-cut.[474] The effectiveness of a warning symbol depends on factors such as the concept it represents, the context in which it appears, depiction quality, prior training and knowledge of the target audience.[475] Several studies have reported that compared with its absence, the inclusion of a pictorial increases safe behaviour, but the degree in which pictorials affect behaviour varies among the studies.[476] Occasionally, no effect has been measured.[477]

The main goal of Friedmann's laboratory experiment was to determine the effects of three independent variables on noticing, reading and following the warning and on recalling the warning information. The independent variables were adding symbols to written warnings, subjects' familiarity with the product and type of hazard.

A total of 144 college students participated in the experiment. Based on the results of a pilot study, a familiar product (liquid drainer) and an unfamiliar

472 Wogalter & Young 1991.

473 Edworthy & Adams 1996, p. 60.

474 Wogalter e.a. 2006, p. 172.

475 Wogalter e.a. 2006, p. 159.

476 See *e.g.* Schneider 1977; Jaynes & Boles 1990; Otsubo 1988 (described in paragraph 3.6.9.8); Wogalter, Kalsher & Racicot 1993; Wogalter e.a. 1991b (described in paragraph 3.6.3); Wogalter, Kalsher & Racicot 1992.

477 Friedmann 1988 (described in paragraph 3.6.9.7).

product (wood cleaner) were selected. The warnings studied were positioned on the back panel of the products' container. The back panel displayed five sections: a warning message located at the top with a promotional paragraph underneath, directions for use, a paragraph specifying the uses of the product and a list of ingredients. The written warning messages contained a signal word followed by information about the nature of the hazard and information about the precautionary action to be taken before using the consumer product. The hazard warned against was eye contact or inhalation. In the conditions of adding symbolic information, the symbols were placed next to the written warning information. Either a proactive or a reactive symbol for each hazard type was used. There were six label conditions for each product. Three different warning labels for the inhalation hazard and three for the eye contact hazard (words only, words and proactive symbol, words and reactive symbol). The subjects were instructed to use the wood cleaner or the liquid drain opener under the guise of another study. The necessary safety goggles or respirator was placed among similar equipment on a cart approximately two feet from the designated work area. Compliance was based on whether the subject put on the goggles or the respirator.

A follow-up questionnaire was administered to determine whether the subject noticed, read the warning and if not why the warning was not heeded, and if the subject recalled the specific content of the warning. Subjective ratings of perceived confidence in using the product safely, hazardousness of the product, likelihood of injury, and severity of injury were also collected on the basis of a 7-point scale.

Across all behavioural measures, there was a steady decline in the number of subjects who first noticed (88%), then read (46%), and finally followed the warning (27%). Compliance levels varied from 8% to 42%. Of the subjects, 49% correctly recalled the hazard and 42% correctly recalled the precautionary action. Unfortunately, no analyses were conducted with regard to the familiarity variable due to the circumstance that subjects were not significantly more familiar with the drain opener than with the wood cleaner. Symbols added to written warning labels did not significantly increase levels of compliance. The subjective ratings did show that the labels containing a symbol were perceived as significantly more dangerous than the product that only contained a written warning. This effect, however, did not translate into a significant increase in behavioural compliance. Furthermore, subjects who read the warning, followed the warning or recalled the warning information perceived the product as significantly more dangerous than those who did not read, follow or recall the warning.

A final interesting result concerns subjects' perception of control. Subjects who read the warning but omitted to follow it, admitted that they did not comply with the warning, because they felt that if they used the product in what they considered to be a safe manner they would not be hurt, even though they perceived the product as hazardous.[478]

3.6.9.8 Hazard Perception and the Severity of Consequences

Hazard perception has been found to be an important determinant of warning compliance. People are more likely to comply with a warning if they perceive the product as dangerous.[479]

Since warning studies indicate that hazard perception is a highly influential factor for warning compliance, the effects of warnings on such perceptions is of interest. Several studies have shown that the mere presence of a warning increases hazard perception. Furthermore, in the event that a warning is given, it is expected that influencing hazard perception will benefit warning effectiveness.

Because hazard perception has generally been found to be subject to systematic biases, it is essential that warnings aim at influencing or correcting possible false beliefs about the hazard associated with the product. When a product is perceived to be less hazardous than it really is, people may act with less precaution than is warranted. Hence, warning designers must be cognizant of the circumstance that consumers' perceptions regarding the hazardousness of the product may be incorrect. As shown earlier, users' belief that they have control over the product risks, a high familiarity with the product (warning), and users' benign experiences with the product have been shown to decrease hazard perception.[480] Since the warning research indicates that people's judgement of hazards associated with consumer products is based largely on the costs of the potentially negative consequences, rather than the likelihood of injury, providing information about the severity of the potential consequences seems helpful to steer consumer behaviour in the appropriate way, especially in situations where the consequences of the hazard can be severe.[481] Secondly, using explicit language to describe warning information has demonstrated to increase hazard perception as receivers' perception of injury severity increases.[482] In addition to explicitness, there is evidence that adding certain warning design features can be of assistance in raising hazard perception so that the expectations about the importance and the seriousness of the hazard are corrected. The variables include signal words, colour, location, explicit and concrete warning symbols and interactivity. The type of information (concern-

---

478 Friedmann 1988.

479 See *e.g.* Otsubo 1988. See also studies that subjectively measured the effect of hazard perception on compliance: Vredenburgh e.a. 2005 (described in paragraph 3.6.6.6); Wogalter e.a. 1991a; Wogalter & Barlow 1990; Friedmann 1988 (described in paragraph 3.6.9.7).

480 Rogers, Lamson & Rousseau 2000, p. 127; Van Duijne 2005.

481 Wogalter & Laughery 2006, p. 904; Riley 2006, p. 292; Wogalter e.a. 1991, p. 82

482 DeJoy 1999a, p. 196, 211; Laughery e.a. 1993a; Wogalter e.a. 1999.

ing the consequences, the hazards and instructions) that is included in a warning has also shown to affect the perceived hazardousness of a product and the perceived effectiveness of a warning.[483] However, studies have also shown no significant effect of these variables on hazard perception.[484]

> The study of Otsubo observed users' behaviour with actual products to examine the effects of pictographs and words and user hazard perceptions on warning effectiveness. Two conditions of perceived danger were used, a high level condition for the electrical saw and a low level of perceived danger condition for the use of a jigsaw. It followed from a preliminary survey that the participants had used the circular saw less frequently, that they felt more likely to be injured by the use of the circular saw and the combined score of the dimensions of likelihood and severity was greater for the circular saw than for the jigsaw. The following warning sign conditions were used:
> (1) words only;
> (2) pictograph only;
> (3) words and pictograph; and
> (4) no warning.
> The warning message was designed to communicate the level of hazard by using the signal word CAUTION, the nature of the hazard (SHARP BLADE), the consequence of the hazard (CAN CUT) and the avoidance of the hazard (WEAR GLOVES). The pictograph represented the same message. The warning label was located on the handle of both saws. The dependent variables that measured warning effectiveness were whether participants saw and read the warning, whether they recalled the cause of the hazard, recalled the consequences and recalled the avoidance of the hazard, how confident the subject was regarding the meaning of the warning label and whether the subject complied with the warning.
> The setting represented a real-world worktable. Various tools, protective equipment such as gloves and pieces of wood were available to the subjects. Subjects were told that they would be expected to use an electric saw and to imagine that they were at home alone and wished to repair the small bookcase using the tools available. The actions of subjects were observed such as whether gloves were donned before use of the saw. When the experimenters saw the subject plugging the tool in the extension cord and pressing the switch to turn

483 Wogalter e.a. 1987, p. 606.
484 DeJoy 1999a, p. 195; DeJoy 1999b, p. 229; Williams & Noyes 2007; DeTurck 2002, p. 361; Vredenburgh & Zackowitz 2006, p. 352.

on the power, they were stopped as a precaution against possible injury. An interview was conducted immediately thereafter to elicit the subject's perception, and recall of the warning.

The findings indicated that subjects noticed, read and complied more with warnings placed on the product perceived to be more dangerous than on the product perceived to be less dangerous: 74% noticed, 52% read and 38% complied with the warning on the circular saw whilst 54% noticed, 25% read and 13% complied with the jigsaw warning. Additional data suggest that people more familiar with the use of the product (the jigsaw) will tend to read, comply and recall the warning less than those less familiar with it. Also people more confident with the use of the product (the jigsaw) will tend to read and comply less than those less confident with its use. The type of warning label showed no significant effect. However, in all conditions with a warning label, an average of 25.5% complied with the warning, with the highest compliance rate of 50% for the condition of words and a pictograph on the circular saw warning. When there was no warning label, 0% wore gloves. The findings support the contention that the use of conspicuously designed and placed warning labels on products will influence people to behave cautiously.[485]

Study 4 of Wogalter e.a. 1999b examined the effect of injury likelihood and severity information on compliance behaviour.

In a laboratory setting, college students had to measure and mix various 'chemicals'. Participants received instructions as to how to perform the task. Compliance was measured by the wearing of gloves during the performance of the chemistry laboratory task. These instructions contained either no warning or a warning. There were five conditions:

(1) low likelihood-low severity: 'Contact with skin can cause mild skin irritation. Wear gloves';

(2) low likelihood-high severity: 'Contact with skin can cause intense skin irritation. Wear gloves';

(3) high likelihood-low severity: 'Contact with skin will cause mild skin irritation. Wear gloves';

(4) high likelihood-high severity: 'Contact with skin will cause intense skin irritation. Wear gloves';

(5) control condition: no warning.

The results showed that among the warning conditions, compliance was highest in the low likelihood-high severity condition (81.3%) and lowest in the low

485 Otsubo 1988.

likelihood-low severity condition (43.8%). Behavioural compliance was 68.8% for high likelihood-low severity; 68.8% for high likelihood-high severity, and 13% for the no warning condition. Thus, the highest effectiveness score was 68.3% and was obtained in the second warning condition. The results confirm that information expressing severe injury has a strong effect on compliance when hazard likelihood is low. There was no difference in compliance when the warning described a higher likelihood. This could be explained by the fact that the participants did not believe they would actually be injured when they behaved contrary to the warning.[486]

3.6.9.9 Explicit Information

Providing explicit information in a warning has been demonstrated to produce higher levels of actual safe behaviour compared with non-explicit language.[487] Furthermore, studies that have measured warning compliance subjectively also report this trend.[488] A plausible explanation is that using explicit language to describe information components of a warning message raises knowledge, hazard perception and subsequently behavioural compliance.[489] The effect of explicitness is especially true for severe hazards. This implies according to the warning research literature that the explicitness of information should increase as the severity of the hazard increases.[490]

In the study of Frantz, 80 subjects used a water-repellent sealer containing one of four labels to determine the effect of the explicitness of precautions and of location on attention to the warning and on compliance during task performance.
The subjects were rather inexperienced with the use of this product. There were four label conditions. The following non-explicit precautions/safety instructions were used:
1) Keep away from open flame or spark;
2) Use in a well-ventilated area;
3) Avoid contact with skin;
4) Avoid contact with eyes.
The explicit versions of the precautions were:
1) Search for and extinguish all flames and remove all sources of ignition;
2) Open windows to vent vapors to outdoors;

486 Wogalter e.a. 1999.
487 See *e.g.* Dingus, Wreggit & Hathaway 1993 (described in paragraph 3.6.6.7); Frantz 1994.
488 See *e.g.* Heaps & Henley 1999; Laughery e.a. 1993a; Laughery e.a. 1991.
489 Laughery & Smith 2006.
490 Laughery & Smith 2006; Edworthy & Adams 1996, p. 68.

3) Wear rubber gloves;
4) Wear protective glasses.

The first label contained warnings and non-explicit precautions and they were located on the side panel of the container separated from the directions for use, which were on the back panel. In the second condition, the precautions were explicit and they were also located on the side panel. In the third condition, the precautions were non-explicit and integrated into the section with the directions for use on the back panel. The final condition contained explicit precautions that were added to the directions for use section. It was hypothesised that the latter condition was the most effective.

Participants were assigned to one of the label conditions and were not told of the real objective of the study. The experiment took place in a kitchen where they were instructed to perform the tasks of unclogging a sink with a drain opener and then applying a coat of water sealant to a wooden plant stand. The water repellent sealer was placed in the cabinet under the sink. Gloves and goggles were adjacent to the product. The wooden plant was on the kitchen counter. Above the sink was a window. A scented candle burned openly approximately 1.5 meters away from the plant stand. Behavioural compliance with each of the four safety precautions (i.e extinguish candle, open window, wear rubber gloves and wear goggles) was measured through hidden camera observation. After the task, subjects were also interviewed and asked about the issues of noticing, reading and complying with the information.

The results showed a significant effect with regard to the location of the precautions. Placing them in the directions for use section (third and fourth condition) significantly increased the number of participants who complied with each of the precautions. The results also showed a significant effect for the explicit version of opening the window and wearing protective glasses. The compliance scores for each precaution when integrated with the instructions (condition 3) were 15%, 20%, 50% and 30% when they were not explicit and 30%, 55%, 95% and 80% when they were described in an explicit manner (condition 4).

In addition, the data of the self-reports showed a relationship between reading and complying with the warning: for subjects who read a precaution the rate of compliance was much higher than for subjects who reported not to have read the warning. The situation of not reading the warning could be viewed as a control condition: this result indicates that the warning was effective as the degree of safe behaviour in the presence of a warning was higher than in its absence. Another important finding is that the factors of location and explicitness interacted, which means that the most effective label condition was the one with the best location and the best presentation of information: the precautions were in the instructions section and in explicit language.[491]

3.6.9.10 Familiarity Belief

Another important factor is the belief of familiarity. In general, greater familiarity and/or experience with a particular or a similar product is associated with a lower likelihood to notice, read, and comply with a warning.[492] However, there is research indicating that there are situations of product use in which familiarity increases compliant behaviour.[493] This may be explained by the circumstance that those product users have more knowledge and experience about the hazards, the consequences and how to avoid the hazards. It is suggested that this effect may be attributed to an increase in perceived hazard.[494]

> Wogalter, Barlow and Murphy examined the effect of the placement of warnings and users' experience on behavioural compliance.
> Twelve participants performed the task of connecting a disk drive to a computer. Compliance behaviour was recorded by the experimenters and consisted of performing the three behaviours of (1) turning off the computer, (2) touching the metal connector on the back of the computer and (3) ejecting the transport disk, before attaching the disk drive to the computer. After the task, participants completed a questionnaire that assessed their previous experience with electronic appliances. Furthermore, seven methods of directing people to warnings in an owner's manual were investigated in order to increase compliance behaviour during the installation of the computer disk drive:
> (1) In the control condition, the disk drive was accompanied by a basic manual of thirteen pages that contained the warning message on pages six and seven that informed users of the electrical hazard and the negative consequences and instructed them to perform the three behaviours. Thus, in this condition, the warnings were embedded in other information.
> (2) In the second condition, the manual was identical except that the warning message was also reprinted redundantly on page two of the manual (which was blank in the basic manual).
> The remaining five conditions were identical to the redundant-warning condition, except that they also included a supplemental directive that was placed

491 Frantz 1994.

492 See *e.g.* Wogalter, Barlow & Murphy 1995 (described in paragraph 3.6.9.10); Otsubo 1988 (described in paragraph 3.6.9.8); Harrell 2003; Goldhaber & DeTurck 1988a; Zeitlin 1994; Lehto & Foley 1991. For studies that measured the effect of familiarity on the *intention* to comply with the warning, see Godfrey e.a. 1983; Godfrey & Laughery 1984.

493 See *e.g.* Ortiz, Resnick & Kengskool 2000; Edworthy e.a. 2004 (described in paragraph 3.6.9.2).

494 Laughery 2006, p. 474.

at various locations. The supplemental directive stated: '«Please Read Page 2 of the Owner's Manual Before Connecting the Equipment»' and was printed in red, bold large letters.

(3) In the third condition, the directive was placed on the top of the shipping box.
(4) In the fourth condition, the directive was located on the cover page of the manual.
(5) In the fifth condition, the directive was placed in an accompanying leaflet.
(6) In the sixth condition, the directive was attached to the cable of the drive.
(7) In the seventh condition, the directive was attached to the front of the drive and covering the disk drive opening.

The results showed a trend of greater compliance with the redundant-warning (only) manual compared to the basic manual, but the difference was not statistically significant. When the supplemental directive was located most proximally to the equipment and required physical interaction with the warning during the task (condition seven), compliance was significantly greater than the basic and redundant-warning (only) manual conditions. With regard to overall compliance, the results showed significantly lower rates for the basic manual control condition compared to all of the other conditions, expect for condition two. The difference in behavioural compliance between the control condition and condition seven was also significant. With respect to the separate compliance behaviours, the compliance rates were highest in the seventh condition (92%, 100%, 83%) and lowest in the control condition (42%, 33%, 33%). The findings of this experiment confirm that well-placed safety information can be useful in alerting users with varying levels of experience and in influencing their behaviour. More specifically, they suggest that people are more likely to notice, read and comply with a warning when it is placed in close proximity of the product, namely, where and when a warning is needed. It must be noted that the high compliance scores of the directive that was attached to the drive can also be explained by its interactivity: the user had to remove it in order to eject the disk.

Furthermore, the results demonstrated that, in general, users with less experience connecting electronic equipment complied more frequently with the warnings than users with greater experience. Additional results indicate that less experienced users were not further influenced by the presence and location of the supplemental directive, but highly experienced users complied significantly more often with the warning when the supplemental directive was placed in the more proximal locations (especially conditions five and seven) than when the warning only appeared in the owner's manual. This latter effect could be

> explained by script theory. It is conceivable that the presence of the supplemental directive introduced an additional, unfamiliar, feature into the task, and as a result interrupted the script-driven behaviour of the experienced users, whilst the script-driven process was automatic and not interrupted when the warning message was placed in a known location, such as in the manual.

3.6.9.11 Costs of Compliance: Time, Effort and Convenience to Carry Out the Behaviour

As discussed earlier, a great deal of research quite consistently shows that the perceived cost of compliance plays an important role in the level of warning effectiveness.[495] Consequently, the implication from the research for facilitating warning effectiveness is that the cost of compliance must be kept within a reasonable level. This entails that the directed behaviour should be easy to perform, the behavioural effort to comply with a warning should be as low as feasible and the comfort of using the product as high as possible.[496] It has been noted in the warning research literature that the other way to steer the decision to comply in a positive direction involves increasing the perceived benefits of warning compliance. This can be done by increasing the perceived effectiveness of the recommended actions to avoid the hazard or by emphasising the level of perceived hazard through warning design features.[497]

> The purpose of the study of Hunn and Dingus was to examine the effects of manipulating three variables on behaviour during the use of a common household cleaning consumer product. The variables were the type and depth of information presented on the warning, the level of interactivity with the warning and the cost of complying with the warning.
> The study involved two information conditions: a label with conventional consumer information and an ANSI warning with the addition of specific consequence information. The specific consequence information contained safety statistics such as 'Household cleaning products are associated with 54.000 injuries per year'. Furthermore, there were two levels of cost of compliance: high cost (no gloves were provided) and low cost (gloves were attached by a rubber band to the bottle). Finally, as regards the interactivity of the subject with the warning label, there were three conditions: a standard product label printed on the spray bottle, a one-time interactive warning attached to the

495 See *e.g.* Wogalter e.a. 1987; Dingus, Hathaway & Hunn 1991; Hunn & Dingus 1992 (described in paragraph 3.6.9.11); Dingus, Wreggit & Hathaway 1993 (described in paragraph 3.6.6.7); Lehto & Foley 1991; Wogalter, McKenna & Allison 1988 (described in paragraph 36.8.2).

496 Lehto 1992, p. 130.

497 DeJoy 1999b, p. 238.

spray bottle and a continuously interactive trigger-block warning on the spray bottle.

The purpose of the study was hidden under the guise of testing the quality of the cleaning product. A total of 356 subjects participated in the experiment. Participants were given the product and were instructed to use it at home for a week and then to return it. They were asked to fill out a questionnaire. Two questions addressed the perceived hazardousness of the product. Whether or not the gloves were used (at least once) was visually tested by examining the gloves when they were returned.

Results indicated that cost of compliance was the only variable that significantly affected the safe use of the product. Over all conditions, 88% used the gloves when they were provided (low cost condition) and 25% said they used their own gloves when they were not included (high cost condition). The compliance rate was highest for the continuously interactive warning that contained specific consequence information and where the gloves were provided (95%), the lowest for the standard label in the high cost condition (11%). Even though the effect of interactivity was not significant, the results showed that in the high cost condition the differences were large between the standard label (11%) and the two interactive ones (34% and 27%). The differences between the one-time interactivity label and the continuously interactive label were small. Another finding is that subjects who were given the gloves perceived the product as more dangerous than subjects who received the product without the gloves.[498]

3.6.9.12 Social Influence of Others

Research also supports the powerful effect of social influence of others on the behaviour of product users.[499] Social influence of others can be exerted by people that are in the immediate vicinity of the warning receiver or by role models that do not appear live, but on print (warning label) or on screen (videotape or television advertisement).

Racicot and Wogalter conducted two experiments to investigate the effectiveness of different kinds of video presentations on the participants' behavioural compliance to the safety directives of the warning (donning mask and gloves). In the first experiment, 36 participants were assigned to one of three conditions. In the first condition, participants were only exposed to a videotape of a static warning sign for 30 seconds. The second condition consisted of the video

498 Hunn & Dingus 1992.

499 See e.g Wogalter e.a. 1987; DeTurck, Rachlin & Young 1994; DeTurck, Chih & Hsu 1999 (described in paragraph 3.6.8.4); Wogalter, Allison & McKenna 1989; Racicot & Wogalter 1995 (described in paragraph 3.6.9.12); Racicot & Wogalter 1992; Chy-Dejoras 1992; Ferrari & Chan 1991.

warning sign that was presented for 10 seconds followed by a shot on the screen of the gloves and a mask for eight seconds, and then followed by a 12-second clip showing a male approaching a table and putting on a mask and a pair of gloves. The third condition was identical to the second, only a voice warning was added during the screen shot of the sign. The warning sign was identical to the one of the chemistry laboratory experiment in Wogalter e.a. 1991b (described in paragraph 3.6.3). After watching the videotape, participants entered the laboratory room and were asked to perform the task which consisted of measuring and mixing the dummy chemicals.

The results indicated that behavioural compliance was significantly higher when participants were exposed to the video of the sign and the role model (92%) and the video of the sign, the role model and the voice warning (100%) compared to the videotaped warning sign alone (50%). The addition of a voice warning to the sign plus role model condition produced no further increase in compliance over the condition without voice. There was no significant effect of adding a voice warning. This might be due to the fact that the social influence of the role model on compliance was already very high.

A second experiment assessed whether a delay between the time of exposure to the video warning and time of performing the task safely would produce a decrease in compliant behaviour. In addition, it also replicated the study mentioned above.

The experiment showed that a delay of several days did not reduce the effectiveness of a video warning. This result suggests that the behavioural change induced by the video is robust over time. Furthermore, the findings again demonstrated a positive influence of video role modelling on warning compliance compared to a static warning sign.[500]

3.6.9.13 Stress

The external factor of stress is also relevant for warning compliance. Even though this hypothesis is not broadly tested in the warning research, it is palpable that stress can weaken warning compliance, as stress has shown to affect information processing and the quality of people's judgements. More specifically, research has indicated that under higher stress peripheral vision narrows and attention becomes restricted as a result of which the available information may not be noticed and used.[501]

500 Racicot & Wogalter 1995.
501 Wogalter e.a. 1998, p. 144.

One study in the warning research involving two chemistry laboratory experiments examined the impact of time stress and location on behavioural compliance.

In the first experiment, 80 undergraduate students performed the common task in which they weighed and measured various 'chemicals'. There were four conditions:

(1) higher stress/posted sign;
(2) higher stress/within instructions;
(3) lower stress/posted sign; and
(4) lower stress/within instructions.

The stress manipulation was a combination of two factors: time pressure and social monitoring by another person. In the high stress condition, participants were given a time limit to complete the task, and the experimenter stood immediately adjacent to the participant, appearing to be monitoring the participant's performance. In the low stress condition, participants were given as much time as they needed to complete the task and the experimenter stood at a distance, out of the participant's field of view. As regards the location, the warning to wear a mask and gloves was either posted as a large sign on a wall directly in front of the participant or it was present in a set of instructions.

The results of the first experiment showed that complying with the warning was significantly higher among participants who were under low stress and exposed to the warning that was placed within the instructions. The questionnaire that was undertaken after the completion of the task confirmed the stress effect, given that participants in the higher stress condition reported that they felt more stress and worry and that the experimenter bothered them. In the low stress condition, they reported that they were feeling less stress. Regarding the effect of location, the results were also significant: participants complied significantly more often when the warning was placed within the instruction sheet of the task than when the warning was placed on a sign. It might be that participants assumed that the sign was not directed to them while the warning in the instructions was perceived as relevant to them. This shows that it is important to place a warning in a location where people are likely to look.

To find out which stress factor had a greater effect on compliance, a second experiment with 80 subjects was conducted to determine the individual effect of time stress or social monitoring on behavioural compliance with a posted sign warning. There were four conditions:

> (1) no time pressure/no social monitoring;
> (2) no time pressure/social monitoring;
> (3) time pressure/no social monitoring; and
> (4) time pressure/social monitoring.
>
> Here, the results showed that time pressure significantly reduced compliance compared with its absence. The experiment failed to show a significant effect of social monitoring on compliance. It was expected that the presence of the experimenter would reduce compliance. However, it produced a small compliance enhancement. This might be explained by the fact that the presence of another person stimulated participants to produce rule-following behaviour. Overall, the hypothesis that greater stress decreases warning compliance is confirmed by the results. Likewise, warning location affects compliance.[502]

3.6.9.14 Mental Work Load

Like stress, a high mental work load which involves performing multiple tasks concurrently, generally has a detrimental effect on the product users' limited processing resources and is likely to decrease compliant behaviour.[503]

> This study examined the effect of the variable of cognitive task load on warning compliance behaviour. A total of 44 participants performed a task in which they installed an external disk drive to a computer. Inside the accompanying owner's manual were a set of specific procedures that were to be followed during the installation to avoid damaging the equipment. These steps included:
> (1) turning off the computer;
> (2) touching the computer's rear metal connector to discharge static electricity;
> (3) ejecting a transport disk from the disk drive.
> The experiment contained 3 conditions. In the low and high task load conditions, two 15-minute cassette tapes with a series of either single digit or double digit math problems were played. For both recordings the math problems were spoken by a female every 15 seconds. Participants were to say aloud the answers to the math problems within the allocated time of 15 seconds whilst simultaneously installing the disk drive to the computer. Participants were told to do the best that they could with the math problems, but to mainly focus on the primary task of installation.

502 Wogalter & Usher 1999.
503 Wogalter & Usher 1999; Duffy, Kalsher & Wogalter 1995 (described in paragraph 3.6.4.8).

A control condition was lacking in the extra task. Compliance was measured by observing whether participants followed the three steps of installation expressed in the owner's manual. Compliances scores were analysed separately for the three precautionary instructions and together.
The results showed a significant reduction in compliance behaviour concerning the instruction of discharging static electricity in the high load condition (25%) compared to the no load condition (88%). The low task load condition also produced less compliance behaviour (63%), and this was significantly different from the high load condition. For the other two steps, the results showed a detrimental, but not significant, effect on compliance. The total compliance measure showed a significant effect of conditions: high task load participants had lower total compliance scores than the no load participants. As regards the additional task, the results demonstrated that the accuracy of solving the maths problems was significantly greater for participants in the low task load than in the high task load, which confirmed the difficulty difference in mental load of this task. These findings suggest that warning effectiveness can be reduced when the mental resources necessary to carry out compliance are being absorbed by other concurrently performed tasks.[504]

3.6.10 A Warning System

Considering the factors and guidelines described above when designing warnings will benefit the effective processing of those warnings. A principle that also facilitates warning effectiveness and that already received attention yet not explicitly, is the approach of a warning system to communicate safety information.

Since there are so many ways in which warning information can be transmitted, it is not surprising that it has been argued in the literature that it is a too narrow view to think that a warning is only a sign or a portion of a product label. There are instances in which a single warning is not sufficient to adequately reach the population to whom it is directed. A *warning system* refers to the variety of media and channels (components) that are used to communicate a warning to the target audience. The components of the warning system are not necessarily identical in terms of content and purpose. Some components can be used to attract attention, whereas others are mainly intended to inform. Another benefit of this system approach is that different components can be used to communicate to different subgroups of users.[505]

504 Wogalter & Usher 1999.

505 Wogalter 2006a, p. 8, 9; Wogalter & Laughery 2006, p. 893.

The most common and preferred way of communication is a printed warning on the product itself and/or on its package. However, there may be circumstances under which additional components are needed to convey information. The use of product manuals is a means to provide visual information on the use and maintenance of the product, but also the hazards associated with product usage. Package inserts are another example of a component of the warning system. Alternative warning label designs have proven to be a helpful method to provide extra information. Verbal statements at the point of sale can also be part of the system. In addition, more than one modality can be used to convey the message. Usually the visual channel is used to transmit the message, but the other human senses (hearing, smell, taste, touch) are able to process warning information as well. Furthermore, many media can be involved as a component, such as a producer's website, brochures, posters, television advertisements, electronic media and so forth.[506]

3.6.11 Testing Warnings

Warning researchers recommend including testing as an integral part of the warning design process to evaluate the warning's effectiveness. Design guidelines can be useful, especially when they are based on empirical evidence from the warning research. However, because products, environments and warning receivers differ in various ways, merely using guidelines and voluntary standards may not be optimal for producing an effective warning.[507]

Testing or evaluation of warnings falls into two main categories. The first concerns formative evaluation. This iterative process occurs whilst the warning is being designed. Mock-ups of the warning design can be tested on participants and then altered on the basis of feedback or results of the measurements. The advantage of this form of design evaluation is that the process is iterative: problems can be defined early in the design process and changes can be made throughout the design process until a final iteration is agreed upon. The second approach is summative evaluation. Summative evaluation involves testing the final warning label after all design activities have been completed. Because summative evaluation requires a completed design and testing within a real-world environment, the costs may pose a disadvantage over formative evaluation.[508] Testing can entail using small groups of product users or experts who give ideas for improvement, but also evaluations by a large group of people. The warning should be tested on participants that are representative of the target audience.[509]

---

506 See *e.g.* paragraph 3.6.3 on the characteristics of the channel.
507 Wogalter 2006a, p. 9.
508 Wogalter, Conzola & Smith-Jackson 2002, p. 225.
509 Wogalter & Laughery 2006, p. 907.

As described in paragraph 3.5, research methods, such as questionnaires that involve subjective ratings and field/lab experiments, are available to measure the extent in which the stages involved in warning processing are achieved. Ideally, all warnings should be tested by measuring behavioural compliance in a real-world context. Measuring the intermediate stages can also offer insights on the possible failures of warning effectiveness. A study can indicate the most favourable warning design, seeing that several experimental warning conditions can be devised to examine which warning design alternative is most effective, for example in attracting attention, being legible, or providing a comprehensible warning. If, for instance, testing shows that less than half of the participants noticed the warning, the study succeeded in pointing out that the warning needs to be enhanced on this point.[510]

3.6.12 *Future Research Directions*

The previous paragraphs have shown that the warning research has made substantial progress, but obviously there is room for more work. The warning research literature has designated areas that need further investigation. The most important ones are mentioned here.

Firstly, it has been argued in the warning literature that future research should focus on achieving a better understanding of the complex warning process by investigating potentially influential variables that have been neglected or have received little focus thus far. Variables such as memory load and self efficacy may have a substantial influence on warning compliance, but have not yet been empirically examined or seldom in the context of product warning effectiveness.[511]

Furthermore, much of the existing research has focused on variables that contribute to noticing, encoding, understanding of a warning and less investigation has been done on factors that influence attitudes and beliefs, motivation and behavioural compliance.[512] Hence, future research should especially be directed towards how warning designs can influence people's attitudes and beliefs and their motivation.[513] Moreover, even though designing studies to collect data on behavioural compliance is challenging, various techniques of measuring behavioural compliance are available and they should be incorporated more often.[514]

A second research goal mentioned by warning researchers is to gain further insight into the relative effect sizes of variables, that is, which variables have the most influence

510 Wogalter, Conzola & Smith-Jackson 2002, p. 226.
511 Rogers, Lamson & Rousseau 2000, p. 133.
512 Rogers, Lamson & Rousseau 2000, p. 132, 115; Lesch 2006, p. 143.
513 Vredenburgh & Zackowitz 2006, p. 353; Riley 2006, p. 296.
514 Kalsher & Williams 2006, p. 329.

on the stages of the warning process and which play a minor role. For example, does colour have a bigger impact on noticing the warning or is it the size of the warning that has the greatest impact?[515]

Thirdly, the interactive process between the variables must be uncovered more deeply. More knowledge is needed on how variables interact with one another, such as familiarity, hazard perception and information seeking.[516]

Another suggestion made by the warning research is that in view of the growing international trade of consumer products, more research should deal with factors related to the diversity in target audience. Having a better understanding of cultural differences, language barriers and illiteracy levels and their effects on warning effectiveness can offer suggestions for effective warning design.[517]

Also, the usefulness of technology calls for research activity that explores the potential of technology in achieving more effective warnings. Compared to static warnings that are abundantly present, dynamic warnings are scarce. Dynamic warnings have the ability to change, for example a change in warning content over time, a change in sound etc. Display technology can provide dynamic warning displays in cars, on TV screens etc. and sensor technology can offer warnings that include sensors that detect hazards. Such technologies are available and are likely to have a positive impact on the noticeability of a warning, and on compliance.[518]

3.6.13 Conclusion

Paragraph 3.6 used the C-HIP model to analyse how warnings are being communicated and processed by individuals as well as to present the tremendous body of warning research.

The model predicts that source and channel can exert an influence on communicating warning information to receivers. Warning studies examining the impact of source information and the type of modality on warning compliance have confirmed this viewpoint. The C-HIP model also shows that once the warning information has reached the receiver, several sequential mental steps need to be successfully completed to elicit behavioural compliance. The receiver stage can be broken down into the substages of attention switch and attention maintenance, comprehension and memory, attitudes and beliefs, and motivation. Behavioural compliance is the end result. This means that if further processing of the warning information within the warning receiver is blocked at a stage, the desired behavioural compliance with the warning will not be achieved. The general research

515 Rogers, Lamson & Rousseau 2000, p. 134; Lesch 2006, p. 143.

516 Rogers, Lamson & Rousseau 2000, p. 134.

517 See paragraph 2.7 on the differences in characteristics of warning receivers.

518 Laughery 2006, p. 476.

finding that there is a decline in the number of people who notice a warning, then read it and finally comply with it within the group of participants corresponds with this.[519] Hence, there is not a direct causal relationship between consumers attending to a warning and their behaviour. Consequently, another prediction that can be derived from the model is that the ultimate impact of exposing consumers to a warning message on behavioural compliance can be relatively small.[520] For example, if 60% of 100 consumers see the warning, 75% of those individuals read and understand it, 60% of those consumers who have understood the warning believe it, 40% are motivated to perform the desired behaviour and 90% actually follow the warning, the probability that the warning will lead to appropriate product use is then approximately 10%. This example illustrates that in the event that the effectiveness of each information processing stage is low, the use of a warning will not be a promising approach for improving product safety. Even if the previous intermediate stages have been successful, the level of behavioural compliance can still be low, because the high costs of complying have produced a bottleneck.

In the discussion about the information processing stages including the stage of behaviour, potentially influential factors pertaining to the effectiveness of each stage and the design implications that followed from the findings were described. Progress has been made by way of research efforts. The studies have indicated factors that can be identified as important, and as a result have contributed to a better understanding of the complex warning process. In general, the factors influencing the effectiveness of the information processing stages and behavioural effectiveness relate to the environment in which the product is used, to human characteristics of product users, and to warning design features. Much of the warning research has focused on design variables such as size, pictorials, and signal words. These factors mainly influence the noticeability and understandability of warnings. However, less research has been done on the influence of external factors that are related to the situation of product use in the environment. Furthermore, relatively few behavioural and survey studies have measured the influence of various attitudes and beliefs on warning effectiveness.

The preceding paragraphs illustrated that different factors can be of relevance at different stages. In addition, one single factor can be important for the successful processing at more than one stage: belief factors can also exert an influence on attention which confirms the prediction of the model that the process is not linear, but that the information processing stages interact. Naturally, factors that influence the earlier information processing stages are indirectly of importance in determining whether the warning will have a positive effect on behaviour. In general, the variables that particularly influence the initial stages are

519 See *e.g.* Friedmann 1988 (described in paragraph 3.6.9.7).

520 Lehto & Miller 1988, p. 227; DeTurck & Goldhaber 1989, p. 348.

expected to be positively correlated with the likelihood of compliance.[521] The list of factors per stage is not exhaustive, but it serves as a means to present the potentially influential variables that have been identified through scientific studies. Consequently, because these factors pertain to the effective processing of the substages and the behavioural effectiveness of product warnings, they must be considered first in the design process of warnings.

To attract the attention of warning receivers the factors of size, brightness and colour contrast, borders, signal words, symbols, location, interactivity, hazard perception, familiarity, information seeking mode, channel and environmental conditions are especially of relevance.

Maintaining the attention of warning receivers depends on similar factors like colour contrast, symbols, hazard perception, the familiarity belief, environmental conditions and location, but additional factors such as letter case, font style, alternative label designs and layout of the warning information are especially relevant for enhancing the encoding process of information.

The comprehensibility of the warning message is mainly determined by variables that pertain to the cognitive abilities of receivers, such as their language skills, reading abilities and the level of knowledge, and to warning variables including signal words, colour, warning symbols and explicit information.

Attitudes and beliefs of warning receivers can strongly influence whether a warning will be effective. The factors that affect processing at this stage are hazard perception, cognitive biases that may affect the people's belief about the hazard (especially availability, overconfidence, suppression and message framing), perceived control, familiarity belief, prior injury experience, personal relevancy, risk-taking style, costs of compliance, perceived effectiveness of recommendations and the belief of self efficacy.

Many variables that affect beliefs and attitudes are also valuable for motivating receivers to produce the appropriate behaviour. Primary variables that influence receivers' motivation are hazard perception, costs of non-compliance, costs of compliance, social influence, stress and mental work load.

Finally, the end stage of behavioural compliance. The main factors that have been shown to promote actual compliant behaviour are location, interactivity, colour, channel, warning symbols, hazard perception, costs of non-compliance, explicit information, the familiarity belief, costs of compliance, social influence, stress and mental work load. The factors of adding information components, adding information about the warning's source, perceived control and people's prior injury experience have influenced behavioural intentions of complying with a warning.

---

521 Laughery 2006, p. 473. See *e.g.* the field demonstration experiment of the water fountain warning in which adding vivid-enhancing features to the warning sign caused a significant increase in behavioural compliance. The vivid-enhancing characteristics consisted of the use of colour, a pictorial and the increase in size of the warning sign, see Wogalter e.a. 1987, p. 609.

From this wealth of influential variables, a shift can be made to identify the variables that have shown to be *most significant* in the success or failure of warnings in capturing attention (noticing and encoding) and providing information to make the decision to comply. In a recent review of the warning research, it was concluded that for the process of noticing and encoding a warning, the variables of size, location, colour contrast, signal words, warning symbols, hazard perception and familiarity are most significant. For warning compliance decisions, the variables that affect noticing and encoding a warning are important as well as the variables of warning symbols, explicit information, familiarity, hazard perception, modelling and lastly, the costs of compliance which factor is a very important consideration in warning design.[522]

In the following paragraphs (paragraphs 3.6.10 and 3.6.11), attention was paid to two important design principles that relate to a warning as a warning system that contains several components that are used to communicate (part of) the warning information. The other principle concerned testing. Testing a (dimension of the) warning on a representative sample of the target audience can yield inside information with respect to its potential effectiveness that could not have been answered on the basis of the mere use of design guidelines.

Lastly, paragraph 3.6.12 called attention to the directions in which future research should develop. Although the earlier paragraphs showed the tremendous amount of research in the area of the effectiveness of warnings, much still needs to be done to get a complete understanding of the factors that influence the processing of warnings and the conditions under which warnings can lead to behavioural compliance.

3.7 Receiver Characteristics and Product Warnings

3.7.1 Introduction

Paragraph 3.6 dealt with the effectiveness of product warnings across the five dimensions and has shown that individual characteristics or person variables play an important role in how product users approach and respond to warnings. In fact, they can affect all stages of cognitive processing.[523] It is recognised by warning researchers that a failure to consider person variables when designing warnings will lead to warnings of which it is expected that they have less effect on behaviour.

There are various receiver characteristics that account for individual differences. Whilst the effects of physical warning design factors have been repeatedly investigated, there is a

522 Laughery 2006.

523 Rogers, Lamson & Rousseau 2000; Smith-Jackson 2006a; Lesch 2006.

paucity of empirical evidence related to the precise impact of different receiver characteristics. This paragraph takes a closer look at the most commonly discussed individual characteristics in the area of warnings, starting with age differences.

3.7.2 Age Differences

With regard to age differences, the warning research has mainly studied the effects of aging.[524] The aging research in general has identified that normal aging is associated with a variety of changes in the visual system and cognitive function.[525] These age-related changes in perception and cognition are also likely to have an impact on an effective processing of warnings. The warning research on aging is limited, but by means of the general aging research, warning design recommendations have been proposed to compensate for the limited abilities of older adults.[526]

Age-related changes in perception relevant to visual warnings concern a reduction in the ability to discriminate between colours (colour vision); a reduced ability to distinguish between adjacent areas that differ in light intensities and patterns that vary in width (contrast sensitivity); an increased susceptibility to glare (glare sensitivity); a decline in the ability to track changes in stimuli (temporal resolution); a decline of the ability to resolve small details (visual acuity) and a reduced ability to select and process relevant information while rejecting irrelevant information (visual search). Because of these, older adults may have a reduced likelihood of noticing and encoding warning information.[527] For example, as a result of impaired acuity, reading text becomes more problematic for older adults.[528] This is especially the case for product warnings since warning information is often in small print. Researchers have suggested ways to increase the amount of space available on product packages and consequently the size of the warning information. The elderly are a primary target group for alternative product labels such as tags and wings. Research findings have consistently shown that older adults prefer labels that have an increased surface area.[529] The younger adults' preference is usually less distinct; they regard a standard label sufficient. The importance of designing a warning that conforms to the needs of the targeted audience can be illustrated by consumer products such as medications. Certain pharmaceuticals are mainly supplied to the elderly. This subpopulation could be

524 For obvious reasons, warnings usually target adults or adolescents. For this reason, most research and design guidelines focus on warnings for adults and not children. Designing warnings for children requires consideration of their unique strengths and limitations. For more information on this topic see Lueder & Rice 2008.

525 See on the topic of aging and information processing Birren, Schaie & Abeles 2006; Fisk 2004.

526 Rousseau, Lamson & Rogers 1998; Mayhorn & Podany 2006.

527 Rousseau, Lamson & Rogers 1998, p. 647.

528 Vanderplas & Vanderplas 1980.

529 See *e.g.* Kalsher, Wogalter & Racicot 1996; Wogalter & Vigilante 2003; Barlow & Wogalter 1991.

at risk as a result of a poorly designed warning that has not taken into account the visual disabilities or other age-related cognitive deficits of older adults. If they have trouble reading the label of pharmaceuticals, misuse of the product will lie in wait.

Changes in the cognitive abilities of people can influence warning processing as well. One of the age-related changes concerns a reduction in the ability to understand complex text and symbols. The warning research has found age differences in the processing of warning symbols, but the results are mixed.[530] Several studies demonstrate that the comprehension levels of older adults are poorer than those of younger adults.[531] Nevertheless, there are studies that show no significant age-related differences for symbol comprehension.[532] Furthermore, research indicates that older adults have a reduced ability to keep information active in memory (working memory capacity). This means that there are limited resources available to store new information in memory or to retrieve previously stored information. For example, when a warning contains several successive instructions that need to be followed, all of the information must maintain active in working memory in order to use the product safely. Because of a decline in working memory capacity, older adults may forget some instructions and as a result the probability of not complying with the warning increases. Though, the influence of working memory on warning effectiveness has not been empirically investigated.[533] As people grow older, their *prospective memory* can also decline. This entails remembering to perform an action in the future.[534] Research suggests that salient cues, such as an interactive warning, may be particularly well-suited to increase the effectiveness of warnings for older adults as they trigger remembering to perform an action.[535]

3.7.3 Gender Differences

Little attention has been devoted to how gender affects the warning process. This may be due to the fact that many consumer products are not gender specific and hence designing gender neutral warnings would be more appropriate. Nevertheless, if the target audience of a specific warning mainly consists of females, research findings on gender differences may assist in designing warnings that take the specific capacities of women into consideration.[536] It appears that warnings are likely to be more effective with females than males, as some studies showed that females were somewhat more likely to look for, read and

530 Hancock e.a. 2004, p. 184.

531 See *e.g.* Lesch 2003; Hancock e.a. 2004; Hancock, Rogers & Fisk 2001; Easterby & Hakiel 1981.

532 See *e.g.* Kline e.a. 1990; Kline & Fuchs 1993.

533 Rousseau, Lamson & Rogers 1998, p. 652.

534 Mayhorn & Podany 2006, p. 358; Hancock e.a. 2006, p. 271.

535 Rousseau, Lamson & Rogers 1998, p. 654.

536 Wogalter, Conzola & Smith-Jackson 2002, p. 225.

comply with warnings.[537] Furthermore, females tend to have higher hazard perceptions.[538] Unfortunately, the warning research on gender differences has not been entirely consistent. Because of the mixed results, it is hard to say whether gender is an important factor for warning effectiveness or how warnings should be designed that target a specific gender.[539]

3.7.4 *Differences in Receiver Competence*

Competence can be defined as possessing the capacity to meet the demands of a particular task. It concerns sensory, physical, and cognitive capabilities that receivers need to have to process warnings and to safely use a product.[540] Receiver characteristics that relate to the senses such as limitations in the ability to see or hear are, unsurprisingly, important for the communication of warnings. Furthermore, it is obvious that one must have the physical ability to handle the product: the instructions of a warning must not be difficult to carry out. The warning literature also notes that consideration needs to be given to cognitive abilities, such as the language abilities, their reading skills and the level of technical knowledge the target audience of a warning possesses. Little research has examined the ideal reading level for warning text, but it is advised to use simple language and to include other ways to communicate the information in order to cope with people with literacy and readability issues.[541]

3.7.5 *Differences in Risk Perception*

A person variable that is imperative for effective warning processing is people's expectations of the risks attached to using a consumer product. It is a consistent finding in the warning research that the more hazardous the product is perceived to be, the more likely the user will look for and read warning information, and the more likely they will comply with warnings.[542] It follows that risk perception varies as does a function of the nature of a product, but also a function of individual differences.[543] For example, findings from observational studies on risk perception in product use show that some product users regard a situation or a particular consumer product dangerous whilst others believe it is

537 See *e.g.* Godfrey e.a. 1983; LaRue & Cohen 1987; Vredenburgh & Cohen 1993; see also Abroms e.a. 2003; Ferrari & Chan 1991.

538 See *e.g.* Vredenburgh & Cohen 1993; Young, Martin & Wogalter 1989; Chen, Gilson & Wang 1999; Friedmann 1988 (described in paragraph 3.6.9.7).

539 Smith-Jackson 2006a, p. 339; Young e.a. 1999, p. 696.

540 Young e.a. 1999, p. 698; Laughery & Young 1991b.

541 Smith-Jackson 2006a, p. 340.

542 See *e.g.* Donner & Brelsford 1988; Friedmann 1988 (described in paragraph 3.6.9.7); Otsubo 1988 (described in paragraph 3.6.9.8).

543 Young e.a. 1999, p. 702.

safe. This variability may be explained by differences in users' understanding of the functioning of the product.[544]

3.7.6 *Personality Differences*

Individual differences in personality can influence a person's expectations about a product and safe usage, which subsequently affects whether a product user is predisposed to comply with warnings. Since not many warning studies have addressed such expectations nor examined their behavioural effect in isolation or simultaneously, it has been noted that there is still a lot yet to be learned.

Personality factors relevant to warnings may be people's belief about their ability to perform a behaviour or task successfully (self efficacy), their belief of whether they can control risks and their attitude towards risk-taking including the related trait of sensation seeking which draws people towards dangerous activities.[545]

3.7.7 *Differences in Familiarity and Experience*

Various studies have investigated the effect of the familiarity with a product on subjective hazard perception and behaviour. Research suggests that people are generally less attentive to warnings as they become more familiar with a product and vice versa, because of their own well-formed expectations that arise from familiarity. Moreover, people who are more familiar with a product report they are less likely to comply with warnings.[546]

3.7.8 *Cultural Differences*

Because warning features that are effective in one culture may not be effective in others, it is important to explore and identify possible cultural differences. Culture can be defined as a collection of values, beliefs, traditions and behaviour patterns shared by a group of people.[547] Several characteristics are important dimensions of culture, such as age, gender, ethnicity and socio-economic status. Due to the broadening of international trade, it becomes increasingly important to effectively communicate safety information to individuals of different languages and cultures.[548] Unhappily, research on cultural differences related to warnings is relatively absent. There are studies that have compared symbol

544 Van Duijne 2005, p. 124, 233.
545 Vredenburgh & Helmick-Rich 2006.
546 Vredenburgh & Zackowitz 2006, p. 347.
547 Smith-Jackson 2006b, p. 364.
548 Wogalter, Conzola & Smith-Jackson 2002, p. 224.

comprehension or hazard perception across cultures. However, thus far, the existing research is not at a level to support the development of effective warning design guidelines for diverse cultures.[549]

From the available findings, the results do indicate that cultures can differ in terms of hazard perception.[550] A recent study compared perceived hazard ratings with regard to the ANSI warning components (colours, symbols, and signal words) of American participants with those of Chinese participants.[551] American students perceived the colour red as most hazardous whilst Chinese students rated orange to connote the highest level of hazard. The researcher argues that this difference could be explained by the fact that in China red is associated with 'happiness' whilst in the United States red represents 'danger' and 'stop'.[552] Nevertheless, there was some degree of similarity across cultures with regard to the hazard perception ratings in response to combinations of the warning components. Moreover, the results showed that Chinese participants consistently provided lower absolute ratings of perceived hazard level than did American participants. This finding may have the following safety implication: given that a warning component received much lower than anticipated perceived hazard ratings, Chinese product users may subsequently produce less caution than necessary to reduce the risk of personal injury. Therefore, a preliminary recommendation could be to use stronger warning components in order to communicate to Chinese people the same hazard level that is communicated to Americans.[553]

3.7.9 *Conclusion*

The available findings in the warning research demonstrate that people differ in the way they deal with consumer products and how they interact with warnings.[554] The effectiveness of warnings is related to the extent to which a warning is compatible with the needs and capabilities of the intended users, and these needs and capabilities vary from person to person.[555] Hence, for reasons of safety, the design of warnings must be user-oriented. This corresponds with the 'know the warning receiver' design principle that has been recommended by warning researchers: the design of the warning should be as closely matched

549 Smith-Jackson 2006b, p. 368.
550 See *e.g.* Martin, Smith-Jackson & Artis 2003; Smith-Jackson & Essuman-Johnson 2002; Lesch e.a. 2009.
551 Lesch e.a. 2009.
552 Lesch e.a. 2009, p. 960.
553 Lesch e.a. 2009, p. 959.
554 Smith-Jackson 2006a.
555 Smith-Jackson 2006a, p. 335.

with its target audience as possible.[556] Employing a warning system that includes several warning components instead of one source of information facilitates this.[557]

Admittedly, in most cases it would not be feasible to design warnings for every individual difference. There are numerous individual differences. Nevertheless, there are instances in which it would indeed be beneficial to consider the relevant research findings on this topic. Many product warnings may be intended for the general population, but they may also be directed at a specific target audience. For example, certain health care products are primarily used by female consumers. By understanding and taking into account women's knowledge and beliefs with regard to the hazards attached to such products rather than applying a gender-neutral approach, the warning's design and its effectiveness can be optimised.[558] Or, when designing products that are typically used by older adults, the use of larger warning text is wise. When the warning is directed to a general target audience that consists of subgroups that differ on relevant receiver characteristics, the use of multiple warning messages helps to sufficiently address the important needs of different people.[559]

It has been recommended to include a representative sample of the target audience to which the warning will apply when designing or evaluating a warning. On the basis of this selection, the essential differences in the target population can be analysed and the implications following from the research findings can be considered.[560]

This paragraph gave an overview of the most common individual differences in the context of product warnings. The most important seem to be risk/hazard perception and product familiarity. Other person variables that have been identified as potentially influential are demographic variables such as age and gender, but also culture, literacy and personality factors.[561] It is worthy to note that it has been argued that the demographic factors of age and gender are not single variables themselves: an older person might be less likely to notice a warning because of vision problems, not because of his or her age per se. In addition, a woman might be more likely to comply with a warning than a male, because of her perception of the hazard or her risk taking style, not because of her gender per se.[562] Gathering knowledge and/or collecting data about the essential characteristics may require time, effort and money, but without such information the effectiveness of the warning design will be reduced.

Empirical knowledge and understanding of the limited abilities of the target audience of the warning is thus necessary in order to design effective warnings. There is, however, a need for research to better understand how individual factors affect the warning process

556 Wogalter & Laughery 2006, p. 907; Laughery & Brelsford 1991.
557 See paragraph 3.6.10.
558 See Smith-Jackson 2006a, p. 339.
559 Wogalter 2006a, p. 8, 9; Wogalter & Laughery 2006, p. 905; Rousseau & Wogalter 2006, p. 154.
560 Wogalter, Conzola & Smith-Jackson 2002, p. 225; Smith-Jackson 2006a, p. 335.
561 Smith-Jackson 2006a.
562 Rogers, Lamson & Rousseau 2000, p. 133.

and to provide guidance for designing and evaluating warnings that effectively address them.[563] For example, given the rise in the aging population worldwide, unfolding a clearer picture of the aging effects on warning processing is becoming increasingly important. Nonetheless, future empirical research is needed to provide conclusive guidelines on how to deal effectively with receiver characteristics.

3.8 HAZARDS THAT NEED A WARNING

3.8.1 *Introduction*

Most of the warning research has focused on how the format and content of effective warnings should be designed. Relatively scarce attention has been paid to when consumers need a warning against product hazards.[564] Nevertheless, because this warning issue corresponds to the similar type of question that is addressed in a legal context, the considerations that have been made by warning researchers in this respect are described below in paragraph 3.8.2. Furthermore, the warning research literature has touched upon a phenomenon related to issuing warnings, which concerns *overusing* warnings. As the term already suggests, overusing warnings can have several negative effects for the processing of warnings in particular and in general and there is general agreement amongst warning researchers that there is a need for a sensitive approach to the use of warnings (paragraph 3.8.3). The final paragraph of paragraph 3.8.4 closes this paragraph.

3.8.2 *Considerations with Respect to what Hazards Need a Warning*

A set of general considerations has been suggested in the warning research literature that can serve as a guidance regarding the question whether a warning for a hazard should be provided. According to the warning research literature, important factors that indicate that warnings are needed are: (1) the existence of a significant hazard; (2) the hazard, consequences and appropriate actions to avoid the hazard are not known by the people who are exposed to the hazard; (3) a reminder warning is needed to assure awareness of the hazard at the proper time of need; and (4) the hazards are not open and obvious.[565]

It follows from the first consideration that the question with reference to what hazards need a warning relates to the (ergonomic) process of identifying the hazards attached to

563 Laughery 2006, p. 475.
564 McCarthy e.a. 1995, p. 2164.
565 Laughery & Hammond 1999, p. 10; Wogalter 2006a, p. 5. See also McCarthy e.a. 1982; McCarthy e.a. 1995; Edworthy & Adams 1996, p. 48.

the product and quantifying them on the basis of the likelihood that they can cause damage and the seriousness of the potential damage.[566] This process is described in more detail in the upcoming paragraph.

It also follows from these considerations that the question respecting what hazards need a warning relates to what information is needed in a warning. Because a warning message can provide information about several hazard-related aspects, the issue with respect to whether a warning against a hazard is needed overlaps to a great extent with the considerations with regard to when an information category should be given. This topic was already covered in the previous paragraph.[567]

3.8.3 Overuse of Warnings

3.8.3.1 Its Meaning

The phenomenon related to issuing warnings that has been reported in the warning research literature that may have a detrimental effect on safety is having a plethora of product warnings in society. Usually this is referred to as 'overusing' warnings, but 'overwarning' and sometimes the term 'warning overload' have also been used to refer to too many warnings in the world.[568]

Frantz e.a. 1999 use the term *overuse* to refer to situations where providing a warning or warnings may serve to reduce hazard communication and/or safety. This is a rather broad definition or an umbrella term which encompasses the situation of having too many warnings that accompany a product, either cluttered together in one warning message or as separate warnings on the given product, as well as the situation of too many products with warnings in general.[569] *Overwarning* has been defined as the extent to which the world is filled with warnings.[570] It seems to me that this term is similar to the second situation of overusing warnings. In addition, it should be noted that overusing warnings and overwarning are not the same as an information overload. An information overload concerns the issue of processing capacity being overwhelmed by the amount of information in a given situation.[571] For example, because a product contains multiple separate warnings or a single extensive one. Hence, it can be said that this concern is covered by the first situation of overusing warnings.

---

566 Frantz, Rhoades & Lehto 1999, p. 293.

567 For more information see paragraph 3.6.6.6. These rules also overlap to some extent with the criteria for prioritisation, which principle is discussed in more detail in paragraph 3.6.5.6.

568 Wogalter & Vigilante 2006, p. 258; Wogalter & Laughery 2006, p. 906; Frantz e.a. 1999.

569 Frantz e.a. 1999.

570 Wogalter & Vigilante 2006, p. 258; Wogalter & Laughery 2006, p. 906; McCormick & Sanders 1992, p. 684.

571 See also paragraph 3.6.5.6 of this chapter.

3.8.3.2 Negative Consequences

The warning research literature discusses a number of counter-productive consequences that can be associated with overusing warnings on a general level and on a concrete level. These are based on theory, practical experience and empirical findings.

One of the frequently mentioned effects of overusing warnings is that, as a result of an ubiquity of warnings in the world, consumers may be less likely to attend to warnings in general. The notion is that if warnings were to be put on every product, people would tune them out.[572] There is concern that, as the number of warnings grows and the prevalence of warnings about low level risks increases, people will increasingly ignore or disregard them. The presence of many warnings about very minor risks in society may negatively affect the credibility and believability of warnings. People might view warnings as false alarms.[573]

Overusing can also negatively influence the effective processing of a warning message on a given product. Too many warnings within one warning message on a product may negatively affect attention to a specific warning.[574] Furthermore, as the number of warnings increases on a label, it is likely that recall for a given warning decreases.[575] This adverse consequence is related to the limitations of the short-term memory. Products that have lengthy warning messages pose a high work load on individuals, who will have increased difficulty with encoding all the information. This concerns an information overload. There is also recent research indicating that the number of warnings on a product can have a negative effect on perceived danger. It seems that a separate warning message is perceived with less risk if it is embedded with other warnings on a label that represent low risks.[576]

Even though there is more research needed that addresses and clarifies these aforementioned effects of overusing warnings, there is total agreement amongst warning researchers that the practice of providing a warning about every conceivable risk associated with a product is an ill-fated and incorrect approach. Warning designers need to be selective in issuing warnings. It has been contended that the decision making process with regard to when to provide warnings, does not only require consideration of the potential positive impact of the warning's design on people, but also of the negative impact that may be related to adding a warning to the product in particular and in general.[577]

---

572 Wogalter & Vigilante 2006, p. 258; Wogalter & Laughery 2006, p. 906; Frantz e.a. 1999, p. 91; Vanilla Research 2007, p. 24.

573 Edworthy & Adams 1996, p. 47; Frantz e.a. 1999; Rogers, Lamson & Rousseau 2000, p. 134; Stewart & Martin 1994, p. 7.

574 Frantz e.a. 1999

575 Rothstein 1985.

576 Chen, Gilson & Mouloua 1997.

577 Frantz e.a. 1999, p. 920.

The qualitative consumer research study of Vanilla Research is illustrative for how warnings can be overused. The case study on product safety used the example of the warnings for a well-known brand of toaster to explore consumer attitudes. It included 12 safety instructions and 26 important safeguards. Many consumers of the study felt that the list was verging on the ridiculous and that few would in practice read them since the core useful content was generally known. They believed that such an exhaustive list was to protect the company for being sued, rather than consumers from being hurt. The consumers who did read the warnings tended to focus on the text that was emboldened or underlined.[578]

3.8.4 Conclusion

This paragraph provided considerations with respect to the issue of when warnings should be provided. Even though this topic is essential for warning design and effectiveness, it has received relatively little attention.

Paragraph 3.8.2 sketched how it can be generally determined whether a warning against a hazard is needed. According to the warning research literature, important considerations that indicate that warnings are needed are: (1) the existence of a significant hazard; (2) the hazard, consequences and appropriate actions to avoid the hazard are not known by the people who are exposed to the hazard; (3) a reminder warning is needed to assure awareness of the hazard at the proper time of need; and (4) the hazards are not open and obvious.

Furthermore, paragraph 3.8.3 paid attention to the phenomenon of overusing warnings that has been addressed by warning researchers. It explained that an overuse of warnings refers to having an abundance of warnings in society. The potential counter-productive consequences that can be associated with overusing warnings were also described. These are based on theory, practical experience and empirical findings. It not only concerns potential problems with regard to warning effectiveness on a concrete level, such as a reduced believability of a warning, but also on a general level, such as reduced attention to warnings in general.

578 Vanilla Research 2007, p. 24.

3.9 The Role of Warnings in the Design Process of Safe Products

3.9.1 *Introduction*

Given the previous discussion, it is safe to say that warnings are no safety panacea. Nevertheless, this does not suggest that warnings are useless. It demonstrates that careful thought is needed with regard to when warnings should be used compared to other design methods that can reduce risk.

In this paragraph, I will discuss the design process of safe products in general on the basis of the ergonomic literature. This includes the role that warnings play in the design process of products with respect to preventing and reducing accidents involving products. The ergonomic literature offers a step-by-step approach with regard to the way in which attention can be paid to safety when designing products. The reason for this discussion is that prior to the decision of whether warnings are needed and how they should be designed, producers, more specifically their designers, must first take other steps to ensure safety through product design.

In the next paragraphs of 3.9 of this chapter, the three basic steps for designing safe products from the literature are discussed. These are in essence: identifying product risks (paragraph 3.9.3), assessing the severity and probability of the risks (paragraph 3.9.4) and determining the design solutions to eliminate or control the risks (paragraph 3.9.5). In paragraph 3.9.5, special attention is first paid to the hazard control hierarchy perspective that shows when warnings are preferred relative to other design methods to control risks. Secondly, it discusses the lessons of Norman's design philosophy and relevant insights from observational research on consumer products that can both be adopted by designers to assist them in designing safe products. The closing paragraph provides a summary of the lessons learned from the ergonomic literature on designing safe products.

3.9.2 *The Design of Safe Products: A Step-by-Step Approach*

The safety of product users is an important consideration in the design of consumer products. Accident statistics and product recall reports indicate that there is still room for improvement. Ergonomics also include other objectives into the design of products, such as the product's performance, cost, aesthetics and user satisfaction.[579]

The discipline of ergonomics emphasises that it is important to pay attention to safety and to the user in the design stage in a systematic way and this must be done sooner rather

579 McCormick & Sanders 1992, p. 4, 661; Helander 2006, p. 14; Dirken 2004.

than later.[580] Nevertheless, there are downsides with regard to paying attention to safety in design. Safety imposes additional requirements on the design and the design process that may add to costs, which costs may lead to decreased profit margins or even loss of market share to competitors with a less safe but cheaper design. Furthermore, the ergonomic objective of safety can conflict with other objectives. For example, it might be that consumers may be unwilling to accept the safer version as it imposes constraints on their use. Because of these challenges, safety must be considered in an early stage, rather than having to make expensive and less user-friendly safety add-ons later.[581]

To enhance the effectiveness of warnings, it has been advocated by ergonomists that warnings should be viewed as an integral part of the design process, which means that consideration of the hazards that may need a warning and the design of warnings should also begin early in the design process when the design is still relatively fluid. Unfortunately, the literature mentions that many warnings are usually developed after the design of the product is fixed, which in turn can negatively affect the quality of warnings. This can be the case in the event that during the design process no account has been taken of the placement of a warning on the product itself, resulting in a warning that is poorly placed and less likely to attract attention.[582]

As noted earlier, the ergonomic literature offers a step-by-step approach to the design of safe products.[583] The first two steps of this approach concern carrying out a hazard analysis and a risk assessment. These are widely accepted tools, especially in the industrial sector.[584] The specific content of a hazard analysis and risk assessment can take various forms. But executing a hazard analysis and a risk assessment generally entails finding the answers to the questions regarding what can happen with the product, what are the consequences thereof, and how likely is it that this will happen. Such an analysis can be done for the evaluation of existing products, new product ideas, concept proposals, and definitive designs and prototypes.[585] The hazard analysis and risk assessment are useful in that they assist designers in their third step of deciding how to ensure safety through the product's design.[586] Hence, the information resulting from the hazard analysis and risk assessment is also valuable for the design of effective warnings.[587]

580 Van Aken 1996, p. 14; Stanton 1998a, p. 9; Norman 2002, p. 156.

581 Hale, Kirwan & Kjellén 2007, p. 308; Stanton 1998a, p. 6.

582 Laughery & Hammond 1999, p. 10; Van Aken 1996, p. 97; Wogalter & Laughery 2006, p. 908.

583 Van Aken 1996, p. 25.

584 Van Aken 1997, p. 87; see also Hale e.a. 1990.

585 Van Aken 1996, p. 15; Van Duijne, Van Aken & Schouten 2008, p. 246.

586 Van Aken 1996, chapters 4 and 5; Bakken 2005, p. 28-20; Van Duijne, Van Aken & Schouten 2008; Wogalter, Conzola & Vigilante 2006, p. 489.

587 Wogalter, Conzola & Vigilante 2006, p. 489.

3.9.3 Step 1: Risk Identification

3.9.3.1 Formulating Possible Accident Scenarios

Before producers can deal with hazards by design, they first need to be identified and estimated. Some hazards can easily be identified in view of their obviousness or because they are generally known. Other hazards are more hidden and require more research effort. A *hazard* can be defined as a potential source of injury or damage. A hazard is always related to a specific characteristic of a product and is defined as such, for example a sharp edge or a rotating part.[588]

Various techniques are available that can be employed to identify and understand hazards posed by a product.[589] One fruitful way is to use a checklist of potential hazard sources as a starting point.[590] Furthermore, it is of importance that possible accident scenarios are contemplated according to the literature. An accident scenario entails an interaction between a person and a product that possesses hazardous characteristics: it describes the activity of the person(s) involved, the hazard(s), the external factors of the situation and the potential injury. Thus, to formulate any relevant accident scenarios, it is useful to make a checklist of factors that may contribute to the occurrence of accidents. On the basis of building up possible combinations of factors pertaining to the product, the user(s), the environment in which it can be used and how the user(s) may handle the product, relevant accident scenarios can be uncovered. It is important in this stage that a broad range of scenarios is identified and described in detail. This means that designers must not only consider their view on how the product should be used, but also take account of the variation in use situations.[591]

It has been argued that expectations about the correct behaviour of relevant product users should not be too high. For example, assuming that product users will take all preventive measures would mean that accident scenarios that may be relevant are being excluded. Hence, worst-case scenarios should be included as well.[592]

---

588 Van Aken 1996, p. 15.

589 Young e.a. 2006b.

590 The potential hazard sources can be divided in nine categories that can subsequently be subcategorised: mechanic hazards (*e.g.* sharp edges, slippery surface, rotating parts, high/low voltage), hazards resulting from extreme temperatures (*e.g.* open flame, hot gases), radiation hazards (*e.g.* ultraviolet radiation), fire and explosion hazards (*e.g.* explosive mixtures, ignition sources), materials and toxity hazards (*e.g.* toxic solid or fluid), product operating hazards (*e.g.* failure to stop, overexertion, warning text and symbols), hazards resulting from product appearance (misleading form), see Van Aken 1996, p. 61; Van Duijne, Van Aken & Schouten 2008, p. 247. For more information, see also the risk matrix that is developed by the Dutch Consumer Safety Institute as a guide for the design of safe products at <www.eisenwijzer.nl>.

591 Hale, Kirwan & Kjellén 2007, p. 315; Norman 2002.

592 Van Duijne, Van Aken & Schouten 2008, p. 247; Van Aken 1996, p. 45-61.

3.9.3.2 Factors Contributing to the Occurrence of Accidents Involving Consumer Products

Formulating possible accident scenarios with varying types of injury is challenging, especially if it concerns the development of a new product or if no accident data are available yet. Even though it is known at a general level that accidents with consumer products are associated with the components of human (including behaviour), the product design, and the environment, it remains difficult to unravel the exact causes. The ergonomic literature provides a starting point. It has put various factors forward that can be regarded as potentially contributing to the occurrence of accidents. These factors are commonly divided into characteristics of the product, the environment, the user, and the actions by users.[593]

User characteristics can be categorised in terms of the sensory, mental and physical characteristics of humans. In addition, human characteristics can be ordered in terms of their temporary condition at the time of the accident, such as fatigue, distraction, being hurried etc., and in more permanent conditions like the individuals' knowledge and personality traits. Physical characteristics concern body dimensions, exertable forces (for squeezing, pushing, pinching etc.), handedness (left or right), people's state of health including diseases and impairments, perceptual-motor skills and performance such as one's reaction speed, visual-manual coordination and people's movements. The functioning of the senses like vision, hearing and touch has also been associated with accidents. Mental characteristics of humans have to do with cognition (*e.g.* knowledge and experience), risk perception, people's attitudes and their motivation to act safely, personality, intelligence and demographical characteristics (age and gender).

The product itself can also play an important role in the occurrence of accidents. The appearance of a product may provoke particular ways of use, not only safe use, but also unsafe use in the event that the product's features do not provide any feedback about its hazards or when the product looks safe when that is not the case. Examples of product features are colour, texture, composition, size, shape, mass, fragility, signs and symbols etc. Factors affecting product failures may also be relevant in the occurrence of accidents, such as the wearing out of the product or a lack of maintenance.

The environment in which the product is used may directly or indirectly have influence on the accident by influencing the performance of the product user. Environmental factors can relate to the physical environment, such as the available surface, noise, weather conditions, but also to the social-economical environment, such as cultural differences in cooking and eating habits or the presence of others.

---

593 Van Aken 1996, p. 46-61; Norris & Wilson 1999, p. 76. See Weegels 1996 for more literature references concerning studies that have investigated the relevancy of different factors for the occurrence of accidents. See also Dirken 2004.

Lastly, the actions by product users can contribute to the occurrence of accidents. Use actions involve people's actual behaviour with regard to a product in terms of manipulations, postures, movements and so on. It has to do with what users can do with a product and how.[594]

It follows that a broad variety of factors can contribute to the occurrence of accidents which hinders a good prediction of the occurrence of an accident. In the literature, extensive attention has been paid to the characteristics of humans to predict how they will handle the product. However, it seems that these characteristics are only remotely related to accidents, as they only remotely predict the variation in product use that can lead to injury.[595] Furthermore, compared with product failures, little appears to be known about the featural and functional characteristics of the product and their interaction with users. The actions by product users have also received little research attention. Research usually addresses this by using self-reports of subjects instead of actually observing what subjects do with the product. This is unfortunate, since people saying how they use products is not the same as observing their actual behaviour. It has been argued that this lack of knowledge gained from research on users' interactions with a specific product and how they perceive its characteristics can be seen as an obstacle for improving product safety.[596]

3.9.3.3 Information Sources on Accidents

In addition to the aforementioned list of factors that can yield basic information that help generate scenarios that describe how a person interacts with a product, the ergonomic literature mentions more sources of information that can help designers predict accidents.[597]

National and international epidemiological data provide a starting point to gather information on accidents with a particular or a similar type of consumer product. They provide statistical data about the numbers and types of injuries that are registered when the persons involved visited institutions for medical treatment.[598] Hence, such data indicate the nature, scope and severity of accidents with a consumer product. Unfortunately, these data may not be complete and only tend to describe accident scenarios in a global way. They do not provide any insight in the way in which products play a role in accidents and the actual circumstances under which the accidents occur. It does not tell anything about

594 Weegels 1996, p. 12-20.

595 Kanis 1998; Weegels 1996, p. 111.

596 Weegels 1996; Van Duijne 2005; Kanis 1998.

597 Van Aken 1996, p. 71-92.

598 On EU level, the Injury Database (IDB) is available that is based on a systematic injury surveillance system collecting accident and injury data from selected emergency departments of Member State hospitals, providing a complement to and integrating existing data sources, such as routine causes of death statistics, hospital discharge registers and data sources specific to injury areas, including home and leisure accidents and accidents at work. See <https://webgate.ec.europa.eu/idb/index.cfm?fuseaction=home&CFID=3840922&CFTOKEN=4547d662957f20ae-5D63A2E4-E903-2A56-48EE36A47A515B31&jsessionid=360847cb65234042a4d2TR>.

whether the person involved was aware of the risks and if so, how the risks were perceived. In addition, the information does not shed any light on how the characteristics of the product have influenced the users' behaviour and contributed to the accident.[599]

Quantitative retrospective methods for collecting accident data are surveys using questionnaires or telephone interviews with people who suffered injuries as a result of the use of a consumer product. Also, design appraisals by experts, such as ergonomists, or representative users may prove helpful in the identification of hazards.[600]

A method to collect more comprehensive data on why and how accidents with consumer products occur is observational accident research. Weegels has introduced a qualitative retrospective approach that is especially attuned to the observation and analysis of accidents involving consumer products. It consists of reconstructions of accidents in an investigation on-site and interviews with the victims. Both are recoded on audio- and videotape. After the interview, measurements are taken of the situation, the product, the hands and forces that the person can exert.[601] The accident data obtained through this approach seems to be of high quality, because it allows subjects to demonstrate to the researcher what actually happened during interaction with a specific product and it allows the researcher to do an open-ended interview combined with a checklist. On the other hand, this method can prove costly and time consuming and the findings may lack some validity because victims may have a distorted picture of the accident as a result of the memory process.[602] Qualitative observational research with consumers that have not been in an accident has also been shown to be a useful tool for collecting comprehensive data. This can be done by utilising Van Duijne's approach that consists of observation and recording subjects who show how they use a particular consumer product in a naturalistic setting, followed by an open-ended interview at their home.[603] Important user characteristics (hand length, grip forces, sensory limitations) are also measured or addressed. The comprehensive findings of such a study that investigates users' interaction with the product and users' (mis)understanding of the risks can be very helpful in improving the safety of the product's design and reducing accidents.[604]

---

599 Van Duijne e.a. 2008, p. 105; Van Duijne, Van Aken & Schouten 2008, p. 247.

600 Page 1998, p. 139, Norris & Wilson 1999; Van Duijne, Van Aken & Schouten 2008, p. 248.

601 Weegels 1996, p. 35-82. See also paragraph 3.5.2.3. Weegels has also proposed counter-factual reasoning as a method for analysing the accident data. When accident data are collected, the next step involves identifying the causes of the accident. Counter-factual reasoning entails the reasoning: if the condition had been absent, would the accident still have happened? If the accident would not have happened in the absence of the condition, then the condition can be considered a contributory factor. On the basis of this reasoning, it becomes clear how factors are related to the accident and which combinations of factors resulted in the accident.

602 Weegels 1996, p. 27, 108; Page 1998, p. 131.

603 Van Duijne 2005, p. 57-80. See also paragraph 3.5.2.3.

604 Van Duijne 2005, p. 58.

3.9.4 *Step 2: Risk Quantification: Injury Severity and Injury Likelihood*

Having identified the hazards and having generated relevant accident scenarios, the second step of the approach to design safe products is that each hazard needs to be quantified in respect of the relative severity of the potential consequences of hazard exposure and the probability of hazard exposure to determine the size of the risk. Because it is often not possible to design out all the hazards that are associated with using a product, designers must make choices and determine what hazards they will address first. An assessment of the risks by specifying the severity of the consequences and the likelihood that an accident scenario may occur benefits designers in this process.[605] The literature provides a helpful way of quantifying, which involves dividing these two parameters into categories.[606]

The parameter of injury severity can, for example, have the following four injury levels on a scale from one to six, but another type of injury scale is also possible:
1. negligible: hazards will not result in injury;
2. marginal: hazards may cause minor injury;
3. critical: hazards may cause severe injury;
4. catastrophic: hazards may cause death.

The dimension of qualitative probability of hazard exposure can contain six categories:
1. impossible: psychically impossible to occur;
2. extremely improbable: probability of occurrence cannot be distinguished from zero, so improbable that it can be assumed that injury will never be experienced;
3. remote: so improbable that it can be assumed that this will not be experienced by a specific individual, but for a group of individuals, it is unlikely to occur but possible;
4. occasional: unlikely to occur for one specific individual but may occur several times for a group of people;
5. reasonably probable: will occur several times for each individual and will occur frequently for a group of individuals;
6. frequent: likely to occur frequently for an individual and continuously for a group of individuals.

On the basis of combinations of these parameters, a matrix can be created that shows the relative urgency to control certain risks. The values must be multiplied. Combinations higher than nine are not permitted and should be controlled. For example, the combination of a marginal level of injury severity (injury level two) multiplied by the probability level

605 Van Aken 1996, p. 63.
606 Bakken 2005; Van Aken 1996, p. 64.

of occasional (level four) is permitted. However, when the probability level is not four, but five, manufacturers should deal with this risk. Note that an index of eight can still require design intervention, even though permissible.[607]

In other words, using the matrix to assess the risks comes down to evaluating whether the risks associated with the use of the product are of an acceptable level or whether they need to be addressed. It is difficult to answer the question of what is acceptable. A technical approach to achieving an acceptable risk level may involve comparing the product risk to risks people are already running, such as natural causes of death. Furthermore, a more or less generally accepted principle is that an acceptable level is achieved provided that it is technically impossible to reduce the risk any further.[608] Other factors that may play a role in this evaluation relate to how product users perceive risks, such as the benefits and the costs of the risks, the severity of the consequences, whether the risk is generally known and whether exposure to it is on a voluntary basis by the product users or not.[609]

3.9.5 *Step 3: Risk Reduction by Design*

3.9.5.1 Hazard Control Hierarchy: General

After having carried out a hazard analysis (step 1) and a risk assessment (step 2), producer should consider which strategies to apply in order to achieve an acceptable level of risk (step 3) according to the ergonomic literature. The safety literature offers a simple, but systematic perspective regarding the way in which hazards should be controlled. This well-accepted perspective is called the *hazard control hierarchy*. It takes account of the limitations of people by prioritising a sequence of hazard control methods from most to least effective in preventing accidents, *i.e.* design, guard, warn.[610]

There are several versions of this perspective, varying in the number of sequential strategies to deal with hazards. For example, safety in the work place is approached differently than the safety of consumer products. In case of occupational safety, employers have a greater range of intervention strategies at their disposal to prevent or minimise the occurrence of accidents. Employers can determine how, how long and when operators work with the product. Employers can exert influence on the people that work with the dangerous products by selecting only workers that have the capability of using the product safely. Furthermore, they can train their personnel how to work safely. Another intervention strategy is that employers, unlike manufacturers, can supervise to ensure that workers

607 Bakken 2005, p. 28-20; cf. Van Aken 1996, p. 64; Van Duijne, Van Aken & Schouten 2008.
608 Van Aken 1996, p. 68.
609 Van Aken 1996, p. 69; Van Duijne, Van Aken & Schouten 2008, p. 251.
610 McCormick & Sanders 1992, p. 681; Wogalter & Laughery 2006, p. 889; Wogalter 2006a, p. 4; Lehto 2006, p. 65; Statler 2005; Van Aken 1996, p. 27-33.

follow safety methods and do not violate safety rules. Enforcement of safety requirements is also a strategy that can be applied by employers.[611] Given that producers are restricted in their strategies to improve safety, it merits setting high safety requirements to product design.[612]

A difficulty that may arise when applying this approach is that the available method to address one hazard, for example elimination, may create a new hazard. A safety solution for one particular hazard may also interfere with controlling another. In addition, usually more than one hazardous characteristic is attached to the product and it is practically impossible to deal with them all. Hence, designers must prioritise the hazards and decide which ones definitely need to be dealt with.

3.9.5.2 First Line of Defence: Design Out

The warning literature makes frequent mention of the hazard control hierarchy model consisting of the methods of design, guard, warn to guide designers in their efforts to develop safer consumer products.

First and foremost, the best line of defence against injury is to design out the hazard. If the hazardous characteristic of the product is eliminated by means of an alternative design, then the likelihood of injury is greatly reduced. For example, if a nonflammable solvent can be used for a cleaning task, then such as solution is preferable to wearing protective equipment and/or warning against using the flammable solvent in a situation where a possible ignition source exists.[613] However, it is seldom possible to design out and eliminate all the hazards associated with the product. Evidently, a hazard cannot be designed out of a product when the product is inherently dangerous. Removing the hazard would not be a feasible solution if it impairs the intended functioning of the product. For example, eliminating the sharp blade of a power tool makes the product useless. Nevertheless, it may be possible to reduce the hazard by lessening its sharpness.

3.9.5.3 Second Line of Defence: Guard

For hazards that cannot be eliminated or effectively minimised by alternative design, the second-best strategy is to guard the hazard or to put a barrier around it so that consumers cannot encounter the hazard. Personal protective equipment, such as gloves or eye, ear and feet protection, separates the product user from the hazard. Other forms of guarding are the dead-man switch on a lawn mower that stops the rotating cutting blade when the handle is released, child resistant caps, locks, but also regulations that require selling alcohol beverages to adults only or wearing a seat belt in the car. If designing out and

611 Lehto & Salvendy 1995, p. 2156; Lehto 2006, p. 65.
612 Van Aken 1996, p. 13.
613 Laughery & Hammond 1999, p. 4.

guarding against the hazard are not possible, warnings or other forms of communication should be considered.

3.9.5.4 Third Line of Defence: Warn

Warning messages are the third line of defence. Because of an increased reliability on human behaviour, warnings on consumer products are a last resort. They are the least desirable of the three approaches for managing hazards associated with consumer products. The consumers' responsibility is to carry out the behaviour expressed in the warning, but they may not see the warning, may not understand or believe it or may not be sufficiently motivated to perform the desired behaviour. Consequently, to ensure the role that warnings play in protecting consumers, they need to be well designed. A final remark is that it must be borne in mind that a product should not be on the market if one method or combinations of these three methods are still not effective in controlling the hazard. If they have already been put on the market, recalling the product is preferred in order to prevent materialisation of a significant hazard.[614]

3.9.5.5 Design Implications from the Hazard Control Hierarchy Model

An implication that follows from the hazard control strategy is that warnings should not be a replacement or substitute for good design or guarding. Rather, they should be viewed as a supplement to the other approaches to safety. In this regard, providing a warning when it is unlikely to be effective and without considering other potentially more promising intervention strategies, can be seen as a misuse of warnings.[615]

The above-mentioned hierarchy and its prioritisation in methods to effectively deal with hazards can be clarified with the distinction between active and passive approaches to injury control. A general principle is that passive approaches to safety are preferred to active approaches. Active approaches in this context refer to situations where some knowledge and/or actions are required on the part of the individual. The passive approach removes or alters the hazard itself and thus provides some level of automatic protection for the individual without requiring any alteration of behaviour. For example, seat belts in cars are a form of an active approach in that the occupant is required to fasten the seat belt. Warnings can be classified as active, behavioural measures of injury control, since the warning receiver must perform self-protective behaviour directed by the warning to avoid possible harm. The use of air bags, on the other hand, is a passive approach.[616]

An approach that has dominated the accident research in the 20th century for a number of years and that does not reflect the idea behind the hazard control hierarchy is the per-

614 Wogalter 2006a, p. 4.
615 Lehto & Salvendy 1995, p. 2162.
616 Laughery & Hammond 1999, p. 5; DeJoy, Cameron & Della 2006, p. 36.

spective of accident proneness. Accident proneness implies that certain individuals are more likely to have accidents than others because they have enduring characteristics, such as personality, that make them unsafe. To prevent and reduce accidents, it was thought that this group of people need to be identified and trained, or even to exclude them (from the work place). Nowadays, this approach is considered misleading. Especially with consumer products, it is not possible to preclude a group of people. People who have experienced accidents, such as factory workers, cannot be seen as accident prone. This tendency to blame individuals for accidents is seemingly counter-productive. Mainly thanks to the ergonomic discipline, it is nowadays realised that human errors are mostly caused or induced by poor design, and the modern approach to accident prevention is to design products so that they fit the limitations and capabilities of users.[617]

In sum, the hazard control hierarchy offers a perspective on how to deal with product hazards in order to prevent and reduce accidents involving consumer products. Using design solution strategies that aim at changing human behaviour, as is the case with warnings, are least preferred. To prevent accidents, the focus should rather be on improving safety via the design of products. After all, it is much easier to redesign products than to redesign the patterns of behaviour of consumers.[618]

3.9.5.6 Design Implications: Norman's Principles of User-Centred Design

Norman also offers an approach to the design of products.[619] This approach aims at influencing the behaviour of users by applying manipulations to the design of products. It consists of several principles to achieve a design that helps users understand and learn how everyday things, like products with knobs and dials, controls and switches, lights and meters, operate.[620] His design philosophy is user-centred: it is a design approach from the user's point of view by considering the needs and interests of the user.

Norman advocates that there are a number of psychological principles that need to be taken into account for the design of products that are understandable and usable. It follows that these principles also relate to the aspect of safety: if product users do not understand the features of a product, unsafe actions can be performed that may result in injury. The main principles are visibility, constraints, affordances, natural mappings and feedback.[621]

The starting point of a successful design is that the product's design must have a good conceptual model. This model is referred to as the design model and concerns the product's conceptualisation that the designer has in mind. The design model must capture the important parts of the operation of the product and must be understandable to the user.

617 Helander 2005, p. 1-3; Van Aken 1996, p. 21; Weegels 1996, p. 20.
618 Adler 1995, p. 84.
619 Norman 2002; see also Van Aken 1996, p. 37-43.
620 Norman 2002, p. 8.
621 Norman 2002, p. 188, 12.

Thus, when users see a product for the first time, it must be clear to them from the appearance of the product how it operates. There must be clues for users that help them learn what actions are possible, what the effect is of user actions and what the current state of a product is. The user's model refers to what the user develops in his mind to explain product use, how he/she perceives what can be done with the product. Ideally, the user's model and the design model are equivalent. The designer must ensure that the image revealed by the product is an accurate translation of the design model. This is important, because the designer can only communicate with the user through the product itself, in other words through its physical appearance, its operation, the way it responds and the manuals and instructions that accompany the product.[622] Hence, to generate good communication, it is vital that the designer creates a clear conceptual model of the product and that he/she succeeds in adequately translating it into the product's design. Users must also correctly interpret the product or otherwise personal injury may lie in wait.[623]

To provide users a good image of the design model, the principle of visibility is fundamental. Relevant parts of a product must be visible to indicate what parts have a function and to indicate the way in which the user must interact with that part of the product so that users know what to do. This means for instance, that the design must show what parts move and the kind of movement that is possible (pushing, pulling and turning). The design must also give visual or auditory feedback to indicate the immediate and obvious effect of an action. When an action is lacking from an apparent result, you may conclude that it was ineffective and repeat it.[624]

The use of affordances is also a relevant aspect for the design of understandable and usable everyday things. Affordances refer to the perceived and actual properties of the product, primarily those fundamental properties that determine just how the product could possibly be used. They provide strong clues to the operations of things: knobs are for turning, slots are for inserting things into etc. When affordances are taken advantage of in design, users know what to do just by looking: the design signals and guides its proper operation and no picture or instruction label is required.[625]

Even though labels and instruction manuals are often necessary, the appropriate use of mappings can also minimise the need for them.[626] Mapping means the relationship between two things, in this case between controls and their movements and the results. Design should exploit the use of good mappings since their effect is immediate understanding; there is no need for learning and remembering. Norman provides the example of stoves that are often frustrating to use because the user does not immediately know what

622 Norman 2002, p. 189.
623 Van Aken 1996, p. 37.
624 Norman 2002, p. 8, 99.
625 Norman 2002, p. 9.
626 Norman 2002, p. 78.

control goes with which burner. Here, the design does not make use of natural mappings. By spatially arranging the controls in the same pattern as the burners, the information is available in the design without the need to label a control. The advantage of mapping is that overloading of the memory is reduced, as the information necessary for performing a task is made externally available.[627]

Constraints (natural, cultural and logical) also have the power to provide information about the proper course of action and reduce the amount of knowledge that must be learned internally. Constraints reduce the number of alternative actions and as a result, can give users the feeling that there is only one possible thing to do.[628] Logical constraints use reasoning to determine the alternatives. An example of a cultural constraint is the learned meaning of turning screws clockwise to tighten and counter clockwise to loosen. Natural or physical constraints concern the physical properties of objects that constrain possible operations. Accordingly, a large peg cannot fit into a small hole and a push bar on a door invites pushing only. They are closely related to real affordances in that they both signal an action. Affordances suggest the range of possibilities; constraints limit the number of alternative actions.[629]

Another principle is to design a product that allows error. Designers should take into account that product users can make errors, everyone does. To reduce them, it is important that the design makes it easy to discover errors and that the design makes it possible to reverse erroneous actions or to make it harder to do what cannot be corrected.[630] Forcing functions (interlocks, lockins and lockouts) are examples of strong constraints that make it easy to discover erroneous behaviour and prevent possible accidents. Nevertheless, they can be a nuisance in normal use. An interlock forces operations to take place in the proper sequence of actions, like the pin of a fire extinguisher. They prevent the accidental use of devices. A lockin keeps an operation active, preventing someone from prematurely stopping it. A lockout feature prevents an event from occurring, or prevents someone from entering a place that is dangerous.[631]

3.9.5.7 Design Implications Derived from Observational Research

Norman 's basic principles can be adopted by designers to assist them in designing usable, understandable and safe products. Lessons in design can also come from insights derived from observational studies.

It has been noted that it can be a major difficulty for designers to predict at the design stage the variety of ways in which their product will be used, especially if accidents with

627 Norman 2002, p. 75.
628 Norman 2002, p. 199.
629 Norman 2002, p. 88.
630 Norman 2002, p. 131.
631 Norman 2002, p. 135.

that or a similar product have not yet occurred. This burden of designers or their restraint to find out seems to be the root of a great number of safety problems.[632] Knowledge of how users interpret the product and its way of use is essential to safe design, as it allows the designer to set appropriate limits to the safe use of the product. When the limits within which the design can operate safely are too restrictive, the design takes little use variation into account and accidents are bound to happen.

Norman notes that failing to consider a broad range of use situations can be the consequence of the designers' expertise. A common pitfall is that designers often think of themselves as typical users. They think they know how people will use it, because they are users too. However, there is a difference between the expertise designers hold in their head and the limited knowledge of consumers. Hence, it is essential that designers not only consider their view on how the product should be used, but also the various ways in which the product can be used by consumers. It is thus of paramount importance that designers acquire information about how product users interact with the product.[633] Especially qualitative observational research can provide rich data. Through observational qualitative research, with subjects who have or have not been involved in an accident, precise information can be gained about users' risk perception and specific user-product interactions that may lead to accidents. Even though product usage can never be fully predicted and even though it is impossible to prevent all accidents by design, qualitative observational research can be effective in guiding designers in their efforts to improve safety by design.[634]

Recently, observational studies have been carried out by Weegels and Van Duijne and their preliminary findings provide general insights that may have design implications.[635]

A most important insight is that product users are not always aware of running the risk of injuring themselves whilst operating it. Awareness of the risks is essential for formulating accurate judgements of the product's risks and for deciding to follow a warning and/or how to use the product.[636] The findings, of studies that selected subjects that were involved in an accident and subjects who were not, showed that a large number of product users had no idea that they were running the risk of getting injured during the (dangerous) usage of their consumer product.[637] Although the researchers acknowledge that these findings should not be seen as a finalised view, but rather as an impetus for further study on risk perception and awareness in consumer product use, the findings do imply a lack of awareness of running the risk of being injured during product use.

632 Hale, Kirwan & Kjellén 2007, p. 310.

633 Hale, Kirwan & Kjellén 2007, p. 313 ff; Norman 2002, p. 155.

634 Note that offering general guidelines derived from observational research to designers can be problematic, as it is difficult to generalise the results that pertain to specific consumer products and because the occurrence of accidents involving consumer products is largely unpredictable, see Weegels 1996, p. 112.

635 Weegels 1996; Weegels & Kanis 2000; Van Duijne 2005; Van Duijne e.a. 2008. See also paragraph 3.6.7.9.

636 Cf. DeJoy 1991.

637 Weegels & Kanis 2000, p. 367; Van Duijne 2005, p. 237, 226.

Based on the evidence of her studies, Van Duijne puts three reasons forward to explain the absence of risk awareness. A main reason why product users overlook particular risks that can lead to injury is that it seems that product users do not fully apply their attentional resources to understand the functional structure of the consumer product and its potential accident mechanisms and jump to a particular way of using the product. Instead of systematically analysing the product, they take short cuts and look for cues in featural and functional product characteristics as regards the way in which to use the product. Secondly, product users may believe that they control the risks during product use and consequently they are unaware of some risks. Lastly, users consider the benefits and costs of their actions. Omitting precautionary actions saves them effort, but without being aware of it, they are running more risk.[638]

The lack of risk awareness can also be related to the circumstance that users are familiar and experienced with the product at the time of an accident and hence, are behaving on autopilot. Using a consumer product on a regular basis may result in paying less attention to product characteristics.[639] People doing their everyday things on a routine basis rarely anticipate accidents in advance: they run the risks, but they do not take them according to Wagenaar.[640] Wagenaar concludes that people who act on a routine basis do not consider the risks and therefore it cannot be said that their way of evaluating risks is wrong and needs to be improved. This general insight of lack of risk awareness is noteworthy in that it does not support the general idea found in the literature, such as in normative decision making models including the value-expectancy theory, that accidents are caused by misperception of risk or risk-taking behaviour. Whilst risk taking behaviour is based on consciously evaluating the available options before accepting the risk and whilst misperceiving the risk can include underestimating the risk, these findings imply that many accidents occur as a result of automated behaviour, which does not involve a consideration of the risk by the individual. Thus, the idea that people engage in risk-taking behaviour after conscious decision making does not completely hold for accidents involving consumer products.[641] Furthermore, this insight implies that trying to raise risk awareness through risk communication like warnings in cases in which the behaviour is routine is not really tenable.[642]

---

638 Van Duijne 2005, p. 237.

639 Weegels 1996, p. 103; Van Duijne 2005, p. 236.

640 Wagenaar 1992, p. 262. This automated behaviour of individuals can be explained by schemata or script theory: people have a hierarchy of pre-arranged schemata for events in their mind with built-in criteria that have evolved on the basis of experience and that controls their behaviour, see Norman 1981. See also paragraph 3.6.7.6.

641 Weegels & Kanis 2000, p. 369; Van Duijne 2005, p. 237.

642 Wagenaar 1992, p. 279; Weegels 1996, p. 103.

The study of Kanis and Weegels observed and interviewed forty-two people that were recent victims of an accident involving consumer products such as kitchen utensils, do-it-yourself products and personal care products. The accidents were reconstructed in an investigation on-site. Subjects were asked to demonstrate what happened. All observed actions and interviews were recorded. The purpose of the study was to explore the influence of featural and functional characteristics of consumer products on risk perception and awareness. The following questions were asked: (1) to what extent they were aware of running the risk of being injured, (2) and if so, to what extent did they change their behaviour, and (3) to what extent did featural and functional product characteristics affect risk perception and awareness and product use. The results showed that two participants were not aware of the risk of injuring themselves while working with their product at the time of the accident. The number one reason for not being aware was that the possibility of an accident did not occur to the subject. This was the case for a user who cut off his finger tip while uncorking a bottle of wine because the neck of a bottle of wine broke off. Another reason given was that the product looked safe. For example, a subject used a children's table as a step to reach for something. The table top broke and wounded the woman's skin badly. The table looked firm from the outside. However, the top was made out of chipboard. This latter finding indicates that a product characteristic may put users on the wrong track, thus giving them a false sense of safety. Furthermore, there were twelve participants who were aware of some risk(s), but not of the risk of this particular accident. The majority of the subjects believed that they were cautious in using their product. The number of subjects that were in fact fully aware of the possibility of an accident was ten. Eight of them knew that the product was not meant to be used the way in which they were using it. These subjects that perceived the risk, did not adjust their use activities because they were in a hurry, they found it too much effort to apply a proper tool or they had others reasons for their behaviour. For example, the user of a planing machine was planing a small piece of wood and he slid into the cutters with the middle fingers of both hands. Although he knew that the machine should not be used for small pieces, he went on because he did not want to waste the wood.[643]

643 Weegels & Kanis 2000.

Another central finding is that subjective risk awareness and perception is triggered by the functional and featural product characteristics.[644] Product users attend selectively to the most salient product characteristics to infer a way of operating the product to get the results they are satisfied with and this is usually done on the basis of a quick inspection. The state of satisfaction hinders new or additional searches of information, which would be required from a safety perspective.[645] Featural product characteristics entail the more permanent properties of a product that are perceivable by the user, such as colour, shape and texture. Functional product characteristics reflect the functional condition of a product, such as a red glow shown by a product being heated.[646] Users interpret product characteristics as signals or cues for how the product is intended to be used and they do not anticipate risks when performing actions that they consider to be intended by the design. If users have misperceived the functioning and effect of protective features or if they have overlooked the functioning of hazardous product characteristics that influence the outcome of their actions, they may incorrectly believe that their actions are not risky.[647] Hence, featural and functional product characteristics can put users on the wrong track. As mentioned above, the appearance of the product can lead to product users being unaware of a specific risk or the way in which the product might harm them. If a product looks safe, people may assume that certain actions can do no harm.[648] A related safety problem arises when functional features of the product do not provide adequate feedback with regard to its (hazardous) condition.[649] If adequate feedback is missing, product users may incorrectly believe that their action was without risk for injury.[650]

From the viewpoint of accident prevention, it follows that it is of relevance to designers to know to what extent product features affect users' risk awareness and perception and consequently, how they influence product use, since these characteristics can be manipulated

644 Product characteristics that triggered risk awareness and a perception of risk in the empirical studies of Van Duijne were sharp powered blades, hot liquids and flammable gas.

645 Van Duijne 2005, p. 226.

646 Weegels & Kanis 2000, p. 366.

647 Van Duijne 2005, p. 238.

648 An illustrative example is the accident involving a person who used a small table, being a piece of children's furniture, as a step to take something off the cupboard. While standing on the table, the table top broke at four of the six screws and tore the subject's chin badly open. The table top was made out of chipboard and was fixed with six screws to the frame. According to the subject, the table gave an impression of firmness because of the thick screws. There was nothing to indicate that the table was in fact fragile, see Weegels 1996, p. 144.

649 Weegels 1996, p. 104; Weegels & Kanis 2000, p. 369.

650 One of the accidents observed and analysed in the study of Weegels concerned a woman who sustained hand injury as a result of touching the hot plate in the kitchen while she was wiping the bread crumbs off the hot plate. When the woman entered the kitchen, she saw the crumbs. She did not know that the plate was still hot. After fifteen minutes, the red glow of the plate disappears, leaving no visible sign that it still retains some heat. The subject said that she had overlooked the small lamp which indicates that one of the plates is still hot. If it had been more obvious that the plate was still hot, then the subject might not have removed the crumbs, see Weegels 1996, p. 158.

by the design of a product. In line with Norman's design philosophy, the design of a product can suggest, invite or prohibit certain user actions by providing signals or usecues, such as affordances.[651] A design implication that follows from the field studies is that emphasising the characteristics of a product that trigger a subjective perception of risk may make product users more alert to risk in usage and behave safely. This is not to say that accentuating risk by product characteristics is a straightforward way to circumvent particular use actions and to trigger safe use. A potential problem is to maintain risk perception. Users may get used to the product's appearance, especially in the event that they use the product on a regular basis. Also, consumers may, from an aesthetic point of view, be less interested in purchasing a product that does not look safe. Moreover, this strategy is based on influencing the actions of users to guide safe behaviour. As discussed above, such a behavioural solution is often less preferred than making a product 'fool proof' by designing out the hazard or using safety guards that shield off hazards from product users.[652] However, the latter solution also does not guarantee safe use. Guarding may result in product users having a sense of safety and hence, even behave more recklessly whilst the risk of getting injured is present. Furthermore, a safety guard that isolates the hazard from people can also annoy users when they hinder task performance. As a result, users may bypass or deactivate them and expose themselves to the hazard.

Another insight is that even if product users are aware of the risk, it does not mean that they will behave safely. Changing their risky behaviour can take time or cause too much effort.[653] Product users may also believe that the precautions are unnecessary because they believe that their way of use is already safe.[654] Consumer products are used to achieve a particular result: powered gardening tools are used to have a tidy lawn or hedge. Whilst aiming at achieving such a result, it seems that product users prefer to minimise their efforts in operating a product: they seek the line of least effort and skip important actions. Furthermore, although safety is regarded highly important to users, the findings indicate that other goals regarding the manner of product use, such as comfort, minimal effort, getting the job done, compete with safety and even seem to be superior to safety. Hence, they motivate them to perform particular risky actions.[655]

> A field study of Van Dujine's examined users' risk perception and awareness with regard to replacing a pierceable cartridge in gas lamps. This product, often

651 Cf. Kanis 1999. Kanis refers to the term usecues when he talks about functional and featural product characteristics that users perceive.

652 Weegels & Kanis 2000, p. 369.

653 Weegels & Kanis 2000, p. 368.

654 Van Duijne 2005, p. 227, 237.

655 Van Duijne e.a. 2008, p. 117, 228.

used at camping sites, can pose serious risks because it contains a highly flammable liquefied butane/propane gas under considerable pressure.

Accident statistics have shown that users are especially at risk during the replacement of the cartridge. To obtain results, 23 users of gas products with disposable cartridges were interviewed at camping sites and were asked to demonstrate the replacement of the cartridge. They all had experience with this particular gas lamp or a similar product, but participants varied in their level of experience. The gas lamp of the study consists of a lamp unit and a holder. The lamp unit has a rubber seal and a pin designed to pierce the cartridge. The lamp unit must be screwed off in order to replace the empty cartridge. After removal of the old cartridge, the new one can be placed in the clamps by pressing the cartridge into the holder. The cartridge has ridges that fix the position of the cartridge in the holder. Lastly, the lamp unit must be screwed on as a result of which the pin pierces the cartridge.

The observations showed that twelve participants replaced the cartridge correctly. They seemed aware of the risk of gas leaking if they did not remove the lamp unit first. The other eleven participants used the product unsafely by omitting to unscrew the lamp unit first when replacing the cartridge. This risky action can cause a gas leak. They were unaware that the cartridge needed to be fixed in the holder before piercing it. They used the product's characteristics, such as the pin, to understand how to replace the cartridge by thinking that the cartridge had to be pierced by the pin. Four of the eleven participants realised that they were replacing the cartridge in the wrong, risky way, and stopped their actions and followed the right procedure. They said that the pressing of the cartridge into the holder could create the risk of spilling gas. The other ones were not aware of their unsafe use. They actually were convinced that they operated in a safe manner. Furthermore, participants reported that they took preventive actions to respond to the risks associated with replacing the cartridge. Most participants explained that they replaced it outside the tent because of the possibility of explosions. Participants also told they would take precautions when someone was smoking or when young children were around. When participants were asked if they had read the instruction manual of their own gas product, nine said they never read the instruction manual and thirteen read it once after having purchased the product. During the task of the study, seven participants requested the manual to learn about the procedure of replacing the cartridge. Nevertheless, three of them still performed the task unsafely. Of the sixteen who did not read the manual, seven users followed the incorrect order, but three participants changed their mind. According to the researcher, this finding indicates that the manual contributed little to partici-

> pants' understanding of how to replace the cartridge. As regards the level of experience, the results suggest that experience with the procedure of replacing the cartridge did not improve the performance of the task. This experience is based on memories from events that happened relatively infrequently: many use their gas lamp only during their holiday. Apparently, this knowledge does not last and even experienced participants needed to inspect the product's characteristics in order to understand the procedure.[656]

The aforementioned insights and Norman's principles as well can be viewed as guidelines for designers in dealing with safety and usability problems. Because of their general nature, the challenge of translating them into adequate design solutions and applying them to the product at hand remains. Resolving issues such as how to put more emphasis on the danger without creating a situation that is too dangerous, or how to give adequate feedback is left to the creative mind of the designer.

Even though it seems obvious that designers can learn a great deal from field studies and the insights, it remains to be seen to what extent they are of use to the design practice. It appears that designers generally pay much attention to designing hardware and less on designing safe use.[657] This claim is supported by the evidence of a recent small-scale redesign study that investigated to what extent twelve designers used information about risk perception and user activities in naturalistic settings for generating ideas to improve the safety and usability of a gas lamp.[658] The study showed that many designers did not read the research findings of the field study very closely and sometimes overlooked or misinterpreted information. They were mainly focused at finding the safety and usability problems that they needed to solve.[659] It seems that designers experience knowledge about user activities as a constraint on their freedom to (re)design a product as this information imposes boundaries to what is possible.[660] Having read the empirical findings of the observational study, eleven designers wanted to solve the safety problems by redesigning the product in such a way that it looked safe. A proposed design solution consisted of hiding the danger of the risky cartridge by means of a casing. Only one designer wanted to influence users' perception of risk by unambiguously expressing and communicating the hazards associated with usage through the design. He believed that providing insight into the risks would raise users' awareness of risk and evoke safe behaviour.[661]

656 Van Duijne 2005, p. 153-184.
657 Hale, Kirwan & Kjellén 2007, p. 314.
658 Van Duijne e.a. 2007; Van Duijne 2005, p. 185-222.
659 Van Duijne 2005, p. 221.
660 Van Duijne 2005, p. 241.
661 Van Duijne 2005, p. 205.

3.9.6 Conclusion

This paragraph was centred on explaining the role of warnings when designing for safety. This entailed providing a general description of the step-by-step approach to the design of safe products, since warnings are design solutions for improving product safety. An important part of the step-by-step approach consists of carrying out a hazard analysis and risk assessment. The approach is a useful systematic way for paying attention to safety in the design process. It is argued in the ergonomic literature that considering product safety early in the design process pays off, not only in terms of improved safety by design, but also in terms of costs, since safety imposes additional requirements on the design and the design process that may add to costs. Executing a hazard and risk assessment generally entails finding the answers to the questions concerning what can happen with the product, what are the consequences thereof, and how likely is it that this will happen? Such an analysis can be done for the evaluation of existing products, new product ideas, concept proposals, and definitive designs and prototypes. The hazard analysis and the subsequent risk assessment are valuable in that they assist designers in the successive step of deciding how to ensure safety through the product's design.

During the design process, industrial designers are faced with difficulties. One of these concerns acquiring sufficient knowledge about the relevant accident scenarios that may lead to injury. It is axiomatic that knowledge about the occurrence and the potential causes of accidents involving consumer products is essential when designing for safety. Because accident situations are complex, it is difficult to have a good theory that explains how accidents occur. A more fruitful approach might be to assess the contributing factors to accidents. The literature offers a starting point with regard to factors that are potentially contributory to the occurrence of accidents involving consumer products. These entail characteristics of the environment (*e.g.* humid environment), the product (narrow opening or toxic hazard), characteristics of the user population (*e.g.* mental and sensory characteristics) and the various ways in which the product can be used. The designer must contemplate possible accident scenarios on the basis of the factors. Conceivable combinations of these factors that can lead to an accident can be distinguished. When generating accident scenarios, it is important that designers do not rely solely on their own conceptual model of how the user should interact with the product, because this may hinder improving the safety in design.

Having different sources of information that help designers consider a variety of accident scenarios is valuable. Accident statistics and ergonomic methods, such as interviews, questionnaires, and observation are available as sources of information during the various stages of the design process. Especially observational research provides rich data for constructing possible accident scenarios. These findings may predict future product use that can culminate into injury, since participants' behaviour explains how they interact with the product and how they perceive the product's characteristics.

Having collected information about the relevant ways in which the product is used and having assessed the risks associated with product usage, the designer has the task of finding ways to prevent or reduce the identified safety problems. There are several design solutions for improving the safety of products. The safety literature offers a basic hierarchy scheme of hazard control as guidance for designers. It prioritizes methods from most to least effective in preventing and reducing accidents. As regards the rank order of these methods, the general rule is that it is far better to design out the hazard than to rely on techniques of behaviour modification. After all, it is easier to redesign products than to redesign the behaviour of consumers. In other words, the product's design must eliminate the hazard and if that is not technically or economically feasible, behavioural solutions that reduce the hazard, such as training, education, supervision and warnings, should be considered. Methods such as training and supervision are particularly helpful to improve safety in industrial settings, but they are more difficult to apply to consumer product safety. The variation in use situations and the variation in user population make improving product safety compared with occupational safety more problematic. The warning literature often cites that warnings should thus be viewed as a method of last resort. Warnings need to be relied on when more fundamental solutions to safety problems are infeasible: they should be viewed as a supplement to the other methods and not as a substitute for good design or guarding.

Norman's design philosophy also offers guidelines for designers to achieve products that are easy to understand and user-friendly. Seeing that understanding how a product operates is relevant for safe use, these design solutions are of interest to safe product design. Designers can communicate with users through the design of the product. Users form mental models of the product on the basis of their interpretation and explanation of how the product works. It is essential that designers take account of these interpretations when designing the product, since they influence behavioural actions. The users' model must correspond with the image of the product that the designer had in mind when he/she developed the design model. However, when the designer fails to provide an explicit conceptual design model due to a lack of clues or when the product's appearance is not an accurate translation of the design model, there is a mismatch that can lead to misunderstanding and frustration by users and even damage. Visibility of the possible actions that can be done with the product and the effects of those operations is thus essential for good communication. A product's design should exploit constraints, affordances and natural mappings, because they can provide good signals with regard to how a product works and what users can do with it. In consequence, such design manipulations can guide good and safe product use. As Norman argues, the design should give the answers to how the product works without the need for warning information and symbols and certainly without the need for trial and error by users.

Theoretical insights derived from accident data and field studies with consumers that have not recently been involved in an accident, can also assist designers in their efforts to set criteria for the design of a safer product. Recent studies of Weegels and Van Duijne have revealed important findings with regard to risk awareness and risk perception in consumer product use. How people perceive risks or hazards (terms that are similar for lay persons) can play a significant role in determining how and why accidents take place, because safe behaviour is linked with perceived hazard. Unfortunately, this has not been frequently studied yet.

The findings indicate that product users use product characteristics to learn how to use the product with satisfaction. This corresponds with Norman's philosophy. When judging the risks, research indicates that people refer to the severity of the potential injury and to the controllability of the perceived risks. Users interpret product characteristics as signals or as cues to find out how to use a product safely. The study also showed that external information, such as an instruction manual, was of minor importance for learning how to operate the product safely. Even though product users may recognise product characteristics that are associated with risk, they may be unaware of the types of use actions that can be risky. One insight that follows from the studies is that product users are not always aware of running the risk of injuring themselves whilst operating everyday (familiar) consumer products. Hence, the idea that people only engage in risky behaviour after conscious decision making does not completely hold for consumer products. This implies that there seems to be little point in trying to raise risk awareness of people who use familiar consumer products by means of warning messages. Users may thus have a false sense of safety if product characteristics are not (sufficiently) present to trigger risk perception and guide safe usage. Although the findings are of a preliminary nature, they do emphasise that gaining understanding of how consumers interpret product characteristics is of value for accident prevention. Manipulating the design so that a product looks safe might not always be the appropriate safety solution. Using product characteristics as cues to help users understand the functioning of the product may involve communicating hazards explicitly in featural and functional product characteristics in order to trigger safe usage of consumer products. This strategy can be favoured over making a product fool proof, that is, using safety guards, if it is expected that users will exhibit careless behaviour due to a misplaced feeling of safety or if the perceived costs of annoyance that are caused by a safety guard outweigh the goal of safe use. Related to this, is the insight that even though product users perceive the risk, they may still be motivated to act in a dangerous fashion. Consumers use their products mainly to achieve a particular result. If this result conflicts with operating the product safely, then it is likely that safety loses.

3.10 Summary

In this closing paragraph, the conclusions of the previous paragraphs are grouped together to give an outline of the chapter. Paragraphs 3.3, 3.6, 3.8, and 3.9 discussed the key warning issues that are relevant for the analysis of European product liability law in the next concluding chapter. The other paragraphs fulfilled a supportive role to better understand the warning research and the warning process.

Definitions and Purposes of Warnings

The first warning issue was dealt with in paragraph 3.3. To answer the question why warn, the chapter began with providing a definition of warnings. This thesis looks at warnings from the standpoint of safety, which means that warnings are regarded as safety communications used to inform people about product hazards and to provide safety instructions so that undesirable consequences are avoided or minimised. On a general level, the ultimate goal of a warning is to improve safety and thus reducing accidents that result from ineffective product warnings. A purpose that follows from this is the behavioural or persuasive function of warnings; warnings must persuade warning receivers to behave according to the warning so that the product is used safely. Furthermore, warnings are a means to communicate (new) safety information so that people can make better-informed choices. In addition, warnings may also serve to cue existing knowledge necessary to perform the product task safely at the time receivers are at risk. Even though consumers may have information or experience in their knowledge base about a warning component, it may not be enough to prevent injuries, because people were not consciously aware of it when they should have been. In such instances, warnings have a reminding function.

Theoretical Models of the Warning Process

The following warning issue covered in this chapter concerned the most important one concerning the identification of the factors that can influence the effective processing of warnings by warning receivers. To better understand the extensive discussion in this paragraph (paragraph 3.6), three subquestions were answered in the preceding paragraphs (paragraphs 3.2, 3.4, 3.5).

First of all, it was necessary to explain how people interact with warning information. Several theoretical models have been proposed by the literature to describe the warning process. These were dealt with in paragraph 3.4. The advantage of using a theoretical framework of the warning process is that it helps to structure the rapidly growing body of warning research. Moreover, it provides a method to help uncover why a warning message is not effective in changing behaviour, to evaluate the effectiveness of warnings and to help designers develop better warnings. The paragraph showed that human information processing models have been popular among researchers. Information processing models of the warning process identify several sequential stages of information processing and show that the stages need to be successfully performed by the human for a warning to change behaviour. These models also imply that warning effectiveness will not be greater than the least effective stage in the sequence. A theoretical view underlying many models of the warning process is communication theory. Warnings are a form of communication intended to communicate safety-related information from a source to a warning receiver by means of a channel and it is thus not surprising to regard source, channel and receiver factors as influential on the warning processing stages. An information processing model of the warning process that has gained much popularity in recent years is the Communication Human Information Processing (C-HIP) model. The C-HIP model provides a framework for showing the stages of information flow from a source to a receiver, whereby the receiver successfully processes the warning information to produce subsequent compliant behaviour. A source (*e.g.* the manufacturer) encodes a message (*i.e.* the product warning) into a channel (*e.g.* visual or auditory) that is transmitted and delivered to a receiver (*i.e.* the product user). If the warning is transmitted and delivered to the receiver, the information must then be mentally processed within the receiver. The receiver stage includes human information processing substages prior to carrying out the desired behavioural response to a warning. These substages are attention switch, attention maintenance, comprehension, beliefs and attitudes, and motivation. In the model, behavioural compliance with the warning is the culmination of these subsequent stages. At each receiver stage, the warning information needs to flow through to the next stage, but it may produce a bottleneck that blocks the flow before the process ends in behaviour. Although the model represents a linear process, the stages interact with each other. Later stages can affect the processing in earlier stages.

Theoretical perspectives and models borrowed from other scientific fields such as the persuasion research literature can also advance the design of effective warnings. In general, these models consider the aspects of attitudes and motivation to explain safety-related behaviour. They stress the importance of personal expectations in determining behaviour.

Another model that conceptualises the warning process is the level of performance model. The primary insight from the perspective of the levels of performance model is that, to reduce human error, warnings should be designed to match the operators' level of performance. Unfortunately, since the design of the most effective warning at each level of performance is fundamentally different, the warning designer must make trade-offs. A warning is likely to have the greatest influence when product users operate at a knowledge-based level. Regrettably, most behaviour occurs at skill- or rule-based level.

Main Disciplines of the Warning Research

The other relevant subquestions that relate to paragraph 3.6 were addressed in paragraph 3.2 and paragraph 3.5. The factors of paragraph 3.6 that influence the effectiveness of warnings have been identified through the warning research. This asks for background information on the main scientific disciplines that are involved in the warning research and the methods warning researchers have employed to achieve the findings on warning effectiveness.

This chapter started with introducing the disciplines that have been studying the design of effective warnings. The majority of the warning studies have been carried out by ergonomists. Human factors or ergonomics is concerned with the interface between people using a product and the environment with a view to optimising performance and pleasure. Its main purpose is to design systems, jobs, products, environments that match the psychical and mental abilities and limitations of people. Ergonomics is an applied science which means that knowledge from various sciences is utilised for the design of consumer products. One of these sciences concerns cognitive psychology. This empirical science is especially relevant for the design of effective warnings, since cognitive psychology is an area of psychology that studies and attempts to explain the basic cognitive processes of the mind and this includes information processing.

Various research methods are available to researchers to obtain results regarding the effective processing of warnings. Paragraph 3.5 was used to provide information on how researchers have measured the effects of a potentially influential factor on one of the intermediate information processing stages or on the end stage of behaviour. The C-HIP model was used to describe the common methods carried out to investigate the processing of visual warning messages at each of the intermediate cognitive stages within warning receivers. The relevant methods used in the warning research to study the behavioural stage were also discussed.

Research can be carried out in the field or in the lab. Relatively few warning studies have taken place in the field. The majority of the warning studies are employed under highly controlled laboratory conditions and are primarily based on measuring the effects of one or more variables on an intermediate information processing stage. The independent variables are the variables that the researcher wants to investigate and that are often related to the design of a warning, such as the warning's size or the presence or absence of symbols, but it can also be a situational factor or a receiver characteristic. These variables are being manipulated in order to assess their impact on the dependent variable like attention to warnings or warning compliance.

The mental activities that occur during the processing of warnings cannot be studied directly because the cognitive processes occur in the mind and are unobservable. Therefore, they are measured indirectly by an indicator that can be seen. The stage of attitudes and beliefs and motivation is primarily assessed on the basis of subjective measures by asking participants directly what they think, for example whether they perceived the product as dangerous. In the other stages, more objective methods are available such as measuring the time to respond to a warning or recall tests. Behaviour in terms of following the warning's directions can be measured directly. Data on behavioural compliance with warnings has been collected in a number of ways. Behavioural compliance can be measured directly or indirectly using subjective or objective measures. Contrary to the internal mental processes of the intermediate processing stages of the C-HIP model, behaviour can be measured directly. This entails observing whether people comply or do not comply with the warning. Observation can be done in the lab under controlled conditions with participants, but also in field settings under less controlled conditions. Studying compliance indirectly concerns measuring the variables that influence the substages of the warning process that occur prior to behavioural compliance. These variables may have an indirect impact on warning compliance. Because it is assumed that behavioural compliance is the result of the successful processing of the subsequent substages within the receiver, studying them can be viewed as indirect measures of warning compliance. Behavioural compliance can also be evaluated through subjective measures. For example, self-reports in which consumers tell whether they performed the behaviour in the past or by measuring behavioural intentions, which means that participants are asked to give their judgement regarding the perceived effectiveness of a warning or to what extent they would be willing to comply with the warning in a particular situation. Research findings in social psychology indicate that there is a causal link between behavioural intentions and behaviour. Thus, by assessing whether receivers of a warning would have the intention to follow the warning's directions, behavioural compliance can be predicted. Nevertheless, saying that you have the intention is not the same as actually complying with a warning. Therefore, it is preferred to examine compliance through observing actual behavioural compliance in a controlled laboratory setting or in a field study.

Notwithstanding that behavioural compliance is the most important stage relatively few studies have measured actual behavioural compliance. Studying behavioural compliance can be difficult for several reasons. The main one is that it is unethical to expose participants of a study to real hazards. As a result, researchers must create a product use situation that is safe. Moreover, it must be under realistic conditions of product use for the participants. They must believe that they are in a hazardous situation and that following the warning will prevent the hazard. Another potential obstacle is that of control. It may not always be possible to control or mimic the conditions that are needed for the situation, especially in field or quasi-field studies. Furthermore, behavioural compliance studies can be time and labour consuming as well as expensive. In such instances, using measures of behavioural intentions are more suited.

The Effectiveness of Product Warnings

In the paragraphs of paragraph 3.6, the stages of source, channel, and the information processing stages of the receiver of the C-HIP model were described together with the corresponding factors that have been identified in the warning research as potentially influential. The C-HIP model predicts that source and channel can exert an influence on communicating warning information to receivers. Warning studies examining the impact of source information and the type of modality on warning compliance confirmed this viewpoint. The C-HIP model also shows that once the warning information has reached the receiver, several sequential mental steps need to be successfully completed to elicit behavioural compliance. This means that if further processing of the warning information within the warning receiver is blocked at a stage, the desired behavioural compliance with the warning will not be achieved. The general research finding that there is a decline in the number of people who notice a warning, then read it and finally comply with it within the group of participants corresponds with this. Hence, there is not a direct causal relationship between consumers attending to a warning and their behaviour. Consequently, another prediction that can be derived from the model is that the ultimate impact of exposing consumers to a warning message on behavioural compliance can be relatively small. For example, if 60% of 100 consumers see the warning, 75% of those individuals read and understand it, 60% of those consumers who have understood the warning believe it, 40% are motivated to perform the desired behaviour and 90% actually follow the warning, the probability that the warning will lead to appropriate product use is then approximately 10%. This example illustrates that in the event that the effectiveness of each information processing stage is low, the use of a warning will not be a promising approach for improving product safety. Even if the previous intermediate stages have been successful, the level of behavioural compliance can still be low, because the high costs of complying have produced a bottleneck.

In the discussion about the information processing stages including the stage of behaviour, it became clear that a myriad of factors can influence the effectiveness of warnings. Potentially influential factors pertaining to the effectiveness of each stage of the C-HIP model were described together with important design implications and guidelines suggested by warning researchers.

Progress has been made by way of research efforts. The studies have indicated factors that can be identified as important, and as a result have contributed to a better understanding of the complex warning process. In general, the factors influencing the effectiveness of the information processing stages and behavioural effectiveness relate to the environment in which the product is used, to human characteristics of product users, and to warning design features. Much of the warning research has focused on design variables such as size, pictorials, and signal words. These factors mainly influence the noticeability and understandability of warnings. However, less research has been done on the influence of external factors that are related to the situation of product use in the environment. Furthermore, relatively few behavioural and survey studies have measured the influence of various attitudes and beliefs on warning effectiveness.

The paragraphs of paragraph 3.6 showed that different factors can be of relevance at different stages. In addition, one single factor can be important for the successful processing at more than one stage: belief factors can also exert an influence on attention which confirms the prediction of the model that the process is not linear, but that the information processing stages interact. Naturally, factors that influence the earlier information processing stages are indirectly of importance in determining whether the warning will have a positive effect on behaviour. In general, the variables that particularly influence the initial stages are expected to be positively correlated with the likelihood of compliance. The list of factors per stage is not exhaustive, but it serves as a means to present the potentially influential variables that have been identified through scientific studies. Consequently, because these factors pertain to the effective processing of the substages and the behavioural effectiveness of product warnings, they must be considered first in the design process of warnings.

To attract the attention of warning receivers the factors of size, brightness and colour contrast, borders, signal words, symbols, location, interactivity, hazard perception, familiarity, information seeking mode, environmental conditions and channel are especially of relevance. Maintaining the attention of warning receivers depends on similar factors like colour contrast, symbols, hazard perception, the familiarity belief, environmental conditions and location, but additional factors such as letter case, font style, alternative label designs and layout of the warning information are especially relevant for enhancing the encoding process of information.

The comprehensibility of the warning message is mainly determined by variables that pertain to the cognitive abilities of receivers, such as their language skills, reading abilities and the level of knowledge, and to warning variables including signal words, colour, warning symbols and explicit information.

Attitudes and beliefs of warning receivers can strongly influence whether a warning will be effective. The factors that affect processing at this stage are hazard perception, cognitive biases that may affect the people's belief about the hazard (especially availability, overconfidence, suppression and message framing), perceived control, the familiarity belief, prior injury experience, personal relevancy, risk-taking style, costs of compliance, perceived effectiveness of recommendations and the belief of self efficacy.

Many variables that affect beliefs and attitudes are also valuable for motivating receivers to produce the appropriate behaviour. Primary variables that influence receivers' motivation are hazard perception, costs of non-compliance, costs of compliance, social influence, stress and mental work load.

The final stage is behavioural compliance with a warning. The main factors that have been shown to promote actual compliant behaviour are location, interactivity, colour, channel, warning symbols, hazard perception, costs of non-compliance, explicit information, the familiarity belief, costs of compliance, social influence, stress and mental work load. The factors of adding information components, adding information about the warning's source, perceived control and people's prior injury experience have influenced behavioural intentions of complying with a warning.

From this wealth of influential variables, a shift can be made to identify the variables that have shown to be most significant in the success or failure of warnings in capturing attention (noticing and encoding) and providing information to make the decision to comply. In a recent review of the warning research, it was concluded that for the process of noticing and encoding a warning, the variables of size, location, colour contrast, signal words, warning symbols, hazard perception and familiarity are most significant. For warning compliance decisions, the variables that affect noticing and encoding a warning are important as well as the variables of warning symbols, explicit information, familiarity, hazard perception, modelling and lastly, the costs of compliance which factor is a very important consideration in warning design.

Receiver Characteristics and Product Warnings

Paragraph 3.6 dealt with the effectiveness of product warnings across the five dimensions and showed that individual characteristics (person variables) play an important role in how product users approach and respond to warnings. Paragraph 3.7 described some of these personal variables (differences in age, gender, receiver competence, risk perception, personality, familiarity, culture) and their implications for warning design in more detail.

Whilst the effects of physical warning design factors have been repeatedly investigated, there is a paucity of empirical evidence related to the precise impact of different receiver characteristics. The available findings in the warning research demonstrate that people differ in the way they deal with consumer products and how they interact with warnings. The design of warnings should be user-oriented as the effectiveness of warnings is related to the extent to which a warning is compatible with the needs and capabilities of the intended users, and these needs and capabilities vary from person to person. This corresponds with the 'know the warning receiver' design principle that has been recommended by warning researchers: the design of the warning should be as closely matched with its target audience as possible. Admittedly, in most cases, it would not be feasible to design warnings for every individual difference. There are numerous individual differences. There are instances though in which it would indeed be beneficial to consider the relevant research findings on this topic. Many product warnings may be intended for the general population, but they may also be directed at a specific target audience. For example, certain health care products are primarily used by female consumers. By understanding and taking into account women's knowledge and beliefs with regard to the hazards attached to such products rather than applying a gender-neutral approach, the warning's design and its effectiveness can be optimised. Or, when designing products that are typically used by older adults, the use of larger warning text is wise. It has been recommended to include a representative sample of the target audience to which the warning will apply when designing or evaluating a warning. On the basis of this selection, the essential differences in the target population can be analysed and the implications following from the research findings can be applied. Individual differences that seem to be most important in the context of product warnings are hazard perception and product familiarity. Other person variables that have been identified as potentially influential are demographic variables such as age and gender, but also culture, literacy and personality factors. It is worthy to note that it has been argued that the demographic factors of age and gender are not single variables themselves: an older person might be less likely to notice a warning because of vision problems, not because of his or her age per se. In addition, a woman might be more likely to comply with a warning than a male, because of her perception of the hazard or her risk taking style, not because of her gender per se. Gathering knowledge and/or collecting data about the essential characteristics may require time, effort and money, but without such information the effectiveness of the warning design will be reduced. In sum, empirical knowledge and understanding of the limited abilities of the target audience of the warning is thus necessary in order to design effective warnings. There is, however, a need for research to better understand how individual factors affect the warning process and to provide guidance for designing and evaluating warnings that effectively address them. For example, given the rise in the aging population worldwide, unfolding a clearer picture of the aging effects on warning processing is becoming increasingly important. Nonetheless, future empirical research is needed to provide conclusive guidelines on how to deal effectively with receiver characteristics.

Hazards that Need a Warning

Paragraph 3.8 provided considerations with respect to the issue of when warnings should be provided to consumers. Even though this topic is essential for warning design and effectiveness, it has received relatively little research attention.

Paragraph 3.8.2 sketched how it can be determined whether a warning against a hazard is needed. According to the warning research literature, important considerations that indicate that warnings are needed are: (1) the existence of a significant hazard; (2) the hazard, consequences and appropriate actions to avoid the hazard are not known by the people who are exposed to the hazard; (3) a reminder warning is needed to assure awareness of the hazard at the proper time of need; and (4) the hazards are not open and obvious.

Furthermore, paragraph 3.8.3 paid attention to the phenomenon of overusing warnings which has been addressed by warning researchers. It explained that an overuse of warnings refers to having an abundance of warnings in society. The potential counter-productive consequences that can be associated with overusing warnings were also described. These are based on theory, practical experience and empirical findings. It not only concerns potential problems with regard to warning effectiveness on a concrete level, such as a reduced believability of a warning, but also on a general level, such as reduced attention to warnings in general.

The Role of Warnings in the Design Process of Safe Products

Paragraph 3.9 centred on discussing the final important warning issue of when warnings need to be provided to prevent and reduce accidents. From the discussion on the factors that influence the behavioural effectiveness of warnings, it became clear that warnings are no safety panacea. Luckily, hazards associated with consumer products can usually be controlled in more than one way. Paragraph 3.9 also generally described the design process of safe products. The reason for this is that prior to the decision of how hazard sources associated with product usage should be approached and whether this includes providing warnings, manufacturers, more specifically their designers, must take other steps to ensure product safety. The first part of the step-by-step approach consists of carrying out a hazard analysis and risk assessment. It is a useful systematic tool for paying attention to safety in the design process. It is argued that considering product safety early in the design process pays off, not only in terms of improved safety by design, but also in terms of costs, since safety imposes additional requirements on the design and the design process that may add to costs. Executing a hazard and risk assessment generally entails finding the answers to the questions regarding what can happen with the product, what are the consequences thereof, and how likely is it that this will happen. Such an analysis can be done for the evaluation of existing products, new product ideas, concept proposals, and definitive designs and prototypes. The hazard analysis and risk assessment is valuable in that it assists designers in the successive step of deciding how to ensure safety through the product's design.

During the design process, industrial designers are faced with difficulties. One of these concerns acquiring sufficient knowledge about the relevant accident scenarios that may lead to injury. It is axiomatic that knowledge about the occurrence and the potential causes of accidents involving consumer products is essential when designing for safety. Because accident situations are complex, it is difficult to have a good theory that explains how accidents occur. A more fruitful approach might be to assess the contributing factors to accidents. The literature offers a starting point with regard to factors that are potentially contributory to the occurrence of accidents involving consumer products. These entail characteristics of the environment (*e.g.* humid environment), the product (narrow opening or toxic hazard), characteristics of the user population (*e.g.* mental and sensory characteristics) and the various ways in which the product can be used. The designer must contemplate possible accident scenarios on the basis of the factors. Conceivable combinations of these factors that can lead to an accident can be distinguished. When generating accident scenarios, it is important that designers do not rely solely on their own conceptual model of how the user should interact with the product, because this may hinder improving the safety in design. Having different sources of information that help designers consider a variety of accident scenarios is valuable. Accident statistics and ergonomic methods, such as interviews, questionnaires, and observation are available as sources of information during the various stages of the design process. Especially observational research provides rich data for constructing possible accident scenarios. These findings may predict future product use that can culminate into injury, since participants' behaviour explains how they interact with the product and how they perceive the product's characteristics.

Having collected information about the relevant ways in which the product is used and having assessed the risks associated with product usage, the designer has the task of finding ways to prevent or reduce the identified safety problems. There are several design solutions for improving the safety of products. The safety literature offers a basic hierarchy scheme of hazard control as guidance for designers. It prioritizes methods from most to least effective in preventing and reducing accidents. As regards the rank order of these methods, the general rule is that it is far better to design out the hazard than to rely on techniques of behaviour modification. After all, it is easier to redesign products than to redesign the behaviour of consumers. In other words, the product's design must eliminate the hazard and if that is not technically or economically feasible, behavioural solutions that reduce the hazard, such as training, education, supervision and warnings, should be considered. Methods such as training and supervision are particularly helpful to improve safety in industrial settings, but they are more difficult to apply to consumer product safety. The variation in use situations and the variation in user population make improving product safety compared with occupational safety more problematic. The warning literature often cites that warnings should thus be viewed as a method of last resort. Warnings need to be relied on when more fundamental solutions to safety problems are infeasible: they should be viewed as a supplement to the other methods and not as a substitute for good design or guarding.

Norman's design philosophy also offers guidelines for designers to achieve products that are easy to understand and user-friendly. Seeing that understanding how a product operates is relevant for safe use, these design solutions are of interest to safe product design.

Designers can communicate with users through the design of the product. Users form mental models of the product on the basis of their interpretation and explanation of how the product works. It is essential that designers take account of these interpretations when designing the product, since they influence behavioural actions. The users' model must correspond with the image of the product that the designer had in mind when he/she developed the design model. However, when the designer fails to provide an explicit conceptual design model due to a lack of clues or when the product's appearance is not an accurate translation of the design model, there is a mismatch that can lead to misunderstanding and frustration by users and even damage. Visibility of the possible actions that can be done with the product and the effects of those operations is thus essential for good communication. A product's design should exploit constraints, affordances and natural mappings, because they can provide good signals with regard to how a product works and what users can do with it. In consequence, such design manipulations can guide good and safe product use. As Norman argues, the design should give the answers to how the product works without the need for warning information and symbols and certainly without the need for trial and error by users.

Theoretical insights derived from accident data and field studies with consumers that have not recently been involved in an accident, can also assist designers in their efforts to set criteria for the design of a safer product. Recent studies of Weegels and Van Duijne have revealed important findings with regard to risk awareness and risk perception in consumer product use. How people perceive risks or hazards (terms that are similar for lay persons) can play a significant role in determining how and why accidents take place, because safe behaviour is linked with perceived hazard. Unfortunately, this has not been frequently studied yet. The findings indicate that product users use product characteristics to learn how to use the product with satisfaction. This corresponds with Norman's philosophy. When judging the risks, research indicates that people refer to the severity of the potential injury and to the controllability of the perceived risks. Users interpret product characteristics as signals or as cues to find out how to use a product safely. The study also showed that external information, such as an instruction manual, was of minor importance for learning how to operate the product safely. Even though product users may recognise product characteristics that are associated with risk, they may be unaware of the types of use actions that can be risky. One insight that follows from the studies is that product users are not always aware of running the risk of injuring themselves whilst operating everyday (familiar) consumer products. Hence, the idea that people only engage in risky behaviour after conscious decision making does not completely hold for consumer products. This implies that there seems to be little point in trying to raise risk awareness of people who use familiar consumer products by means of warning messages. Users may thus have a false sense of safety if product characteristics are not (sufficiently) present to trigger risk perception and guide safe usage. Although the findings are of a preliminary nature, they do emphasise that gaining understanding of how consumers interpret product characteristics is of value for accident prevention. Manipulating the design so that a product looks safe might not always be the appropriate safety solution. Using product characteristics as cues to help users understand the functioning of the product may involve communicating hazards explicitly in featural and functional product characteristics in order to trigger safe usage of consumer products. This strategy can be favoured over making a product fool proof, that is, using safety guards, if it is expected that users will exhibit careless behaviour due to a misplaced feeling of safety or if the perceived costs of annoyance that are caused by a safety guard outweigh the goal of safe use. Related to this, is the insight that even though product users perceive the risk, they may still be motivated to act in a dangerous fashion. Consumers use their products mainly to achieve a particular result. If this result conflicts with operating the product safely, then it is likely that safety loses.

4 Warning Defectiveness: Lessons Learned from Cognitive Psychology and Ergonomics

4.1 Introduction

The previous two chapters discussed the topic of product warnings from two different perspectives. Chapter 2 discussed product warnings in the context of European product liability law. Chapter 3 gave a comprehensive analysis of important warning topics with regard to how humans interact with warnings and products. Chapter 3 showed that design implications and design principles pertaining to effective warnings and safe products are at hand as a result of the considerable body of empirical warning research and reviews thereof.[1] Not only producers, public law policy-makers, researchers and safety specialists, but also European product liability law can learn from these insights.[2]

This chapter embodies the heart of this book, as it analyses the following warning issues from a legal context:

- What is a product warning? (paragraph 4.2);
- Why warn? (paragraph 4.3);
- What risks need no warning? (paragraph 4.4);
- When should consumers be warned in relation to other design solutions? (paragraph 4.5);
- How should consumers be warned? (paragraph 4.6).

The legal analysis is fuelled by the insights that have been collected in the previous chapter.[3] The insights are used to provide detailed supporting explanations for why European product liability law holds its view on the warning issues mentioned above. Furthermore, they are used to get a better understanding of the warning issues in European product liability law and in consequence the insights contribute to suggesting improvements for how European product liability law should deal with them. A toolkit of recommendations is provided for courts and litigants with regard to how to deal best with these warning

---

1 See *e.g.* Wogalter & Laughery 2006; Laughery 2006; Argo & Main 2004; Wogalter, Conzola & Smith-Jackson 2002; Rogers, Lamson & Rousseau 2000; Cox III e.a. 1997; Stewart & Martin 1994; DeJoy 1989; Lehto & Miller 1986.

2 See Lehto & Miller 1988.

3 Note that in chapter 1 consideration has been given of the potential limitations attached to the value and use of the findings of the warning research literature in a legal setting.

issues within the defectiveness test of the Directive's liability regime. My recommendations are presented alongside the discussion of the above-mentioned warning issues.

4.2 What is a Warning? Legal Analysis of the Legal Meaning of the Term 'Product Warning'

4.2.1 Introduction

At the outset of this concluding chapter, it is necessary to agree upon what constitutes a product warning according to European product liability law, since interpretation difficulties may be present. A range of terms are being utilised in relation to product information and product liability: duty to warn, failure to warn, warning defect, instruction fault, instruction defect, marketing defect, presentation of the product and so forth. Their names differ, but their meanings can overlap. This can give cause for confusion. And how do product warnings fit in? Clarification of what constitutes a product warning in a legal sense and what is meant by a warning defect can contribute to the assessment of warnings under the Directive's liability regime.

This paragraph is organised as follows. To disentangle the legal meaning of product warnings, paragraph 4.2.2 starts with an introduction of the different types of product information within the context of European product liability. In paragraph 4.2.3, a suggestion is made with regard to what should fall within the definition of a product warning by reference to the insights from the warning research literature. The next paragraph (paragraph 4.2.4) addresses the meaning of defectiveness in relation to warnings and suggests rearranging the warning defect category into product information defects. After discussing the legal meaning of the term 'product warning' and 'warning defect', paragraph 4.2.5 describes my recommendation to distinguish between 'warning defect' and 'duty to warn'. Finally, paragraph 4.2.6 provides a summary of the recommendations that were made in the previous paragraphs.

4.2.2 Disentangling the Place of Product Warnings in Product Liability

4.2.2.1 Categories of Product Information

Product information is a term that covers different forms of information, such as text or symbols, and various distinctions can be devised to categorise product information.[4] Product information can for instance be classified on the basis of when the information

4 Cf. Van Aken 1996, p. 93. See also CEN/CENELEC Guide 11 Product information relevant to consumers – Guidelines for standard developers 2006 for categories of product information.

is most useful to provide: before purchase (*e.g.* price, product qualities, such as the level of energy consumption[5] or food ingredients[6]), or after sale and before use (*e.g.* user instructions), or during use (*e.g.* product recall information); whether the information relates to safety (warning messages) or not (information on maintenance and cleaning); whether it is supplied with the product (on the label, on the package or a loose owner's manual) or not (*e.g.* via advertisements in newspapers and on TV, websites).

Naturally, there is no strict division between these categories; they can overlap (see figure 2). Product information that does not relate to safety (*e.g.* information about the specific qualities and characteristics of the product) is usually provided or needed before the purchase of the product. Similarly, information that relates to the safety aspects of a product generally becomes relevant after having purchased the product and before using it. Note though that in some instances safety information is in fact needed to make a well-informed purchase decision, such as the warning on the package of toys that informs buyers about the suitability and safety of a toy for their child.

---

5 See Proposal of the Commission of 13 November 2008 for a Directive of the European Parliament and of the Council on the indication by labelling and standard product information of the consumption of energy and other resources by energy-related products 2008, a proposal on labelling of energy-related products, such as household appliances. The aims of the Directive include contribution to the empowerment of consumers by providing them with useful and comparable information on the use of energy so that they can make better choices.

6 See Proposal of the Commission of 30 January 2008 for a Regulation of the European Parliament and of the Council on the provision of food information to consumers 2008, a proposal that introduces certain general principles regarding the provision of food information. One of the objectives is to contribute to consumer protection by providing appropriate information that enables consumers to make informed, safe, healthy and sustainable food choices.

Figure 2 Categories of product information

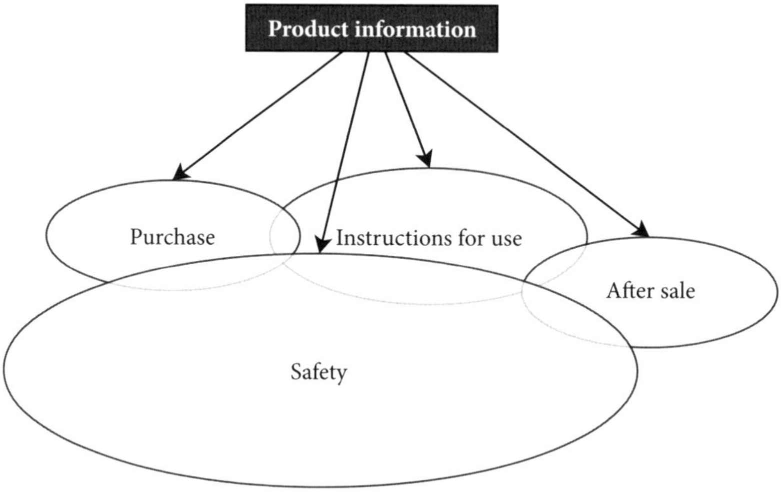

4.2.2.2 Product Information and Product Liability

The Directive uses the term 'presentation of the product' in relation to product information. This is also a broad term. Accordingly, Member States have interpreted this term widely by arguing that the packaging, the advertisements or other promotion material, and also warnings and instructions for use fall within this meaning of this factor. Some countries have even made explicit references in their implementing laws as regards to what falls within the scope of this factor.[7]

This expression does, however, not mean to imply that all product information is of relevance for the assessment of defectiveness; only in the event that it influences the safety expectations of the general public in such a way that the product does not provide the level of safety they are entitled to expect. More specifically, the Directive stipulates that just those risks that led to damage with respect to personal injuries or death and to property are recoverable.[8]

It thus seems that only product information that can influence the safety expectations of consumers needs to be considered in the judgement of defect. One can think of advertisements or other promotion material in which positive statements are made about how safe the product is or advertisements that emphasise a specific product characteristic.[9] This

---

7 *E.g.* the English and Italian implementing laws (see Campbell 2007, p. 189, 308).

8 Article 9 stipulates that commercial loss and damage to the defective product itself fall outside the scope of the Directive.

9 See *e.g.* the Italian case discussed in Rajneri 2005, p. 71. In this case, the victim suffered personal injuries because the front fork of his mountain bike broke whilst he was cycling up a mountain. The Tribunal held the manufacturer liable because the product was advertised as an off-road bicycle. Consequently, the victim's

distinction in product information still covers a lot and it raises the question whether product information that can influence consumers' safety expectations can be equated with the term 'safety information', namely, information related to the safety aspects of a product. At first glance, it would seem that these descriptions are interchangeable. Although this is often the case, there are situations possible in which they do not overlap. Think of quality marks. Such marks can have nothing to do with the safety of the product. However, its appearance on the package of a product may lead consumers to think that it does indeed relate to safety and that the product is safer than it actually is. If personal damage has occurred as a result of this information, it should be possible to commence a legal action under the provisions of the Directive.

Safety information that can certainly play a role in determining defectiveness is in the form of product warnings. Claims related to warnings are a common form of litigation in product liability cases. Popular assertions in this respect are that the absence of a warning rendered the product defective, or that defectiveness was caused by the presence of an inadequate warning. The meaning and interpretation of product warnings is discussed in more detail below.

4.2.3 *Recommendation: Wide Legal Interpretation of the Term 'Product Warning'*

4.2.3.1 General

The term 'warning' is frequently used in discussions about the defectiveness of the product or the duty of due care of producers, but what does it mean? Is there a single legal definition of product warnings? And what is its correct interpretation? Is it an umbrella term for information that is related to the safety aspects of a product, including the safety instructions (*i.e.* a wide interpretation)? Or does it have a narrow scope and does the word 'warning' only refer to informing people that there is a hazard?

This unwanted vagueness with regard to interpretation raises questions like: when information about the hazard is present, but the associated safety precautions to prevent injury are absent, should the injured party allege that the product is defective due to the absence of safety instructions, or should he/she allege that the *warning* is inadequate because it misses important information about the safety instructions?[10] If European product liability law would be based on a clear distinction between visual warnings (only

use of the bike on an inaccessible mountain road could be considered normal use. Another example concerns the early advertisements that promoted the use of cigarettes and suggested that smoking was safe.

10 As for fault-based liability, should the injured party claim that the producer failed to instruct or that the producer violated his duty to warn adequately? See also the third recommendation which explains that fault-based terminology in relation to warnings should be separated from strict liability as much as possible.

conveying information about the nature of the hazard) and safety instructions for use (that explain what to do or not to do in order to avoid damage), then the lack of a warning has to be classified as a warning defect as a result of this distinction (or as a failure to warn under fault-based liability), and the absence of the necessary safety instructions, such as in the above-mentioned example, as an instruction defect (or a failure to instruct under fault-based liability). On the other hand, if the definition of a product warning would be interpreted widely under the law, then it follows that the warning itself can be inadequate and defective, because there is already warning information available, but the information is incomplete.

4.2.3.2 Warnings as Safety Communications

In view of what is said above, there is in my opinion a need for more transparency in European product liability law as regards the scope of the legal term 'product warning' or 'warn'. Neither the text of the Directive, nor a decision of the ECJ defines what constitutes a product warning in legal terms. The precise content of these terms is surrounded by vagueness, which in consequence can complicate the legal assessment of warnings.

A good source of inspiration for finding a proper legal definition concerns the warning research literature. Warning researchers have expressed that it is not easy to come up with an appropriate definition of a warning, given the different types of warnings that abound in the world. As shown in paragraph 3.3, several definitions have nonetheless been suggested by them. One definition in particular appeals to me. This definition of warnings that is provided by Wogalter and Laughery is simple yet gets straight to the core.[11]

With my supplement of the word 'relevant', product warnings should subsequently be defined as follows:

> *A product warning is a safety communication; it is intended to provide relevant information about the product hazard so that undesirable consequences can be avoided or minimised.*

Against this background, the upcoming paragraphs of paragraph 4.2.3 discuss in more detail the aforementioned recommendation to legally define warnings in such a way. On the basis of this definition, two types of warnings can be distinguished that fall within the scope of this definition (paragraph 4.2.3.3). Secondly, this definition entails that three different information categories should be communicated within a visual product warning (paragraph 4.2.3.4). These are information about the (nature of the) hazard, the potential consequences and instruction information. In other words, it has a wide interpretation.

11 See Wogalter 2006a, p. 3; Wogalter & Laughery 2006, p. 889, 891.

The remaining part (paragraphs 4.2.3.5-4.2.3.8) is devoted to explaining the reasons for proposing the wide legal definition of product warnings.

4.2.3.3 Types of Warnings: Safe Use Warnings and Purchase Warnings

In reference to the aforementioned legal definition, two types of warnings can be distinguished. Self-protection against a product's hazard usually comes down to the decision of heeding a given warning. However in some cases, the choice in reality entails the decision not to use the product at all if one does not want to expose oneself to the risk of getting harmed. In this regard, it may be useful to distinguish (though they are not necessarily separate) two types of warnings.[12]

The first type of warnings is present when warnings communicate information that allows users to handle the product in a safe manner; damage resulting from the risk is in principle avoided by complying with the warning. 'Handling' the product safely means that users need to refrain from or produce certain behavioural actions to avoid the risk. In consequence, these warnings can be called *safe use warnings*.

> A half-litre bottle of a fizzy drink containing the statement: 'Warning. Contents under pressure. Cap may blow off causing eye or other serious injury, point away from face and people, especially while opening.' belongs to the first type of warnings, because it informs users how to use the product in a manner without the risk of getting harmed.
> Examples of this type of warnings are also overly present in instruction manuals of electrical appliances. Look inside your kettle's or toaster's user manual – if you have kept it at all – and you are bound to find information like 'Warning. To prevent electric shock unplug before cleaning. Unplug from outlet when not in use and before cleaning. Allow to cool before putting on or taking off parts.' and 'Do not use the appliance if the plug, the main cord, the base or the appliance itself is damaged'.

The second type concerns warnings that allow consumers to decide on the basis of the information whether to use or consume the product or not. Because the risk is unavoidable during use – even if one is awfully circumspect during use –, complete protection from harm boils down to not using the product at all. Even though the risk cannot be avoided during use, the safety level of such a product has usually been accepted by the public because of the product's benefits.

12 In addition, another type of warnings, namely recall warnings or after sale warnings, can be distinguished. These warn consumers of the risks discovered after having put the product into circulation. These warnings also relate to information about hazards to avoid accidents. Nevertheless, they are of a different nature and require a different format and content than normal safe use warnings.

This type of warnings can best be linked to products that warn of inherent, unavoidable hazards or risks during intended product use, such as alcohol or cigarettes.[13] In addition, it also concerns products with warnings of risks that are only unavoidable for a specific group of people, such as allergic consumers, children, diabetics and pregnant women informing them that the product is not suitable for them because of their skills, age, gender, sensitivity and so forth. Since avoidance of the risk means not using the product, this type of warning information is especially important for the purchase decision of potential users. This explains why such warnings are defined here as *purchase warnings*

> Package inserts of drugs generally warn of possible side effects that may occur as a result of using the drug, such as the risk information that blood clots are the most common serious side effect of birth control pills and that the use of birth control pills have a higher incidence of blood clots.
> Warnings against the risk of cross-contamination on pre-packed food labels such as the statement 'This product is made on a production line that also handles sesame' and 'This product may contain traces of nuts' can also be qualified as purchase warnings, as allergic consumers use this information for deciding in light of the risks and benefits of the product whether to expose themselves to the risk and to consume it or not.

The US Restatement of the Law (Third), Torts also draws a distinction between types of warnings and sets forth a definition of informed choice warnings and risk reduction warnings.[14] I believe that the distinction made above is for a large part compatible with this one. Risk reduction warnings can be equated with safe use warnings and purchase warnings largely overlap informed choice warnings. A difference that immediately emerges is that purchase warnings not only include informed choice warnings, such as those that accompany pharmaceuticals, but also warnings that inform that a product use is not suitable for certain persons, because this specific group cannot avoid the risk during use, not even with utmost care. Another possible difference may lie in my narrow interpretation of 'product use' and 'unavoidable'. According to the US Restatement of the Law (Third),

---

13 Damage resulting from such risks has been defined in the European legal literature as system damage, see also paragraph 2.4.4.

14 US Restatement of the Law (Third), Torts: Products Liability, § 2(c), comment i (American Law Institute 1998, p. 30). The Restatement explains that in addition to warnings that inform users and consumers of product risks so that they can reduce the risk of harm by appropriate conduct during use or consumption (risk reduction warnings), there are informed choice warnings. Informed choice warnings are needed to inform users and consumers of non-obvious and not generally known risks that unavoidably inhere in using or consuming the product. They allow the consumer or user to avoid the particular risk by making an informed decision not to purchase or to use the product at all and hence not to encounter the risk. See also Owen 2008, p. 613; Henderson & Twerski 2000-2001, p. 15.

Torts, 'informed choice' warnings are mainly reserved for toxic agents and pharmaceuticals. I prefer to use the term 'informed choice' warnings only for products that unavoidably expose users to an inherent risk during normal use, that is, use as intended by the product, excluding the risks that can happen during behavioural conduct that can be typed as 'misuses'. Those risks are truly inherent because exposure to the risk whilst using the product as intended is unavoidable. For example, kitchen knives have an inherent risk of cutting the user, but when consumers use it as intended, the risk of harm is in principle avoided. Only when one is exposed to a moment of carelessness, the knife can cut you. This behavioural action should in this regard be considered a (foreseeable) 'misuse'. It may be more problematic for cleansing agents. Because direct contact with the hazard can happen easily and is not far off from normal use, the risk increases. However, such products are not intended to be used on bare skin. Damage to the skin can be avoided by wearing rubber gloves. This is normally recommended in the warning message. So, I believe such a warning is a safe use warning and not a purchase warning or an informed choice warning, because even though the risk is inherent, users can prevent materialisation of the risk by being careful.

As noted by commentators such as Henderson & Twerski, it is important to consider whether a distinction in types of warnings aids courts in adjudicating warning defect claims. Accepting a distinction in types of warnings raises the question whether the types demand a similar or dissimilar treatment by courts with regard to the assessment of the adequacy of the warnings. It must be remarked that although the types of warnings described here are distinct, I believe that both communications can be considered warnings since, by their very nature, they contribute to safety. Admittedly, this is less obvious for purchase warnings, but it is argued here that minimising the use of products carrying a purchase warning provides some level of protection against damage. Nonetheless, because they differ, it can be asserted that the assessment of the adequacy of the warning should be in light of the type of warning. The actual assessment is explained in further detail in paragraph 4.6.

4.2.3.4 Warnings Generally Communicate Three Important Information Categories

On the basis of the legal definition given in paragraph 4.2.3.2, the following is also recommended. European product liability law should in my opinion treat a visual product warning of a specific hazard as a *message* that – in principle – communicates three information categories about that specific hazard (see figure 3).

A product warning or a duty to warn against a hazard owns its existence to informing users about the nature of the hazard(s) attached to the product. Obviously, in many instances, warnings indeed contain this component of hazard information. But, there is more information related to that hazard that should in my opinion generally form part of a visual warning. These categories concern information about the consequences of the hazard, such as the severity of the hazard (consequence information) and information about how to avoid or reduce the hazard (instruction information).[15] Thus, having regard to the aforesaid definition of warnings as safety communications, these three information categories give meaning to the expression of 'relevant' information.

Because warning studies generally indicate the finding that severity of the possible harm plays a greater role than the injury probability in the formation of hazard perception for most consumer products, this latter information category has not been included as a fourth relevant category. Researchers have for this reason focused less on determining to what extent this information category can influence warning compliance. Further research may provide a clearer answer. Nonetheless, this information category should not be set aside as it can still be needed with respect to certain types of products. For example, for products that are substantially different from products that are encountered on a regular basis, information about the probability of experiencing a negative outcome can positively steer hazard perception and their decision to comply.[16]

15 Wogalter & Laughery 2006, p. 898.

16 See paragraph 3.6.7.2.

Figure 3 Safety information

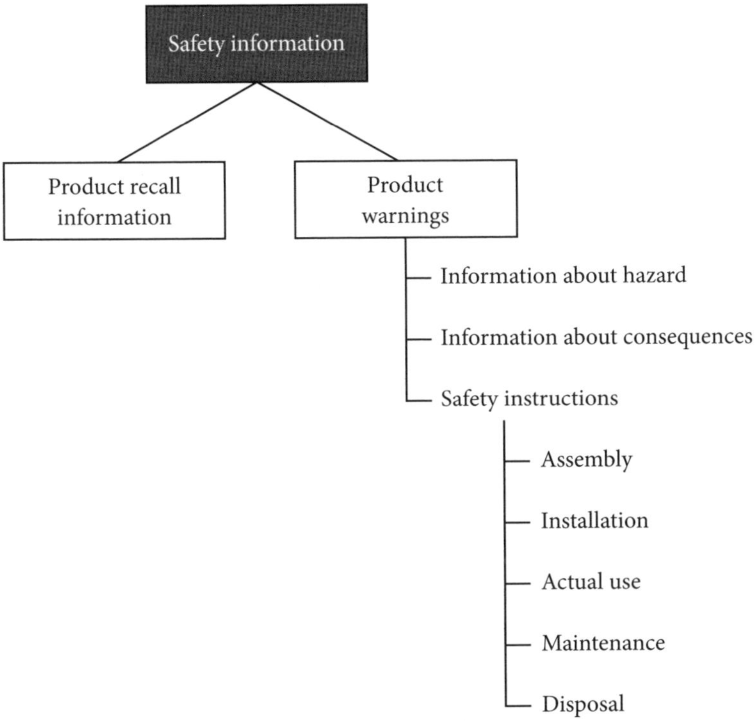

Because the definition of product warnings has a wide scope, it simultaneously says something about the adequacy of printed warnings. After all, the definition entails a requirement with regard to the informational content of a warning. Usually this is communicated in text, but images such as symbols and pictorials can also be useful. Bear in mind that this definition says nothing about the way in which the content should be expressed or how the outward appearance of the warning should be designed.[17]

Another comment is that it is important to realise that instructions for use are not equal to safety instructions for use. As has been pointed out sharply by warning researchers, instructions may or may not concern safety (see figure 3).[18] I assume that academics meant to refer to the safety instructions for use when they write about instructions for use, since these generally influence the safety expectations of people and are thus relevant for determining defectiveness.

17 See paragraph 3.6.6.6 that explains that specificity of information exists at two levels: the number of information categories that are present in a warning and the way in which the information is expressed. See also paragraph 4.6.

18 Wogalter & Laughery 2006, p. 890.

What's more is that instructions may or may not concern actual use of the product (see figure 2). Often instructions relate to how to operate a product, but not all. Products experience various stages during their life cycle and use is only one of them. A basic life cycle of a product can be divided into the stages of creation, distribution, use and disposal.[19] The stage of use can further be divided into actual use of the product by consumers, but also the assembly, installation and maintenance (repair and cleaning) of the product. 'Use' is thus interpreted widely here. The GPSD 's definition of 'safe product' also takes account of the several stages of a product's life cycle.[20] In view of these stages, product hazards do not only have to relate to the actual use or operation of the product. Safety instructions for use can also be associated with avoiding hazards during assembly, installation, and maintenance.[21]

> Suppose you have bought cut flowers. These are usually accompanied by a packaged powder that needs to be mixed with water. It reduces the number of harmful bacteria in the water and increases the probability that one can enjoy the flowers for a couple of days. The package can contain instructions for use that tell you how to cut the stems of the flowers (on the diagonal, not with scissors) and how to use the powder (with ice-cold or lukewarm water). These instructions have nothing to do with safety. Nevertheless, when the package would contain the information that people need to be careful of the thorns when cutting the flowers, it relates to safety and it can be considered a safety instruction for use. Note that this is not a plea for the necessity of giving this safety instruction: it concerns an immediately obvious hazard that needs no warning.

As a general legal rule warnings need to contain these three information categories, but not always. The warning research literature provides important exceptions to the rule of including the three aforementioned information categories in a warning. An information

19 The stage of creation can be divided into design, manufacturing, and packaging; the stage of distribution into transportation, storage, distribution; the stage of use into assembly, use, cleaning and repair; the stage of disposal into disassembly, transport and waste disposal. See Van Aken 1996, p. 25.

20 The main text of Article 2(b) GPSD makes reference to the conditions of use, which can include putting the product into service, installation and maintenance of the product. Point (i) of the article stipulates to consider the instructions for assembly, installation and maintenance. Point (iii) also mentions to take into account the instructions for use and disposal. This may imply that instructions with regard to putting into service, assembly, installation and maintenance fall within the scope of instructions for use and that instructions with regard to disposal form a separate category. Irrespective of whether disposal falls under the scope of the stage of use or whether it is viewed as a separate stage next to use, a distinction in stages emphasises the relevance of identifying the hazards in all stages of the product life cycle.

21 Cf. CEN/CENELEC Guide 11 Product information relevant to consumers – Guidelines for standard developers 2006, p. 8; Grubb & Howells 2007, p. 727.

category can be omitted: (1) when the information category is obvious; (2) when such information already exists in the target audience and a reminder warning is not necessary; or (3) when the content can be readily inferred from the other information component(s) given.[22] These exceptions should also be applied to warnings in product liability cases. Thus, this legal approach to the definition of warnings supports the view that warnings should only contain hazard-related information that is of specific relevance to the target audience of a warning.[23]

> The following examples illustrate the exceptions. It can be argued that the instruction information 'Avoid contact with skin and eyes' on a bottle of shampoo is sufficient for consumers to know that there is a risk of harm if used internally. Explicit reference to the presence of the danger of irritation seems unnecessary, because this information can be inferred from the precaution. In a similar vein, it can also be asserted that it is general knowledge that shampoo contains ingredients that can be irritating to your eyes.
> The warning research literature uses the example of 'Wet floor'. Although this warning only contains hazard information, it can be considered enough because most people already know what the consequences can be and what actions are needed to avoid the danger.

The recommended wide definition does in all probability not bring about a shift in the core issues that need to be assessed by a court. It does, however, take away any semantic vagueness that may be present with respect to whether a claim should be formulated as defectiveness due to the absence of a warning or due to the inadequacy of the provided warning. The lack of clarity is caused by the misleading situation in which there is warning information present, but necessary information to deal with the hazard safely is missing. The wide definition of warnings demonstrates that a warning is more than just a communication of a hazard. And as a result, the absence of one of the information categories with regard to a specific hazard should be viewed as defectiveness based on warning inadequacy. A lack of any information related to a hazard should therefore be regarded as defectiveness due to the absence of a warning message. Obviously, a claim can also be founded on warning inadequacy by alleging that the provided information is not sufficiently noticeable or understandable and so forth.

22 Laughery & Smith 2006, p. 421; Wogalter & Laughery 2006, p. 898. The concepts of obviousness and knowledge are discussed in more detail in paragraphs 4.4.6 and 4.4.7.

23 See also Article 5(1) GPSD.

4.2.3.5 Implication of the Warning Research Literature

The main reason underlying the approach – that visual product warnings should legally be viewed as a message consisting of three basic information categories unless one or more can be omitted – is that findings of warning studies have indicated that these information categories are essential for making informed safety decisions about a hazard.[24] Because of this, warning researchers have advocated that the content of a warning message should typically contain hazard, consequences and instructional information. Including these categories generally makes a warning comprehensible and subsequently enhances the likelihood that warnings are complied with.[25] American non-mandatory standards, such as the ANSI Z535 series, also provide this requirement with regard to the content, though the standard's underlying premise for providing this information seems to be more based on promoting a single uniform graphic system for providing safety information rather than promoting the design of effective warnings.[26]

4.2.3.6 Guidance for Producers

A second explanation for this approach to warnings is that it provides guidance to producers with regard to what hazard-related information should normally be present in a warning according to European product liability law. This reduces the probability that producers make incorrect assumptions about the knowledge level of the target audience. It has been reported that, precisely because warning designers are experts, they have a tendency to overestimate the knowledge level of warning receivers.[27] Hence, proceeding on the legal standpoint that warnings are messages that generally need to contain the three basic information categories reduces the probability that manufacturers leave out essential information.

4.2.3.7 Brevity and Testing

Another reason for this viewpoint towards warnings is that it may encourage producers to take into serious consideration several principles that are significant for warning design. These design principles are brevity of a warning message, testing a warning and employing a warning system.[28]

The reason for assuming this effect is that the approach to warnings not only emphasises communication of the essential information categories, it also necessitates producers to leave information out of the warning when it is not relevant to the target audience. The

24 See *e.g.* Wogalter e.a. 1987; Vredenburgh e.a. 2005.

25 Laughery & Smith 2006; Wogalter 2006a, p. 5; Leonard, Otani & Wogalter 1999, p. 154; Wogalter & Laughery 2006, p. 898.

26 Young e.a. 2006a, p. 450.

27 See paragraph 3.6.6.2.

28 These principles were discussed in the previous chapter and here in more detail in the section "How should consumers be warned?" of this chapter.

proposed approach and European product liability law both recognise that there are different types of warning information such as hazard information and instruction information. However, the starting points from which a warning is viewed differ. By requiring that – in principle – warning messages should contain these three information categories, the presence of the information categories is originally treated as a given. Producers do not need to focus on what types of information should be disclosed anymore. Consequently, their activities can shift towards analysing what information categories can be omitted. Asking this question is vital for the design of effective warnings, since *brevity of a warning* is designated as a key design principle of warnings.[29]

Brevity is important, because warnings that contain a flood of information can have a negative impact on effective information processing.[30] If European product liability law would view warnings as messages that should always contain the three information categories, then this could encourage producers to produce warnings that have excessive information. Moreover, it could discourage them from contemplating what the appropriate informational content of a warning should be. This would certainly not facilitate the effectiveness of warnings. Therefore, by emphasising that superfluous warning information needs to be omitted in view of the principle of brevity, it is expected that producers are triggered to investigate whether an exception is applicable to the warning at hand. This investigation aims at finding out how much hazard-related knowledge the target audience possesses, if an information component is obvious or whether information can be inferred from the other information categories. This can be done by testing the comprehensibility of the warning design on a representative sample of the target audience. This principle was described in paragraph 3.6.11. Testing is recommended as it can provide answers to whether a producer's assumptions about the competence of the target audience were correct and thus whether the producer can rightly omit certain information.

4.2.3.8 Brevity and Employing a Warning System

The approach of viewing warnings as messages also aims at encouraging producers to consider other means to transmit the warning information effectively. In the event that producers are confronted with making a trade-off between brevity and including all relevant information that may lead to an information overload, producers should consider to divert to additional means to communicate the hazard-related information. This relates to the view that exists in the warning research of using a warning system. As discussed in paragraph 3.6.10, a warning system refers to the different means and formats that can be used to transmit a warning message. It mirrors the notion that a warning message is more than just the traditionally used printed warning on the product's label.

29 See paragraph 3.6.6.6 of the previous chapter.

30 This potential problem of an information overload is discussed in more detail in paragraph 3.6.5.6.

Multiple modalities (*e.g.* visual and auditory warnings) and media (*e.g.* product labels, television commercials, media campaigns) can be used to communicate the information. If there is a lot of information that needs to be transmitted, the warning on the label can be accompanied with one or more additional means of communication such as a package insert, a manual, a digital medium like a DVD. Research findings generally indicate that presenting warnings in two modalities (such as a visual warning combined with an oral warning) is better than one modality.[31] Furthermore, including an alternative product label to a product, such as tags, has shown to be a solution to deal with limited space on product labels of small containers.[32]

4.2.4 *Recommendation: Defectiveness in Relation to Product Warnings*

The part 'warning' within 'warning defect' or 'warn' within 'duty to warn' has been dissected and its legal meaning has been illuminated above. It refers to all warning information that is relevant about the hazard, including the safety instructions to avoid the hazard. My second call for more transparency relates to the legal scope of the words 'defect' and 'duty' within the word combinations of 'warning defect' and 'duty to warn'.

To assess whether a producer can be held liable for a product that caused damage, it has been customary to undertake the approach of pointing out the alleged cause of the negligent conduct of the producer or the deficiency in the product. This has resulted in a traditional categorisation of product defects that has provided the parties involved, judges and scholars a handy tool to get a better grip on the liability assessment, both under fault-based and strict product liability. The type of defects most referred to is the tripartite consisting of design defects, manufacturing (or production) defects and warning or instruction defects.[33]

If you take a closer look at the European product liability literature, several definitions circulate, which suggests that a consensus on how to title this category of defects is lacking.[34] For example, in Grubb & Howells 2007, Howells refers to the category of failure to warn/instruct in addition to design defects and manufacturing defects, and does not use the expression of warning defects.[35] Miller & Goldberg 2004 speak of defects in warnings or instructions. Sometimes warnings are boxed in the category of marketing defects. The

31 See paragraphs 3.6.3 and 3.6.9.6.

32 See paragraphs 3.6.5.4 and 3.6.9.3.

33 After-sale defects or post-marketing defects can be considered a fourth category. These defects arise after the marketing of the product. This category is not of relevance to the liability regime of the Directive since Article 7(a) requires that the product was already defective at the time of putting it into circulation.

34 It must be noted that even though different names circulate with regard to the defect category relating to product warnings, vagueness is only of temporary concern seeing that it fades away after dealing with the actual content of the matter.

35 See also Stoppa 1992.

US Restatement of the Law (Third), Torts and the American literature commonly speak of warning defects or defects in instructions for use.[36]

I prefer the use of one umbrella term in European product liability law that encompasses all situations that relate to warning information and damage, and that is the term *warning defect*.[37] From a linguistic viewpoint, this term is more neutral and indicates to cover both the scenario of the absence of a warning and the presence of a warning, whilst the expression of 'failure to warn' can be interpreted to mean that only the situation in which the absence of a warning caused damage is covered. Note that because my recommended legal definition of warnings is wide, there is no need to distinguish between defects in warnings or defects in instructions for use.

As regards the scope of the term 'warning defect', it is logical to assign a wide scope to the term 'defect' within 'warning defect', because it is immediately associated with the question of the ways in which consumer products in relation to warnings can be unsafe. Warning defects refer to two scenarios: (1) the absence of a warning and (2) the presence of a warning. The former concerns situations in which the absence of a warning allegedly rendered the product defective. The latter concerns situations in which the warning that has been provided was inadequately designed and seemingly caused the defectiveness of the product. In addition, a third category of warning defects can be distinguished in respect of scenario 2. This category concerns the situation in which a warning has been provided as a risk-reduction measure, whilst the use of this precautionary measure can be viewed as inappropriate given the level of risk involved. This category reflects that warnings form part of the design of a product and that warnings can thus be viewed as one of the design methods that aim at risk reduction. It emphasises that producers should not misuse warnings, they should be treated as a last-resort measure to reduce risk. Usually such a situation is brought under the scope of design defects, which is logical, because the absence of a safer adequate design feature is the core matter of discussion rather than the content and form of the warning. Hence, to better express the hierarchical order with regard to design methods for risk reduction, I prefer to denominate these situations and title them as 'producer's misuse of a warning'. This warning defect type is discussed in more detail later on in paragraph 4.5.

Moreover, as shown above, various types of product information can play a role in determining product defectiveness. As a result, other defect categories besides warnings can be made in relation to product information. To provide a useful tool to decision makers in their process of discovering the defective nature of product information, I advise referring to *product information defects* next to design defects and manufacturing defects.

36 American Law Institute 1998, p. 14. See *e.g.* Owen 2008.

37 It could also be possible to refer to product information defects, which can be subdivided into marketing defects and warning defects, and possibly other defects that relate to product information but not to warning information and marketing information.

These product information defects would then include warning defects, marketing defects and other possible information defects that influence the safety expectations of the public at large (see figure 4 below).

Figure 4 Product defects

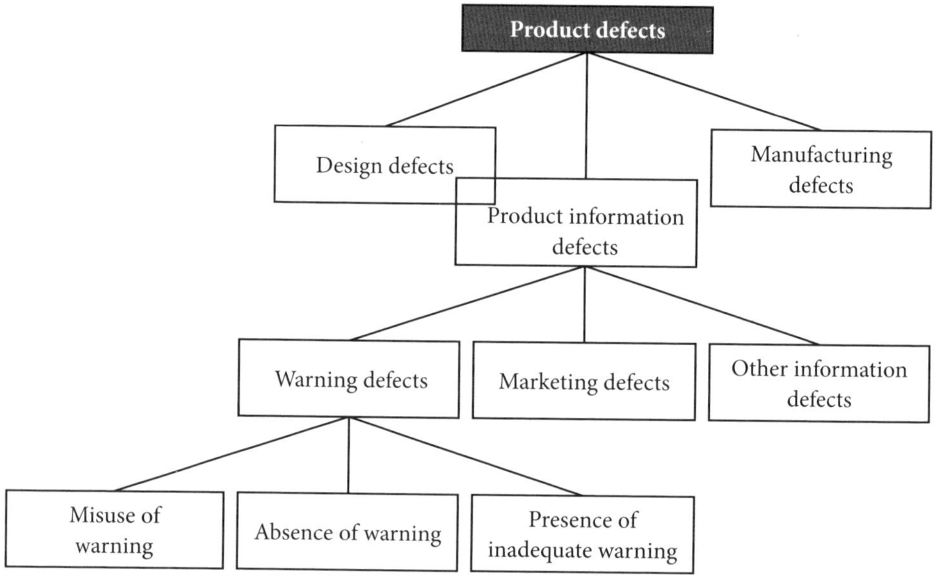

4.2.5 *Recommendation: Maintain Linguistic Distinction between Warning Defect and Duty to Warn*

A final comment with regard to the use of these terms relates to the theoretical and linguistic difference between warning defect and duty to warn. Theoretically, the terms of fault, duty to warn and failure to warn have their origin in fault-based liability. They refer to acts and omissions of the producer and what he did wrong, for that is the cornerstone of fault-based liability. On the other hand, warning defect, or instruction defect and marketing defect are terms that belong in a regime of liability without fault like the Directive, as defect typically concerns the characteristics of the product, such as the presence or absence of a warning message.

In spite of this theoretical divergence, it seems that these terms are regularly used interchangeably by courts and in literature.[38] The reason for not adhering to this distinction is that in practice the assessment of warnings under negligence and the regime of the Directive does not really differ from each other.[39] It is difficult and perhaps even impossible, to determine whether a product is defective because of the absence of a warning or the presence of an inadequate warning, without taking into account whether the producer could and should have provided a warning or whether he could and should have designed a better warning or implemented a better design solution.

Apart from the question whether a clearer distinction needs to be made between the liability standards of the two theories, it is argued here that it is at least desirable to maintain this distinction between the liability theories in language as much as possible. The underlying reason is that the provisions of the liability regime of the Directive and fault-based liability diverge on several aspects and that should be emphasised by staying true to the language of the ground of liability.

So, using the expression whether the producer has a duty to warn under the provisions of the Directive should be avoided; courts and parties involved should formulate this in terms of whether the absence of a warning was the cause of the defectiveness. Nevertheless, within the assessment of whether a warning was needed or whether the printed warning was sufficiently adequate, courts can take into account if the harm could and should have been avoided by the producer.

> The Dutch *Betonmortel* case is used to illustrate the recommendations made above. In this case, the producer of concrete fluid only provided information about the nature of the hazard (in text, *i.e.* the word 'irritant', and a symbol of a cross). Information about the safety instructions for use (*e.g.* wearing protective clothing that avoids contact between the liquid and the skin) and information about the consequences (severe skin burns) were lacking.[40]
> First of all, it can be criticised that the District Court used language of negligence by referring to producer's duty to warn in the context of the standard of defectiveness of the Directive laid down in Article 6:186 DCC. The court ruled that in this specific case the producer did not comply with his *duty to warn*, at least he did not comply with his duty to warn adequately. Regrettably, the claimant also used terms that stem from fault-based product liability. The injured party alleged that the defendant *failed to warn* against alkaline and the

38 See *e.g.* [2002] *EWHC* 490 (*Bogle v. Mc Donald's Restaurants*); [2000] *PIQR* 95 (*Worsely v. Tambrands Ltd*); Rb. Zwolle 24 april 2002, *Praktijkgids* 2002, 5921 (*Mini-tampon*); Hof 's-Hertogenbosch 15 mei 2007, *LJN* 2007, BA6838 (*Nagelstyling*); Hof Arnhem 14 oktober 2003, *NJF* 2004, 46 (*Datafan*).

39 See paragraph 2.4.1.

40 See paragraph 2.4.5.

irritating characteristics of concrete fluid and then argued that in the event that the use of concrete fluid entails that certain precautions need to be taken, the defendant should have informed him of this.
Secondly, the description of the claim by the injured party can be criticised in view of the proposed legal definition of warnings. There was warning information present about the nature of the hazard. Therefore it cannot be treated as defectiveness due to the absence of a warning, or in fault-based language as a failure to warn. It does seem that the court tried to correct this by adding that the defendant did not comply with his *duty to warn adequately* instead of the claimant's formulation that the defendant failed to meet his obligation to warn. Here, it concerns a failure to instruct had he interpreted warnings in a narrow sense. Pursuant to the desired definition of product warnings proposed above, the injured party should have claimed that the product is defective due to the presence of an *inadequate warning message*. The inadequacy consisted of the absence of safety instructions for use. There was too little information available for him to use the product safely. He needed instruction information, for example in the form of the statement 'Use impermeable clothing to protect your hands and legs against burns'. Because of his low knowledge level with regard of this hazard, which relates to him being an unprofessional buyer of the product, it was not possible for him to infer the necessary information from the information that was given.

4.2.6 Conclusion

Three recommendations were made in this paragraph that relate to how product warnings should be interpreted legally and how the term should be used in relation to defectiveness under European product liability law.

In sum, 4.2.3 proposed to legally define visual warnings as safety communications: they are intended to communicate relevant information about the hazard so that undesirable consequences can be avoided or minimised. 'Relevant' refers to conveying different important hazard-related information categories. Warning messages should generally contain the basic information regarding the type of hazard, the consequences of the hazard and the safety instructions, unless one or two information categories can be omitted. This shows that the term 'warning' is interpreted widely here (see figure 3). The exceptions with regard to omitting an information component are that the information component is obvious, generally known or can be inferred from another category.

Four main reasons were given to explain why European product liability law should interpret visual product warnings widely. A key reason is that the research findings and literature on the effectiveness of warnings support the notion that the content of each warning should contain these information categories. These are in principle prerequisite for enabling users to protect themselves against the hazard. Secondly, it provides clarity to injured parties with respect to which starting point to choose when they want to claim compensation. In the event that one of the categories is missing, the adequacy of the warning message has been possibly undermined and the claim should consequently be based on product defectiveness due to the presence of an inadequate warning. When no warning information at all is given, it should be argued that the absence of a warning message renders the product defective. Thirdly, this definition guides producers with respect to the informational content of a warning. Lastly, it may trigger producers to consider testing the content of the warning on comprehensibility to ascertain whether one or two information categories need to be omitted and whether other means to communicate the information need to be included. Producers may also be encouraged to employ a warning system to ensure that the warning information that is relevant and considerable in amount can be communicated effectively. These activities are important because they can secure that the principle of brevity of a warning message is considered sufficiently.

Furthermore, it was recommended in paragraph 4.2.4 to use the categories of product information defects, design defects and manufacturing defects under the Directive, since warning information is not the only form of product information that has the ability to influence safety expectations, which embodies the defectiveness test under the Directive. Product information defects subsequently consist of warning defects, marketing defects and other potential product information defects that can influence safety expectations. I also recommended using the term 'warning defects' to refer to defects in relation to product warnings. Warning defects can then be subdivided into product defectiveness as a result of (1) the absence of a warning; (2) the presence of an inadequate warning; and (3) the inappropriate use of a warning as a precautionary measure whilst a higher level of safety was needed (see figure 4).

The final recommendation in paragraph 4.2.5 advocated for maintaining the linguistic distinction between 'warning defect' and 'duty to warn', since these terms refer to distinct legal theories.

4.3 Why Warn? Legal Rationale of Product Warnings

4.3.1 *Introduction*

Closely related to the discussion of what the term 'product warning' means in European product liability law is the issue of why product warnings are necessary according to law. The basic question of why warn is the second warning issue that is examined here. Before discussing the main questions of when to warn and how to warn, it is fundamental to take a closer look at what is the underlying rationale of warning of potential product hazards and for requiring them in European product liability law. There is some information available on how EU law views product warnings, but not a great deal. This paragraph focuses on what would be a plausible underpinning for setting out product liability requirements that pertain to product warnings. In consequence thereof, it is recommended here that European product liability law should focus on the legal goal and subgoals of product warnings, because this exercise is eventually helpful for framing the legal test of warning adequacy.

In this paragraph I will deal with the following. First, I identify the most plausible legal rationale for product warning duties (paragraph 4.3.2), then I connect legal rationale with the empiricism by asserting that the preventive aspirations of the law are well conceived: effective warning does actually reduce injury rates and therefore there is firm legal justification for warning duties in product liability law (paragraph 4.3.3). In paragraph 4.3.4, I discuss the goal and subgoals that product warnings should fulfil according to European product liability law. This recommendation is derived from the implications that can be drawn from the C-HIP model. Finally, paragraph 4.3.5 provides the closing part of this paragraph.

4.3.2 *The Legal Rationale of Warning Requirements*

4.3.2.1 Safety and Accident Prevention

Earlier, we saw that warnings can be defined as 'safety communications', as they communicate important information that relates to the hazard so that undesirable consequences can be avoided or reduced. Given the aforementioned definition of warnings as safety communications in paragraph 4.2, it is a trite observation that the law has burdened producers with duties concerning these 'safety communications'. It has been an established fact for decennia in product liability that producers have some responsibility to warn consumers of the hazards attached to their products. The legal responsibility primarily lies with producers, because they design, manufacture and put the products into circulation. They receive the benefits and they should consequently also carry the burden. Furthermore, producers are experts in their field. They have (or ought to have) essential knowledge with regard to the risks attached to their products, compared to product users who have no or less information about the risks.

The legal duty to warn descends from the general duty of due care under fault-based liability. It has long since been reflected in tort law that verbal and visual warnings are in essence means to protect potential victims against damage (personal or property). They form a part of a product that can influence the safety expectations of consumers and consequently, they influence the safety level of the product. Good warnings and instructions can render an otherwise unsafe product safe, whilst inadequate warnings and poor instructions may be reasons why a product is deemed unsafe.[41] Hence, they are treated as 'precautionary' measures, actions that need to be taken in order to prevent something dangerous or unpleasant from happening. Usually, there are several precautionary measures that can be taken to deal with a certain risk in a dangerous situation. Courts treat warnings as one of these precautionary measures on the basis of which care can be exercised in respect of the risk. The extent and content of the duty to warn is developed under case law.

Although both the legislative documents at EU level and the legal literature are far from clear-cut on the rationale for duties to warn in product liability, it seems that the underlying principle of treating warnings as precautionary measures in European product liability law is accident prevention or reduction. Communicating warning information enables potential victims to protect themselves against the harm so that unnecessary accidents that might otherwise occur without the presence of the warning can be prevented or reduced. In other words, by accompanying the product with safety information an acceptable level of product safety can be achieved, which safety level would not have been that high in the absence of the warning.

4.3.2.2 Autonomy and Informed Choice

Requiring product warnings by law has also been associated with individual autonomy and having an informed choice. Informed choice is tied to the legally accepted notion of free will, which refers to the individual autonomy to have a choice in what you do, in this case the choice whether or not to protect yourself against the danger warned against. The informed decision of self-protection usually comes down to following the warnings' safety instructions. However in some cases, the choice entails the decision not to use the product at all if one wants to avoid the danger.[42]

From the viewpoint of individual autonomy, it can be asserted that requiring warnings by law contributes to a society in which people can take responsibility for their own decisions. Warnings enable consumers to freely asses and manage the product risks. Hence, if those individuals who have been educated with adequate warning information still want to behave in a risky way, they ought to have the freedom to do that. In such cases, liability is however usually denied. On the basis hereof, it has been argued in the legal literature that by accompanying products with adequate warning information that allows the con-

41 Grubb & Howells 2007, p. 357; Hodges 1993b, p. 103.

42 See paragraph 4.2.3.3 which deals with the distinction between two types of warnings.

sumer to decide whether to expose himself to the hazard, the responsibility for safe use shifts from producer to consumer.[43]

Informed decision making can, in turn, be connected with the 'right to know' of the risks attached to products. It can be said that EU public law that mandates labelling obligations, *e.g.* with regard to chemical information, comes from the idea that EU citizens have a right to know about certain risks and therefore need to be provided with information so that they can protect themselves. This right has not been expressly acknowledged as such in the field of EU product liability law, but traces of its underlying thought could be found in case law where victims allege that the producer is liable because without a warning of the risk, the product should be considered defective. Presumably, a reason for not accepting this 'right to know' as a true consumer right that needs to be protected by tort law on an overall basis, is that it can and should not always prevail. It can conflict with the tort rule that a tortfeasor is not obliged to give up dangerous activities, such as the marketing of a product, merely because there are risks involved. Risks are to a certain degree acceptable in society. Acts and omissions are wrongful, only when the risk exceeds the socially accepted level of safety.[44] If the probability of the risk is trivial, taking precautionary measures, including warnings, is generally not obliged and a finding of liability is as a result absent. This means that there may be cases where consumers in fact do not have the right to receive information, *e.g.* because of the absence of a significant product risk. This limit to product liability is in my opinion justified, especially in view of the efficacy of warnings. Paragraph 4.4, discusses this matter in more detail.

4.3.2.3 Conclusion

It follows from the above that, even though legislative documents, judges and academics have hardly deliberated on the legal rationale(s) of warning duties, there are plausible, interrelated reasons that explain why European product liability obliges producers to provide adequate warnings. I believe that the main answer to the question of why warn lies in promoting safety and preventing/reducing accidents. The legal treatment of product warnings as one of the types of precautionary measures that can protect consumers from harm shows that European product liability considers protection against damage, and the resulting consequence of accident prevention or accident reduction, as the *main* underpinning for laying down warning requirements in product liability law. This treatment places the emphasis on safety and the prevention/reduction of accidents caused by the absence of an adequate warning.

---

43 See Howells, Janssen & Schulze 2005; Owen 2008, p. 585; Hodges 2005, p. 117; Viscusi 1995-1996, p. 625; Giesen 2005, p. 53.

44 Cf. Van Dam 2006, p. 199.

The other two explanations of individual autonomy and 'right to know' are interrelated with the main rationale. I consider them of secondary importance, because in the end both are concerned with contributing to product safety. Their angles differ however. Although the informed decisions pertain to product safety, the rationale of autonomy and informed choice places the emphasis on making autonomous decisions on the basis of the warning information and on consumers' responsibility for taking these decisions and their ensuing consequences. It is a logical choice to follow a warning, since acting in accordance with a warning corresponds to safe behaviour. But if under certain circumstances this does not suit you (*e.g.* because the prescribed way of usage seems too unpleasant or because the benefits of using the product weigh more heavily), you can behave differently. Furthermore, the explanation of having a 'right to know' relates to the explanation of autonomy and informed choice and subsequently to the main rationale, because just such a right allows people to make choices about self-protection.

In conclusion, a basic principle in European product liability is that, since warnings are treated as precautionary measures, accompanying products with warnings increases the safety level of products and they can even make a product acceptably safe, which level would thus be lower in the absence of warnings. European product liability law therefore requires that consumers need to be given warnings. This is not a 'right' that should always be confirmed in court. Product liability cases have shown that this depends on the circumstances of the case. It is therefore important to be mindful of the difference between consumers' right to know and producers' obligation to provide warnings. Moreover, European product liability also imposes the general requirement that the hazard-related information in a warning is adequate so that consumers are in fact able to protect themselves against the harm. The protection usually entails following the warnings' user instructions that enable consumers to avoid the risk during usage. The protection may also lie in the informed choice not to use the product anymore or to accept the risk and use it less frequently because of the unavoidable exposure during normal use. Precisely because protection partly depends on the behavioural actions of consumers, European product liability recognises that consumers also have a duty to take adequate actions to protect themselves. European product liability stipulates that negligent conduct on the part of the claimant can reduce producer's obligation to pay compensation or even worse, bar liability completely.[45] It can therefore be said that in principle adequate warnings lead to a shift in responsibility for safety from producer to consumer.

45 See *e.g.* the legal doctrine of contributory negligence in tort law, and also Article 8(2) of the Product Liability Directive.

4.3.3 The Added Value of Warnings

By treating warnings as precautionary measures to protect the safety interests of others, product liability law proceeds from the presumption that warnings can have a positive effect on behaviour. Before getting into a discussion on the functions that, in my viewpoint, product warnings should have in product liability law, it is important to find out if the underlying rationale of accident prevention can truly be upheld by empiricism. This finding may otherwise urge reconsideration of labelling accident prevention as the underlying legal rationale.

A first major lesson that can be learned from the empirical research on warnings is that product warnings can indeed change behaviour.[46] They can bring an added value to the situation to which they refer.[47] Meta-analyses of warning studies have shown that product warnings significantly increase the amount of safe behaviour.[48] In other words, the research findings generally indicate that the presence of a warning is better in terms of safety than providing no warning at all.[49] Bear in mind that this does not mean to say that there are no studies that have reported no behavioural effects or even adverse effects.[50] It also says little about how many people are expected to comply with warnings. This matter is discussed in more detail in paragraph 4.5.2.

It follows from the above that the behavioural presumption upheld in liability law can be affirmed by the overall findings of the warning studies. On the basis of this insight, it can thus be said that warnings are in principle useful protection mechanisms. Empirical evidence warrants their use as precautionary measures in European tort law.

---

46 Silver & Braun 1999; Kalsher & Williams 2006. See paragraph 3.6.9.1.

47 See paragraph 3.5.4.1.

48 Cox III e.a. 1997; Argo & Main 2004.

49 Silver & Braun 1999 provide an overview of studies that have examined the effects of the presence versus absence of warnings on behavioural intentions and compliance, see also Kalsher & Williams 2006, p. 315. Interestingly, an early quantitative and qualitative review of the warning research up to 1984 yielded the conclusion that there is virtually no reason to anticipate that warnings positively impact safety. On the basis of their findings, the authors questioned the usefulness of warnings as mechanisms to increase safety and reasoned that warnings are ineffective (McCarthy e.a. 1984). This conclusion has been criticised by others, who primarily contended that the conclusion was based more on an absence of empirical evidence than on the existence of non-supportive data (DeJoy 1989; see also Cox III e.a. 1997, p. 195; Young & Lovvoll 1999, p. 44; Wogalter, Fontenelle & Laughery 1985; Godfrey, Rothstein & Laughery 1985). Since then, various additional empirical studies with control groups have been conducted and they have demonstrated that warnings can modify behaviour and thus can be useful as safety mechanisms. Many of the experiments utilised the incidental exposure experimental paradigm consisting of the chemistry demonstration task which was introduced by Wogalter in 1987. Others used a different laboratory setting or have taken place in the field.

50 See *e.g.* Cox III e.a. 1997, p. 199.

4.3.4 *Recommendation: Distinction in Goal and Subgoals of Warnings*

Since we now have established that (1) warnings are legally required because they can prevent accidents and that (2) this presumption is justified because they *can* change behaviour and provide protection, it is essential for European product liability law to be clear about the legal goal that product warnings seek to achieve. Because only then it is possible to get a good answer to the successive question of what requirements need to be met in order to reach that goal. This information is in turn helpful to product liability as it can be guiding for constructing the legal test of warning adequacy.

The discussion on the legal rationale showed that accident prevention can be viewed as the main reason for having warning requirements in European product liability law. In a similar vein, if warnings are legally required because of their contribution to accident prevention/reduction, then it can be said that this forms the main *goal* or *function* of providing product warnings. As mentioned in paragraph 3.3.2, the warning research literature also generally agrees that warnings are ultimately intended to reduce or prevent injuries and other adverse consequences that might otherwise occur without their presence.[51] It considers this a 'function' or a 'purpose' of warnings on a general level. Furthermore, at a more concrete level, the researchers have stated that there are three additional functions/purposes that can be distinguished. These are: (1) informing product users about the hazards so that they can make better, more informed decisions about safety issues; (2) reminding product users who may already know the information but are not consciously aware of the hazard at the appropriate time; and (3) influencing or modifying behaviour by promoting compliance with the warning. The latter one is a concrete version of the former ultimate purpose of accident prevention.[52]

In view of what is just said, I recommend appointing accident prevention as the ultimate goal of warnings in European product liability law. That is *the* goal, and not so much informing. Rather, the information transmitted by a warning is the 'vehicle' to reach the destination of accident prevention; it is the instrument for achieving accident prevention (see figure 5).

51 In the warning research literature, researchers generally speak of 'purposes'. I assume that there is no reason to suggest that this cannot be equated with the word 'goals'.

52 Wogalter 2006a, p. 3; Laughery 2006, p. 469; Kalsher & Williams 2006, p. 313.

Figure 5 Ultimate legal goal of warnings

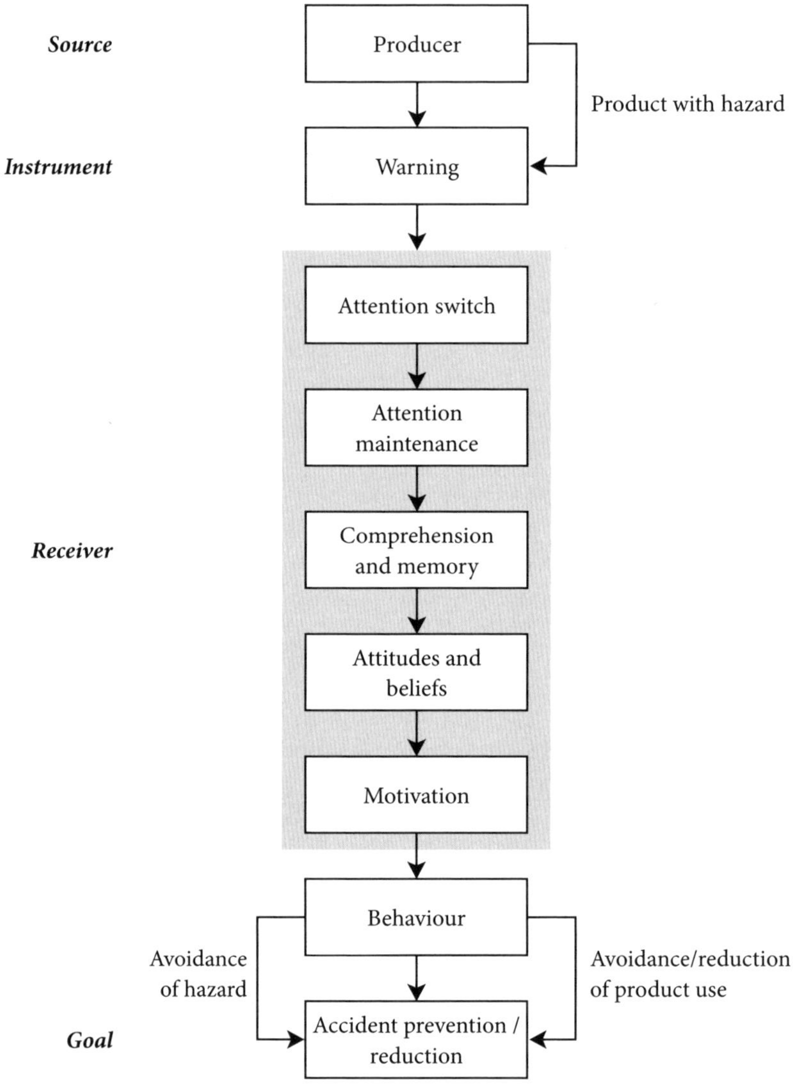

The warning information does need to satisfy a number of requirements before it can have an impact on the behaviour of people. These requirements can be found in the human information processing stages, as depicted by the C-HIP model.[53] As told in the previous chapter, the C-HIP model explains that behavioural compliance with a warning is the

53 For more information of this model see paragraph 3.4.2.2.

culmination of a sequence of several information processing stages within warning receivers. The information processing stages must all be successfully completed for the warning to end in safe behaviour. This implies that the behavioural effectiveness of a warning is determined by the success of each stage of the model. Hence, the information processing stages of the warning process can be viewed as requirements for effective warnings. For this reason, I consider each information processing stage as a *subgoal* that needs to be achieved in order to reach the ultimate goal of accident prevention. So, this means that a warning must be salient; legible, comprehensible and memorable, concur with important beliefs and attitudes and it must motivate so that people will actually behave in accordance with the warning (see figure 6 below).[54]

Having regard to the information processing stages of the C-HIP model, I argue here that the concept of distinguishing subgoals is useful for the legal test of warning adequacy and should play a vital role in determining when to shift the full legal responsibility for safety from producer to consumer. Even though European product liability law commonly treats warnings as protective measures, it appears that it places more emphasis on the warning's subgoal of providing comprehensible information and less on its ultimate goal of achieving safety. It presumes that by giving comprehensible information to warning receivers, product safety is sufficiently achieved. However, as shown by the C-HIP model and figure 6, providing a comprehensible warning is not sufficient for actual behavioural compliance with the warning. This raises the fundamental question concerning when can warnings be considered legally adequate so that the full responsibility for safety shifts from producer to consumer. Should the legal turning point on the line of responsibility for safety be situated at the subgoal of providing comprehensible information as now seems to be the case?[55] Or should it be at the next subgoal of warning information that concurs with important attitudes and beliefs,[56] or even the one after that? In other words, should the responsibility for safety fully shift from producer to consumers and is the legal test of warning adequacy subsequently fulfilled when the information can be considered to be comprehensible? This is discussed in more detail in paragraph 4.6, which deals with the main warning issue 'How should consumers be warned?'.

54 Furthermore, if one takes a closer look at the various stages, this parallel can also be drawn between the stages and the concrete purposes mentioned above by warning researchers. The purposes of informing and reminding correspond with the result of the stage of comprehension and memory. In other words, before a warning can inform and remind people, it needs to be seen (stage of attention switch), read (stage of attention maintenance). The other subgoal of influencing emphasises the subsequent intermediate stages between comprehension and behaviour. It is about persuading and motivating people to act safely. It can thus be argued that all information stages represent subgoals attached to product warnings, since all of them need to be successfully completed to get to the next stage and to ultimately end in safe behaviour.

55 As portrayed in the shift 1 of legal responsibility between producer and consumer of figure 6. The line consists of blocks that represent the subgoals illustrated above. The break represents the shift.

56 As portrayed in the shift 2 of legal responsibility between producer and consumer of figure 6. The line consists of blocks that represent the subgoals illustrated above.

Figure 6 Legal subgoals of warnings

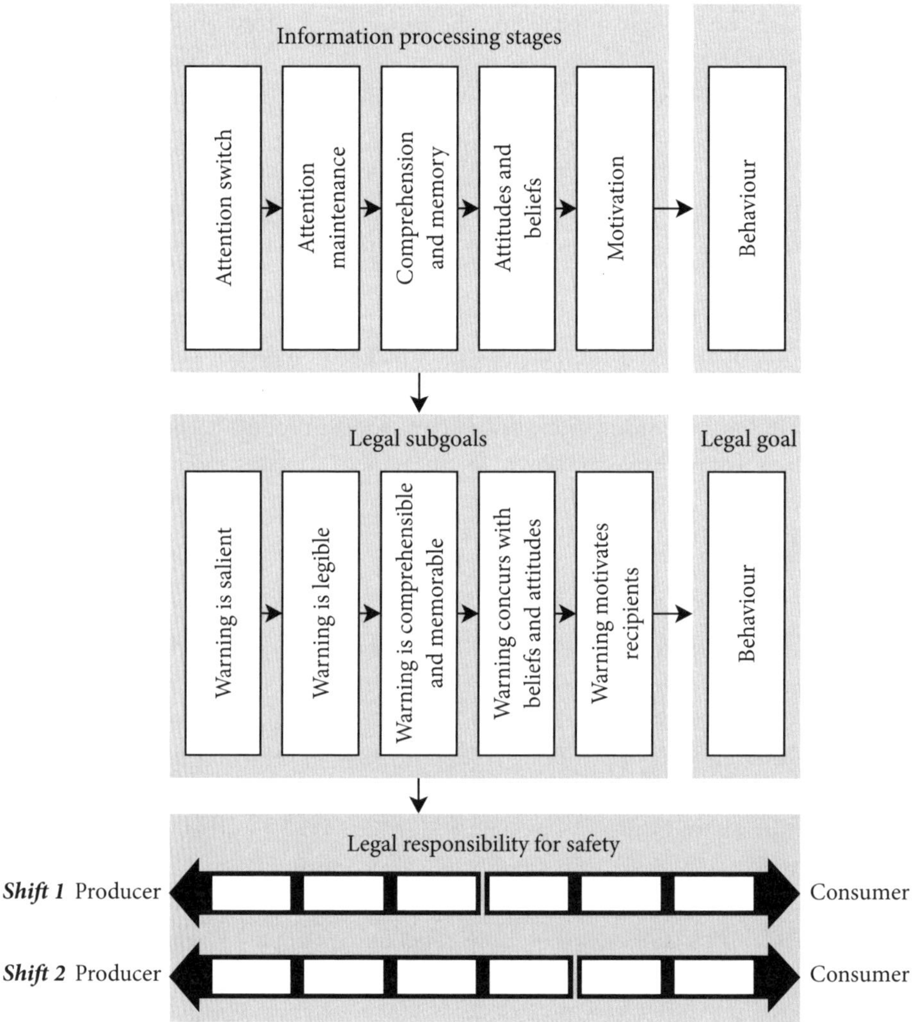

4.3.5 *Conclusion*

For a good understanding with respect to which requirements product liability should lay down for warnings, it is important to have a clear picture of why warnings are required by law in the first place. Unfortunately, there is a dearth on legislation, case law and legal literature that extensively discusses the underlying rationale and goal of imposing warning duties on producers. Nonetheless, paragraph 4.3.2 showed there is considerable support for the argument that the main underlying rationale for requiring warning duties in product liability concerns preventing or reducing the number of accidents that are caused by the absence of an adequate product warning. The reason for this argument is that warnings have traditionally been treated as precautionary measures to protect the interests of potential victims from damage. As concluded in paragraph 4.3.3, an important lesson that can be learned from the warning research studies in this respect is that this treatment is justified. Warnings researchers have done a number of studies that have led up to the general conclusion that warnings can indeed influence behaviour in a positive way. Empirical evidence thus warrants their use as precautionary measures in European product liability law.

Paragraph 4.3.4 showed that it is not difficult to see the link between the questions why warn and what is the goal of a warning according to law. In accordance with the consensus amongst warning researchers, it was recommended here to consider accident prevention or reduction as the ultimate goal of product warnings in European product liability law. Warnings should be viewed as means or instruments intended to accomplish prevention or reduction of unnecessary accidents that happen as a result of the presence of inadequate warnings. Equally, the information embedded in a warning is the 'vehicle' to reach the destination of accident prevention (see figure 5). For warnings to be adequately processed and end in safe behaviour, this information must meet a sequence of conditions. This can be explained by the C-HIP model. The C-HIP model describes behavioural compliance with a warning as the culmination of a sequence of several information processing stages within warning receivers. The information processing stages must all be successfully completed for the warning to end in safe behaviour. Hence, the information processing stages can be considered requirements for effective warnings. For this reason, I recommended in this paragraph to consider each information processing stage as a subgoal that needs to be achieved in order to reach the goal of accident prevention.

It was also argued in paragraph 4.3.4 that it is helpful for European product liability law to focus on what are the main goal and subgoals of product warnings according to law, since this exercise can uncover insights that can be useful for the legal test of the adequacy of warnings (see figure 6). Because the information processing stages can be viewed as subgoals, they can also be considered to give meaning to the determination of when warnings should be considered 'legally adequate' as a result of which the legal responsibility for safety shifts from producer to consumer. This raises the question at what point on the legal line of responsibility for safety there is a shift from producer to consumer. Even though product liability law commonly treats warnings as protective measures, it appears that European courts place much emphasis on reaching the subgoal of giving a comprehensible warning. Such a legal treatment presumes that by giving comprehensible information to warning receivers, product safety is sufficiently achieved. However, as shown by the C-HIP model and figure 6, more cognitive steps within the receiver or subgoals need to be completed to reach safe behaviour. It can thus be argued that these final two subgoals are currently not or not sufficiently considered by law. This deficiency may call for a re-evaluation of the legal test of warning adequacy. This is done in more detail in paragraph 4.6.

4.4 What Risks Need No Warning? Legal Analysis of the Limits to Defectiveness with Respect to the Absence of a Warning

4.4.1 Introduction

The next important warning issue that is addressed here concerns the question regarding what risks should a producer warn against according to European product liability and what not. This question is related to a common allegation in product liability, that is, the claim that the absence of a warning rendered the product defective, hereinafter also referred to as a 'no-warning claim'.

A basic starting principle of European product liability is that the absence of a warning does not by definition render a product defective. The reason for this is twofold. Evidently, not every single risk associated with using the product needs a warning, such as risks that are the result of abuse of the product or that were undiscoverable at the time the product was put into circulation. Secondly, a reason that may be less familiar in the legal literature yet important is that a warning is unnecessary because the risk has already been adequately controlled by other design solutions.

Authors have contended that the emphasis on the presentation of the product as a relevant circumstance for the defectiveness test of the Directive, can lead to an increased reliance on warnings by producers and even by courts.[57] As far as I know, the accuracy of this assertion has never been empirically tested, but one cannot escape the fact that there is an abundance of warnings that accompany consumer products. Naturally, this can be explained by the increase of information obligations under the EU consumer policy, but it may well be so that the implementation of the Directive's liability regime is also responsible. Furthermore, the fact that today's society has high expectations about the safety level of products may also contribute to this.

As was discussed in paragraph 3.8, ergonomists have warned about a proliferation of warnings and the potential detrimental effect it can have on safety. A number of potential problems associated with overusing warnings have been reported in the warning research literature. These concerns have also been mentioned by some in the legal literature. 'Overusing' refers to overloading consumers with too many warnings that accompany a

---

57 Clark 1989, p. 96; Miller & Goldberg 2004, p. 438; Hodges 1993b, p. 115.

given product or too many products with a warning compared to products that have none. Because of these problems, the warning research literature has advocated caution in the indiscriminate use of warnings. It is therefore of significance that European product liability does not elicit producers to 'overuse' warnings.

Two approaches are suggested in this chapter for European product liability law to reduce the number of unnecessary warnings and to subsequently deal with the negative consequences associated with 'overusing' warnings. My first suggestion is made in this chapter and concerns the recommendation for courts to apply limits to defectiveness for cases in which claimants allege that the lack of a warning pertaining to the materialised risk rendered the product defective.[58] By applying these limits and thereby denying defectiveness for the absence of warnings for these categories of risks, European product liability shifts the responsibility for safety from producer to consumer. The proposed limits are helpful to many decision makers in the legal arena. Even though some limits have been built into the Directive, interpretation difficulties with regard to these limits, together with previous case law, can cause uncertainty, which may result in producers having to warn anyhow and courts having to rule in favour of consumers whilst the warning was actually unnecessary. Consequently, the limits assist Member State courts in applying a consistent approach with regard to determining whether the absence of a warning leads to establishing defectiveness. The limits also help producers understand which risks need no warning. Thirdly, the limits can guide claimants with anticipating the successfulness of proceeding on the basis of a no-warning claim.

The second approach that aids in reducing the number of superfluous warnings is to legally counter the inappropriate use of warnings as a precautionary measure to control the risk. The recommendation to adopt the last-resort-measure principle of warnings under the Directive aims at decreasing the number of inappropriately used warnings, since this principle provides producers with a reason to implement design alternatives that are more effective in providing protection than warnings. This decrease, in turn, is likely to facilitate a reduction of unnecessary warnings in Europe. This second approach is dealt with in paragraph 4.5.

The structure of this paragraph is as follows: paragraph 4.4.2 starts with the tool of a risk assessment. As was discussed in paragraph 3.9, a risk assessment forms a key element in the step-by-step approach to the design of safe products, since this tool can assist in the decision making process of whether the risks associated with the product need to be reduced by design or not. This includes the decision whether warnings are needed, since warnings can be viewed as design measures that reduce a risk.[59] On the basis of this tool, hazards

58 Note that even though the product is not considered defective because of the absence of a warning, that does not automatically preclude a finding of defectiveness on other grounds, see paragraph 4.5.6.

59 See paragraph 3.8.

can be identified, accident scenarios can be predicted, and the injury probability and injury severity can be estimated. In consequence thereof, it is possible to make a selection of the risks that need to be managed by design and risks that do not. This tool can thus be helpful to producers for evaluating whether a warning is preferred when designing the product, and for courts in the aftermath of an accident whether it is legally justified to leave a warning out. This paragraph also pays attention to the (renewed) gained prominence of this tool in the product safety policy of the EU. After that, paragraph 4.4.3 sets forth the limits to defectiveness for no-warning claims. Five categories of risks are discussed in detail in paragraphs 4.4.4-4.4.8, of which I recommend that they do not need to carry a product warning. These are risks that have an insignificant size, risks arising from unreasonably expected use, obvious risks, generally known risks[60] and risks that were present yet undiscoverable at the time the product was put into circulation. The ways in which the steps of a risk assessment can be helpful to producers and courts is described in more detail here. It must be noted that these categories regularly overlap; a product risk that materialised in a certain case can fall within the scope of more than one category. The closing paragraph 4.4.9 provides a summary of the previous discussion.

4.4.2 *The Tool of a Risk Assessment*

4.4.2.1 European Developments

The decision of whether providing warnings is superfluous and thus not needed can be facilitated by means of carrying out a risk assessment. As was discussed in paragraph 3.9, the ergonomic literature advises to apply a systematic step-by-step approach to the design of safe products. A risk assessment forms a key element in this approach. The purpose of a risk assessment in consumer product safety is to provide a systematic judgement on the level of risk, which result can assist in determining whether design measures that reduce the risk are needed.[61]

It appears that these ergonomic considerations have not remained unnoticed in the legal arena of consumer safety. In recent years, it has been acknowledged in the European product safety policy that risk assessment forms an essential part in achieving a high level of consumer protection. Recent EU legislation with regard to European safety of food and non-food consumer products, such as the Machinery Directive, the REACH Regulation, the GFL and other important documents with regard to product safety, such as the ISO/IEC Guide 51 and the European Commission's decision with respect to the notification procedures of the GPSD take this tool into account. Some of these are discussed below.

60 A warning for generally known risks can be required under the limited circumstances in which a reminder is needed.

61 Rider e.a. 2009, p. 76.

In spite of the increasing popularity of this tool, subject experts from regulatory bodies and industry noted that there is a great level of discrepancy in nomenclature and divergence in approaches to risk assessment.[62] Initiatives, such as the working group EuroSafe Working Group on Risk Assessment (WGRA) and the working group to improve the RAPEX Guidelines on Risk Assessment, have been taken to deal with these problems.[63]

Machinery Directive

An important piece of legislation that illustrates the importance of a product risk assessment concerns the newly revised Directive 2006/42/EC, hereafter called Machinery Directive.[64] This Directive is mainly for industrial machinery, but consumer products can also fall under its scope.[65] Article 5 of the Machinery Directive stipulates that before placing machinery on the market, the producer must ensure that it satisfies the relevant health and safety requirements set out in Annex I.[66] It introduces the obligation for producers to carry out a risk assessment and a risk reduction.[67] Four general principles are laid down at the beginning of Annex I. The first principle describes the process of a risk assessment (first four dashes) and a risk reduction (fifth dash):

> GENERAL PRINCIPLES
> 1. The manufacturer of machinery or his authorised representative must ensure that a risk assessment is carried out in order to determine the health and safety

62 Rider e.a. 2009, p. 73.

63 EuroSafe stands for the European Association for Injury Prevention and Safety Promotion. EuroSafe is a non-governmental organisation, representing organisations and individuals working to prevent injury and to promote safety. This includes policies and actions for promoting child safety, consumer safety, safety for seniors, safety of vulnerable road users, safety in sports and the prevention of violence and self harm. One of the working areas in the consumer safety programme concerns the WGRA. An important partner in this network is the European Commission. For more information see <www.eurosafe.eu.com>. RAPEX refers to the Community rapid alert system for non-food consumer products.

64 Directive 2006/42/EC of the European Parliament and of the Council of 17 May 2006 on machinery, and amending Directive 95/16/EC (*OJ* 2006, L 157/24). The provisions of the new Machinery Directive became applicable on 29 December 2009. See also European Commission Enterprise and Industry 2010.

65 See *e.g.* for a reference to consumers recitals 3 and 15 and Article 1(k). Electrical and electronic products such as household appliances intended for domestic use are excluded from the scope of the Directive, insofar as they are covered by Directive 2006/95/EC of the European Parliament and of the Council of 12 December 2006 on the harmonisation of the laws of Member States relating to electrical equipment designed for use within certain voltage limits (*OJ* 2006, L 374/10) (hereafter called 'Low Voltage Directive'). This means that electrical gardening machinery or electrical power tools intended for construction and repair work in the home are not concerned by this exclusion and are subject to the Machinery Directive: European Commission Enterprise and Industry 2010, p. 53.

66 Article 5 also states that the manufacturer must ensure that there is a technical file available. According to Annex VII, this file includes documentation on the risk assessment and risk reduction.

67 See recital 23 to the preamble of the Machinery Directive. Article 5 also states that the manufacturer must ensure that there is a technical file available. According to Annex VII, this file includes documentation on the risk assessment and risk reduction.

requirements which apply to the machinery. The machinery must then be designed and constructed taking into account the results of the risk assessment. By the iterative process of risk assessment and risk reduction referred to above, the manufacturer or his authorised representative shall:
- determine the limits of the machinery, which include the intended use and any reasonably foreseeable misuse thereof,
- identify the hazards that can be generated by the machinery and the associated hazardous situations,
- estimate the risks, taking into account the severity of the possible injury or damage to health and the probability of its occurrence,
- evaluate the risks, with a view to determining whether risk reduction is required, in accordance with the objective of this Directive,
- eliminate the hazards or reduce the risks associated with these hazards by application of protective measures, in the order of priority established in section 1.1.2(b).

2. The obligations laid down by the essential health and safety requirements only apply when the corresponding hazard exists for the machinery in question when it is used under the conditions foreseen by the manufacturer or his authorised representative or in foreseeable abnormal situations. In any event, the principles of safety integration referred to in section 1.1.2 and the obligations concerning marking of machinery and instructions referred to in sections 1.7.3 and 1.7.4 apply.
3. The essential health and safety requirements laid down in this Annex are mandatory. However, taking into account the state of the art, it may not be possible to meet the objectives set by them. In that event, the machinery must, as far as possible, be designed and constructed with the purpose of approaching these objectives.
4. This Annex is organised in several parts. The first one has a general scope and is applicable to all kinds of machinery. The other parts refer to certain kinds of more specific hazards. Nevertheless, it is essential to examine the whole of this Annex in order to be sure of meeting all the relevant essential requirements. When machinery is being designed, the requirements of the general part and the requirements of one or more of the other parts shall be taken into account, depending on the results of the risk assessment carried out in accordance with point 1 of these General Principles.

Viewing the above-described steps, it becomes clear that the ISO/IEC Guide 51 has been a source of inspiration for laying down safety principles with regard to the design of safe machinery. This Guide provides standard writers with guidelines for the inclusion of safety aspects in standards and its underlying principles are used as a basis for European product safety standards and legislation with regard to non-food products.[68] The ISO/IEC Guide 51 defines risk assessment for non-food products as the overall process comprising a risk analysis and a risk evaluation. A risk analysis concerns the systematic use of available information to identify hazards and to estimate the risk. A risk evaluation is defined as the procedure based on the risk analysis to determine whether the tolerable risk has been achieved. Figure 7 depicts the process.

68 ISO/IEC 1999.

Figure 7 The iterative process of risk assessment and risk reduction, derived from ISO-IEC Guide 51 Safety aspects – Guidelines for their inclusion in standards 1999, p. 4

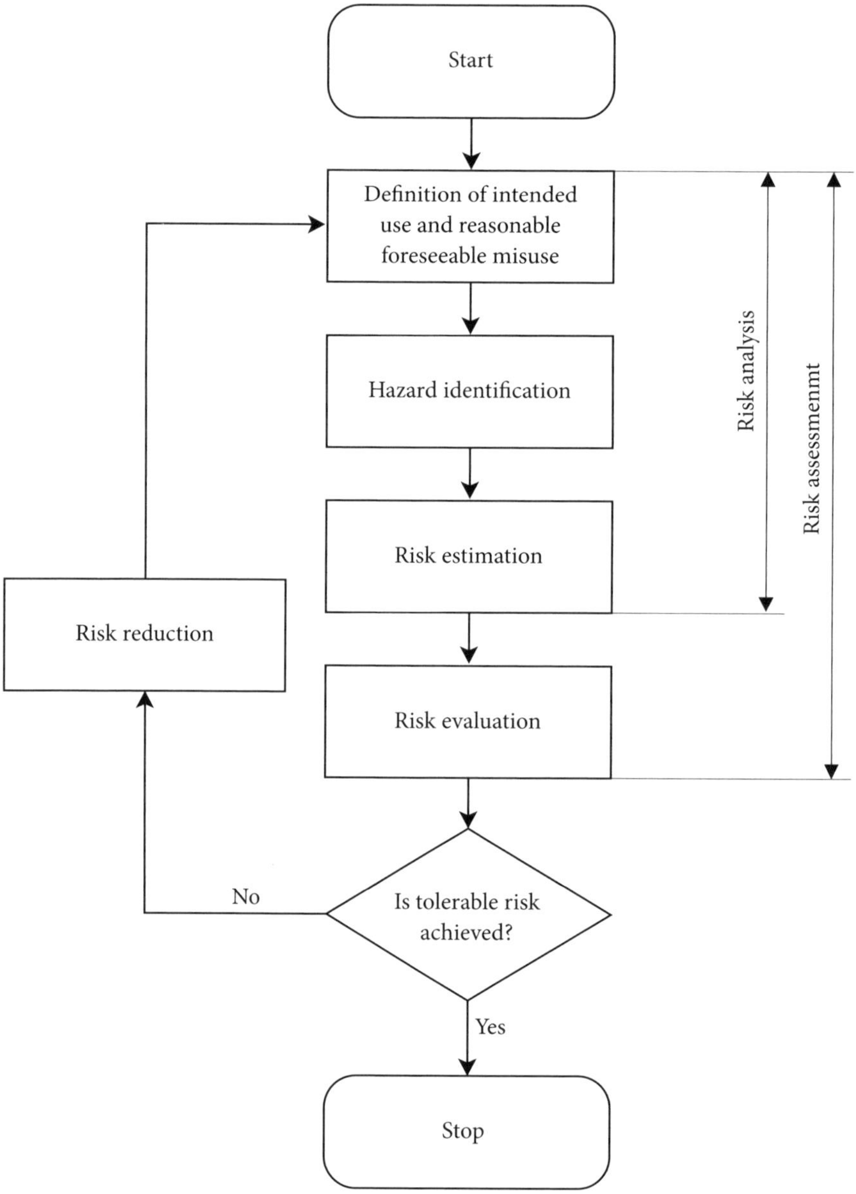

RAPEX Risk Assessment Guidelines

Another important document that exemplifies that risk assessment is high on the agenda in the context of European product safety is the European Commission's decision with respect to the notification procedures of the GPSD. In December 2009, the Commission issued the RAPEX Guidelines, a non-legislative decision, that lays down guidelines that facilitate the operation of the RAPEX notification procedure of Article 12 GPSD and the notification procedure of Article 11 GPSD.[69] The authorities of Member States are the addressees of the Commission's decision.[70] RAPEX involves the system of rapid exchange of information between the Commission and the Member States in the event of a *serious* risk in relation to non-food consumer products.[71] Article 12 GPSD establishes the notification procedure for the exchange of information in case of serious risks. The authorities of the Member States are obliged to notify the Commission through RAPEX of the measures, described in Article 8 GPSD, that are taken to prevent, restrict, or impose specific conditions on the possible marketing or use of consumer products that pose a serious risk to the health and safety of consumers and which risk also has a cross-border effect.[72] Article 11 GPSD involves the mandatory notification procedure with regard to measures adopted by Member States that restrict the marketing of consumer products posing a non-serious risk.[73]

Before an authority of a Member State decides to submit a notification, it is obliged to perform the appropriate risk assessment in order to assess whether a product to be notified poses a serious or a non-serious risk. Because risks can be assessed with many different methods which can lead to diverging outcomes, one of the objectives of the RAPEX Guidelines was to set out a transparent and practicable risk assessment method. Part IV of the RAPEX Guidelines lays down risk assessment guidelines for consumer products.[74] These RAPEX Risk Assessment Guidelines are the result of a revision. Because of difficulties with applying the old guidelines in practice, the European Commission called for a working group of Member State experts with the goal to improve the RAPEX Risk Assessment Guidelines on how to draft a risk assessment under the GPSD. Even though risk assessment remains a complex activity, the new risk assessment guidelines provide more clarity; the

69 Commission Decision 2010/15/EU of 16 December 2009 laying down guidelines for the management of the Community Rapid Information System 'RAPEX' established under Article 12 and of the notification procedure established under Article 11 of Directive 2001/95/EC (*OJ* 2010, L 22/1).

70 Article 3 of the RAPEX Guidelines.

71 RAPEX helps to prevent and restrict the supply of products posing a serious risk to consumer health and safety and facilitates the monitoring of the effectiveness and consistency of market surveillance and enforcement activities in the Member States; see p. 7 of the RAPEX Guidelines.

72 See also p. 8 of the RAPEX Guidelines.

73 See also p. 22 of the RAPEX Guidelines.

74 See p. 33-64 of the RAPEX Guidelines.

elements of the risk assessment have been clearly separated, step-by-step instructions and standardised lists and tables are given to the risk assessors.[75]

Risk assessment is defined by the RAPEX Risk Assessment Guidelines as the procedure for identifying and assessing hazards, consisting of three steps: (1) identification of the seriousness of a hazard; (2) determination of the probability that a consumer will be injured by that hazard; and (3) combination of the hazard with the probability.[76] In addition, pursuant to the Risk Assessment Guidelines, a *hazard* means a source of danger involving the chance of being injured or harmed. A means of quantifying a hazard in a risk assessment is the severity of the possible injury or harm.[77] A *risk* means the balanced combination of a hazard and the probability that damage will occur. Risk describes neither the hazard, nor the probability, but both at the same time.[78]

Its definition of a risk assessment resembles for a large part the latter definition of the Machinery Directive and the ISO/IEC Guide 51 (see figure 7), but the procedure is less extensive. It does not include the step of a risk evaluation. This step appears to be largely incorporated in table 4 of the RAPEX Risk Assessment Guidelines (see figure 8). This table combines four levels of severity of injury and eight levels of probability of damage and illustrates the degree of risk level that results from the combinations. The risk levels are serious risk, high risk, medium risk and low risk. It can be said that these risk levels embody an evaluation of the risks on the basis of which follow-up actions can be considered.[79] The step of taking follow-up actions aims at reducing or eliminating the risk. The RAPEX Risk

75 For more information, see the documents on the website: <http://ec.europa.eu/consumers/safety/committees/index_en.htm#rawg>.

76 See p. 64.

77 See p. 64 of the RAPEX Guidelines. Likewise, Annex I of the Machinery Directive defines 'hazard' as a potential source of injury or damage to health. The ISO/IEC Guide 51 defines it as a potential source of harm. See also paragraph 3.9.3. The RAPEX Guidelines (p. 56-60) also provide a table of groups of product hazards that can be used as guidance to identify the specific hazards intrinsic to a non-food product. Product hazards can relate to size, shape and surface, potential energy, kinetic energy, electrical energy, extreme temperature, radiation, fire and explosion, toxicity, microbiological contamination, and product operating hazards.

78 See p. 64 of the RAPEX Guidelines. Annex I of the Machinery Directive defines 'risk' as a combination of the probability and the degree of an injury or damage to health that can arise in a hazardous situation. The ISO/IEC Guide 51 defines risk as the combination of the probability of occurrence of harm and the severity of that harm (and harm as physical injury or damage to the health of people, or damage to property or the environment). The injury that a hazard can cause can have different levels of severity, from superficial temporary injuries to permanent losses or fatality. The RAPEX Risk Assessment Guidelines distinguishes between four levels. Table 3 (p. 61-63) provides a list of examples of types of injuries that can differ in level of severity. Among these types of injury are: laceration/cut, bruising, concussion, entrapment/pinching, sprain, dislocation, fracture, crushing, amputation etc.

79 The previous chapter (3.9.4) also discussed a way to quantify and evaluate the level of risk. This matrix describes four injury levels and six levels of qualitative probability. On the basis of combining these two parameters (multiplying the values of the levels), a matrix can be created that shows the relative urgency to control a risk. A difference between the two tables is that the RAPEX table has more levels of probability, which makes the measurement more precise.

Assessment Guidelines refer to this procedure as risk management, which corresponds with the step of risk reduction in the model of the Machinery Directive and the ISO/IEC Guide.

Figure 8 Table 4 of the RAPEX Risk Assessment Guidelines[80]

Probability of damage during the foreseeable lifetime of the product		Severity of injury			
		1	2	3	4
High	> 50 %	H	S	S	S
	> 1/10	M	S	S	S
	> 1/100	M	S	S	S
	> 1/1000	L	H	S	S
	> 1/10 000	L	M	H	S
	> 1/100 000	L	L	M	H
	> 1/1 000 000	L	L	L	M
Low	< 1/1 000 000	L	L	L	L

S - Serious Risk
H - High Risk
M - Medium Risk
L - Low Risk

4.4.2.2 Insights for Courts and Producers

The tool of a risk assessment, such as laid down in the Machinery Directive, ISO/IEC Guide 51 and the RAPEX Guidelines, is of significance for all parties concerned with product safety. It must be stated here courts are no risk assessors. That does not mean however, that this tool cannot be of assistance to courts when dealing with product liability issues. It can provide insights to Member State courts when they need to judge if giving a warning was necessary to render the product not defective, as it shows how producers can design safe products, how a risk assessment for consumer products can be properly prepared and what basic principles or steps need to be taken into account by producers.

As has become apparent from the discussion above, the Member State authorities are the addressees of the RAPEX Risk Assessment Guidelines. Even so, the content of this decision of the EC still cannot provide useful information to producers and Member State courts. The RAPEX Risk Assessment Guidelines provide understanding of how – according to the European Commission – Member State authorities should assess risks to evaluate

80 See p. 64.

whether a particular consumer product should be restricted or prevented from the European market. Although these regulators start a risk assessment to determine the need for action with regard to products that have been put on the market, there seems to be no reason to argue that this risk assessment method cannot be applied by producers to evaluate product designs before putting them into circulation, since they have the legal responsibility to design safe products. The RAPEX Risk Assessment Guidelines offer a transparent and practicable method of risk assessment that has been accepted on EU level. The risk assessment method is build up in small manageable steps that are described in detail. The RAPEX Risk Assessment Guidelines describe what points have to be taken into account and what questions have to be asked when preparing a risk assessment. It is admitted that there is no 'one size fits all' approach to risk assessment. Certain products need specific guidelines on how to prepare a risk assessment, for instance machinery, cosmetics and chemicals.[81] Nonetheless, the basic principles of a risk assessment remain the same. In this regard, producers can also take account of the basic principles of a risk assessment identified by the EuroSafe WGRA. [82]

In addition, the RAPEX Risk Assessment Guidelines offer some guidance with respect to making the succeeding decision of whether risk-reduction measures need to be adopted, and if so what measures.[83] The principle underlying the decision is that of proportionality. The European Commission states that serious risks may include withdrawal from the market, but lower levels of risk normally lead to less rigorous measures (*e.g.* adding/changing a warning).[84] The Commission remarks that in the end, there is no automatic link from risk to action. However, factors that can be regarded as relevant for this decision are the degree of exposure of the population to the product as a whole (compared to the individual risk), the severity of the injury (fatal) even if the likelihood of such accidents is extremely unlikely, the availability and difficulty of a design change, public perception of the risk, the type of product and the effectiveness of the action.

Furthermore, even though the table (figure 8) only classifies the level of risk and does not give a definite answer about whether risk reduction is necessary, this classification can still be considered useful because of its implicit evaluation. What is interesting in this respect is that the view with regard to the evaluation of a risk appears to have been altered during the revision process of the RAPEX Guidelines. A draft of the Risk Assessment Guidelines from 2007 also contained a table combining the four levels of severity of injury

81 See p. 39 of the RAPEX Guidelines. There are REACH guidance documents that provide supplementary information to the regulation and that help to fulfil the obligations, for example with regard to the chemical safety assessment.

82 It introduces the elements which should be incorporated in any risk assessment, since according to these experts, their model on risk assessment seems most suitable to deal with consumer products, see Rider e.a. 2009.

83 See p. 47 of the RAPEX Guidelines.

84 See Article 8(b) and (c) GPSD.

and the eight levels of probability of damage, but with a difference.[85] The categories representing the risk levels that resulted from the combinations diverged. The risk levels were serious risk, moderate risk, low risk and acceptable risk. So, the new risk level of low now corresponds with the old risk level of acceptable, the new risk level of medium with low, high with moderate and serious with serious respectively. In consequence, it could be argued that the evaluation of whether a risk can be considered acceptable (and requires no risk-reduction measures) is raised to a higher level. Redefining a low risk into a medium risk indicates that this degree of risk now urges consideration of whether follow-up actions to reduce the risk need to be taken, whilst taking action would have seem to be less likely when it was defined as a low risk. In addition, the evaluation can also suggest to courts that risks that fall within the level of a low risk, can be considered acceptable and need no further risk-reduction measures.[86]

4.4.3 *Recommendation: Limits to Defectiveness for No-Warning Claims*

4.4.3.1 Five Risk Categories

Below, in paragraphs 4.4.4-4.4.8, the five risk categories of which I recommend that they do not need to carry a product warning are discussed in depth. Hence, they embody limits to defectiveness with respect to claims that centre on defectiveness as a result of the absence of a warning. The risk categories are:

(1) insignificant size of risks;
(2) risks arising from unreasonably expected use;
(3) obvious risks;
(4) generally known risks;
(5) risks undiscoverable at time of putting the product into circulation.

The limits to defectiveness are helpful to many in several ways. First, they provide European courts with tools that allow them to assess in a consistent manner whether warnings are not required under the defectiveness test. Secondly, the limits make these conditions under which liability is denied more transparent to the interested parties. Producers need to have knowledge of how courts will decide, so that they are confident in not using a warning for every risk. These limits thus provide producers with useful information for anticipating no-warning claims in the future, or in the aftermath of an accident for successfully counteracting the victims' no-warning claims. Furthermore, the limits can guide claimants in alleging the appropriate cause of the defect. Hence, the limits teach them that in the event

85 For more information see the website: <http://ec.europa.eu/consumers/safety/committees/ra_guidelines_workshop11122007.pdf>.

86 See also paragraph 4.4.4.

of the presence of an obvious risk, it would be wiser to argue that the absence of a safety device/guard rendered the product defective rather than the absence of a warning.

As you can see, these risk categories have been linked to a certain quality of a risk. Bear in mind, however, that the categories normally overlap; a risk can fall within the scope of more than one category. Risks arising from unreasonably expected use are usually generally known or obvious.

If a danger is obvious, it usually implies that the behavioural action leading up to damage, is also obvious or generally known and is in consequence often related to unreasonably expected use. Such a risk can thus be considered both obvious and related to unreasonably expected use that does not need a warning. Similarly, risks that are obvious are usually also well-known. Moreover, the insignificant size of the risk plays a key role in determining whether a warning can be abandoned, as all categories are related to a judgement of the size of the risk at hand.

I believe that within these categories, two groups can be distinguished. First, some risks may not lead to defectiveness on the basis of a lack of a warning, but may be held defective on the basis of another alleged product defect, such as the absence of a more effective design solution. Obvious risks and generally known risks fall within this group. The second group of risks concerns risks that completely fall outside the scope of liability. In the event that a risk is of insignificant size, arises from unreasonably expected use, or is undiscoverable at the time the product was put into circulation, it is argued here that no alternative design method in whatever form is required by the Directive. Accordingly, it may be more appropriate to speak of limits to liability rather than limits to defectiveness with regard to the latter group of risk categories.

4.4.3.2 Rationales for Not Requiring Warnings: General Considerations

The rationale for not requiring a warning for these risks is threefold. First of all, these limits are in line with the Directive's policy goal of creating an appropriate balance between the interests of producers and consumers. By setting these liability limits and thereby denying defectiveness for the absence of warnings of these risks, European product liability shifts the responsibility for avoiding this risk from producer to consumer. Even though from the viewpoint of consumer safety, it could be argued that all risks associated with using the product need to be controlled by producers, no matter how remote, unexpected, known and/or obvious, this reasoning is unjust and not endorsed by the EPLD. As pointed out by Burton J, the test of the Directive does not require an absolute level of safety, nor embodies an absolute liability for any injury caused by the harmful characteristic.[87] Furthermore, the responsibility of consumers is a concept that is part of the Directive's liability system. Although it is difficult to determine this relative safety level that consumers are

87 [2001] 3 All ER 289 (*A v. National Blood Authority*), para. 31.

entitled to expect of products, it follows that the Directive is grounded on striking a balance between protecting consumers from unnecessary harm and ensuring a market that offers a wide range of relatively safe products that are affordable.[88] Hence, these limits might also be viewed as incentives from European product liability law for individuals to take reasonable care for their own safety and for the safety of those around them. The absence of a warning does not automatically lead to compensation. To be successful, the circumstances of the concrete case must indicate that adding a warning would have made a difference.

Moreover, it is advanced here that these limits should be applied by courts as they are likely to discourage producers to use warnings that are actually superfluous as a shield against liability. This effect subsequently contributes to reducing the negative problems associated with overusing warnings.

> The research study on product safety of Vanilla Research[89] was supplemented by desk research and interviews with more than 60 stakeholders across government, business and relevant interest groups. The interim report notes that the discussion with manufacturers on their attitudes to information and risk found significant differences. An interesting finding is that a retailer of electrical appliances suggested that it was now essential to incorporate as many warnings as possible to protect themselves from legal liability. Manufacturers also felt that enforcement and the courts often interpreted legislation in the favour of the consumer. It seemed to them that little consideration was given to the extent that consumers may have brought risks upon themselves through irresponsible behaviour. The research also showed that two other manufacturers, of generally higher priced goods, were less concerned about liability and made a conscious effort to keep the number of warnings down to a reasonable minimum.[90]

4.4.3.3 Rationales for Not Requiring Warnings: Framework for Defectiveness for 'No Warning' Claims

Before discussing the risk categories that do not need a warning in more detail, this paragraph pays attention to how I believe European courts should assess whether the absence of the warning rendered the product defective in terms of the EPLD.

The EPLD prescribes that defectiveness is determined on the basis of establishing whether the product met the safety expectations the public is entitled to have, taking all circumstances into account. As shown in chapter 2, it has been argued in the legal literature that the introduction of the Directive's liability system without fault does not really change

88 See recital 2 EPLD and the Green paper: European Commission 1999.
89 Vanilla Research 2007, p. 23-25.
90 Better Regulation Executive & National Consumer Council July 2007, p. 19.

the way in which liability is established with respect to warning claims.[91] Case law suggests that many European courts use the same method and circumstances to determine whether a warning was required under the Directive as under fault-based liability. Under fault-based liability, courts use a risk-utility analysis to determine wrongful conduct, which means that they evaluate the failure to adopt a warning by balancing the burden (*i.e.* cost) of adopting the warning to avoid the risk against the size of the risk (*i.e.* combination of severity of the harm and the probability).

Although the EPLD requires a consumer expectations test to determine whether the product is defective due to the absence of a warning, courts usually and often implicitly engage in a balancing process where the benefits of providing a warning and the costs of providing a warning are also taken into account. This means that defectiveness as a result of the absence of a warning is denied because providing the warning can be considered not to increase the safety of the product to an acceptable level, as the costs are expected to be less than the safety benefits.[92]

This assessment is in line with how in my opinion warnings should be viewed in European product liability law. As recommended in paragraph 4.3, the rationale of requiring warnings in law is that warnings can be considered to have an added safety value. Equally, if it can be said that giving a warning has no or only little safety value in light of the concrete circumstances of a case, then I believe that this should be decisive for rejecting defectiveness for a no-warning claim. In other words, with regard to the test to determine defectiveness for no-warning claims it should be permitted to include risk-utility factors that balance the costs of providing the warning that is missing against the utility of having the warning present. If it can be expected that there are no (substantial) safety utility/benefits or if the costs of providing the warning can be considered to outweigh the benefits attached to providing a warning, European product liability should not require the presence of a warning under the defectiveness test.

Chapter two showed an indefinite list of circumstances that were considered relevant by European courts for determining whether a warning was required. These circumstances entail: the intended and reasonably expected use of the product; the time that the product was put into circulation; probability that a product danger emerges; the degree of harm arising from that, the nature of the product hazard; the burden/cost of providing a warning or an alternative design; an obvious risk; general knowledge of the risk; the social acceptability of the risk; the degree of certainty with respect to the health hazards associated with product use; (non)compliance with public product safety provisions or voluntary standards; the utility of the product. Although these circumstances can be distilled from case law,

91 *E.g.* Stapleton 1994; Stoppa 1992; Miller & Goldberg 2004, p. 354, 417; Lord Griffiths, De Val & Dormer 1987-1988. For the Netherlands, see Snijders 1984; Dommering-Van Rongen 2000, p. 32; Van Dam 2000, p. 289.

92 Cf. Owen 2008, p. 72.

courts are seldom explicit in how they balance them.[93] Nonetheless, rulings, especially those that involve obvious, generally known risks and small risks, indeed imply that courts take account of the risk-utility factors when determining whether the absence of a warning rendered the product defective under the EPLD. For example, it seems that the underlying reason for rejecting a warning for an obvious risk is that the warning does not make the product safer in respect of providing no warning, as the obviousness of the risk itself already warns users and thus enables consumers to avoid injury. From this it follows that a warning for an obvious risk is not necessary because the minimal utility of adding a printed warning is less than the costs of giving it. The same can be submitted for generally known risks. As for low risks, it would not be surprising to see that many courts would rule that in view of the severity of the injuries suffered a warning should have been provided. This conclusion would point to some sort of balancing exercise undertaken by courts.

However, because misconceptions with respect to assessing the cost component of providing warnings are lurking, courts tend to rule in favour of the presence of a warning whilst the warning should have actually been regarded unnecessary. It appears that the cost component is often underestimated or overlooked by courts. First of all, the cost of adding a warning may be overlooked because courts have a tendency to place high value on the safety utility of a warning precisely because it is a precautionary measure that is usually more practical to implement than other design changes. Furthermore, in the aftermath of an accident, courts may be inclined to focus more on the hazard (in terms of the severity of the injury), rather than on the level of risk (in terms of combination of both the hazard and the probability of consumers being injured).[94] Thirdly, courts can underestimate the costs by structurally viewing the costs of providing a warning as being financially low, because it only requires some extra printing costs in their view. This mistakenly suggests that all risks should have a warning, no matter how small. A fourth fallacy of courts is that they apparently view the costs only from a financial viewpoint, and ignore the social costs of adding a warning. These costs relate to the potential negative effects for warning recipients associated with overusing warnings on a concrete level and in general. Hence, adding a warning cannot only negatively affect the processing of the other warnings on the product, but also the impact warnings have in general.

In view of the rationales of the limits to defectiveness for no-warning claims given above, it is of importance that courts take account of these misconceptions and exercise care when assessing the costs. First, courts must understand that designing an adequate

93 An exception may be the Dutch *Nagelstyling* case discussed in chapter 2, where the Court of Appeal stated that the fact that the nature and/or concentration of the amount of acryl in the defendant's products pose a lesser risk and the circumstance that there is a small number of cases in which the risk will manifest during product use do not lead to the conclusion that no warning against the risks associated with the use of acryl should be provided at all. The Court of Appeal also took into consideration that a warning is not a major measure and that such a measure can prevent far-reaching adverse consequences in a fairly simply way.

94 See p. 38 of the RAPEX Guidelines.

warning may not be so cheap after all, since it requires investigation of how people process warnings. This can be done on the basis of becoming knowledgeable of the scientific literature in this area, but it may also include acquiring expert opinions or testing a prototype warning on a representative user group to better evaluate the warning design. Besides these research costs, the financial costs related to accompanying all products with a warning should of course also be considered. Secondly, to better assess the costs, there must be judicial recognition of the view that the burden of providing a warning should not be assessed solely financially. The negative effects associated with an overuse of warnings are social costs that should also be considered as part of the cost component of adding a warning to the product.[95] Given that the assessment involves consideration of factors relating to the benefits and costs associated with warnings, I argue that courts must take (more) consideration of the potential negative effects of adding warnings, especially when the risk in question falls within an above-mentioned risk category. The problems related to an overuse of warnings are not new to the legal literature. Some authors have voiced their concerns about product liability law creating incentives for producers to warn of every hazard.[96] Admittedly, there are difficulties with assigning weight to this type of cost. Yet, this problem can also occur with other factors that need to be balanced, as the assessment is, at its heart, a normative analysis. Even though the negative effects of adding a warning may be especially difficult to measure, they should have bearing. They should tip the scales of the balance in favour of requiring no warning when courts have doubts about whether the costs can be considered higher than the benefits. Hence, these social costs ought to be used as a push in order to adhere to the limits to defectiveness. For example, courts should explicitly refer to the potential negative effects that can occur to the other warnings present if it is to be expected that such effect(s) can occur. If there is already a long list of warnings available, courts may convincingly argue that adding another is likely to negatively influence the processing of the existing warnings, such as a reduced credibility of the more serious risks or a decreased ability to attract and hold attention to all the warning information. If the circumstances do not indicate any effect(s) on the other warnings of the product, courts can in general point out a reduced effectiveness of warnings in general when products are increasingly carrying warnings.

---

95 Twerski e.a. 1975-1976, p. 517; Henderson & Twerski 1990, p. 296; Jankowski 1995, p. 298.

96 The hazards of overwarning/overusing warnings have been noticed in the legal literature, especially by American scholars, such as Twerski e.a. 1975-1976, p. 513; Henderson & Twerski 1990, p. 296; Noah 1994, p. 374; Schwartz & Driver 1983, p. 60; Owen 2008, p. 610, but also by European scholars, such as Grubb & Howells 2007, p. 375; Giesen 2005, p. 52.

4.4.4 *Insignificant Size of Risks*

The first limit entails that warnings of risks that have an insignificant size should not be required under European product liability law. The viewpoint that not every risk requires care has long been recognised in European civil liability law. Risks are to a certain degree acceptable in society. As was noted above, under fault-based liability, the need to take precautionary measures, such as warnings, is balanced against the size of the risk. A common rule of thumb is that the greater the risk, the higher the level of care.[97] Similarly, small risks require small precautions. It is settled case law in the Netherlands that an act or omission of someone is not wrongful by the mere fact that there is a possibility that damage to people can occur as a result of this conduct and that the risk is materialised. The same can be said for the acts and omissions of producers.[98] By the same token, it has been noted in the academic literature that a threshold of seriousness of the risk needs to be passed before European product liability obliges producers to take protective measures, such as warnings.[99] Hence, product risks that have not been controlled by producers can still be acceptable.

This limit is of special importance to the effectiveness of warnings since it can contribute to reducing the number of unnecessary warnings. There is general agreement among warning researchers that warnings about trivial risks are unwelcome as they are likely to have negative effects on the effectiveness of warnings. They have argued that it is fair to assume that consumers would become inured to all warnings if every possible risk carries a warning. On a concrete level, adding a warning of a trivial risk can for example negatively influence the credibility of other warnings that accompany the product, or cause the warning label to convey too much information to process at once.[100]

Even though the size of the risk associated with the product can be viewed as low, producers may still be inclined to provide a warning. This form of 'defensive warning' to minimise liability can plausibly be explained by the producers' uncertainty with regard to what a court would decide in a product liability case. There are courts that tend to believe that adopting a warning is relatively costless, and these are likely to have little difficulty in incurring liability for defectiveness because a warning is missing.

It follows from the above that it is imperative that European courts carefully consider the cost component of adding a warning when the risk in question is small. The premise adopted by courts that warnings are relatively costless suggests that all potential risks, no matter how remote, should carry a warning. This should not be generally held in European product liability law. Especially when courts are challenged to deal with a small risk, it is

97 See *e.g.* Van Dam 2006.
98 See paragraph 2.2.
99 Grubb & Howells 2007, p. 14, 375; Miller & Goldberg 2004, p. 462; Hodges 1993b, p. 104.
100 Edworthy & Adams 1996, p. 48.

paramount that they not only take account of the financial costs of adding an extra warning to the product, but also the social costs that relate to the negative effects of overusing warnings.[101] By considering these social costs too, it is expected that courts thoroughly decide whether the warning will be of use or not. Furthermore, European product liability law should not signal to producers that providing warnings is an easy and cheap task. The cost of providing a warning entails more than just putting printed text about the presence of a risk on the product or in an accompanying manual. The warning research literature shows that the design of a good warning is a complex task, as it requires expert knowledge of the environment in which the product is going to be used, the characteristics of the target audience, including their limited abilities to process information, and the hazard itself. Furthermore, in order to design good warnings, producers need to take account of the state-of-the-art scientific warning literature that deals with how people process warnings and how the design of warnings can influence this process.[102]

Unfortunately, the defectiveness criterion of the Directive has its shortcomings with regard to its precise interpretation. It could follow from a strict application of the defectiveness test that every risk, irrespective of its size, must carry a warning, as this enables consumers to adjust their safety expectations. In other words, that in the absence of a warning consumers are entitled to expect absolute safety with regard to the risk. Seeing that warnings can influence people's perception of risk attached to the product, the absence of any warning might lead consumers to think that using the product is without risk. I believe, however, that this should in general not be an expectation that consumers are entitled to have with regard to the safety of consumer products.[103] Risks cannot only be communicated by printed product warnings, but also by the product's design itself, its appearance. Moreover, there is always a degree of risk present when using consumer products that have been designed by mankind and the public ought to recognise this, because if the defectiveness requirement pushed producers to warn of every single risk, especially the ones of insignificant size, people would become inundated with product warnings. This evidently would nullify the real impact that warnings ought to have on people. Given that superfluous warnings need to be avoided, such an interpretation of the defectiveness test is therefore unacceptable.[104]

---

101 See Twerski e.a. 1975-1976, p. 517.

102 Frantz, Rhoades & Lehto 1999, p. 292; Wogalter, Conzola & Vigilante 2006.

103 It seems that blood products are an exception; they do need to be absolutely safe. In the Dutch *HIV* case, the claimant had contracted HIV as a result of a blood transfusion. The infection had not been detected, because it was in the three-month 'window period'. The District Court decided that the general public may expect that blood products are 100% free of HIV in the Netherlands, taking into account the vital importance in blood products and the fact that in principle no alternatives exist. Even though there is a slight risk of infection it is, in the opinion of the court, not a generally known fact.

104 The GPSD's definition of a 'safe product' might also be problematic. Pursuant to this definition, a product has an acceptable level of safety if it *does not present any* risk or only the *minimum* risks compatible with a product's use. Although the GPSD does not explain when a risk can be considered acceptable, referring to

There may be another potential obstacle when dealing with small product risks. It follows from the steps of a risk assessment, described in paragraph 4.4.2 and depicted by figure 7, that it is important that courts are not tempted into accepting defectiveness because of the seriousness of the injury suffered by the victim, whilst the probability that the injury occurs is minimal. In other words, courts must be mindful to focus on the combination that the hazard will occur (*i.e.* the risk) when assessing the costs or burden of providing a warning instead of merely looking at the severity of the hazard that is intrinsic to the product. An injury scenario leading to a less severe injury may be much more likely than an injury scenario ending up in death, and the less severe injury may therefore result in a higher risk.[105] Hence, a lesson learned from the risk assessment procedures described above is that a hazard assessment can lead to a different outcome than a risk assessment. Consequently, the type of assessment adopted by a court is of essence as it can influence the judicial outcome. Furthermore, the table presented by the European Commission in the RAPEX Risk Assessment Guidelines (see figure 8) can serve as a guiding principle for Member State courts when determining whether the risk in a concrete case is of such an insignificant size that no warning is required, as this table distinguishes between four risk levels (serious, high, medium and low). It could be asserted that risks that fall within the level of a low risk generally require no warning.

A special note with respect to warnings of small risks is that such a warning can be required because it addresses the needs of a vulnerable group of people of which it is foreseeable that they can be harmed by the product's hazardous properties. This may entail users, but also bystanders. The GPSD provides a standard level of protection to these consumers, because it explicitly stipulates that within the meaning of a 'safe product' the different types of consumers that are at risk during use should be taken into account.[106] Because often only a small percentage of the population can be considered vulnerable in some respect, *e.g.* people who suffer from a certain disease, the risk inevitably has a size of which it generally can be questioned whether it necessitates a warning. A separate risk assessment that only focuses on the size of the risk for these types of consumers can reveal a high risk, *e.g.* because they have less ability to recognise the hazard or because many are particularly sensitive to the hazard and experience severe injuries. For this reason, the RAPEX Risk Assessment Guidelines view identification of the consumer(s) as an essential step within the preparation of a risk assessment and this Commission Decision advises to

---

no or only minimum risks implies that the standard has been set high. From a product safety viewpoint, this can be viewed as sensible. However, from a warnings-related perspective, this is less fortunate, because this standard might confuse producers to think that they need to provide warnings for very minor risk to meet the safety requirement of the GPSD. This potential obstacle strengthens the significance of reflecting the viewpoint in European product liability that some risks are so small that they need not be controlled by producers, not by design and guarding nor through warnings.

105 See p. 38 of the RAPEX Guidelines.

106 Article 2(b)(iv) GPSD; see also Article 14, para. 4(c) GFL.

make risk assessments for different consumer types.[107] Even though a user may not be considered the intended user of the product by the producer, circumstances may still require the presence of a warning that informs 'special' consumers that the product can be harmful to them or not suitable for them. Such a type of warning was termed a purchase warning in paragraph 4.2. Although the Directive concentrates on meeting the expectations of the general public, it is hard to imagine that it would not protect the safety interests of special categories of consumers that form a minority.[108] The recognition of vulnerable consumers in EU public law offers a vital indication that these categories should not be ignored within the scope of European product liability law. In this respect, it can be said that warnings are generally a suitable means to take account of the interests of a foreseeable small part of the target audience; such warnings do not restrict the use of the product by the majority of the target audience of the product and are often less expensive than other design changes.

4.4.5 *Risks Arising from Unreasonably Expected Use*

The second shield against liability relates to the way in which products are used. Because of their hazardous properties, many products are capable of causing damage, but whether the product hazard actually materialises depends on the manner in which people interact with the product.[109] It is recommended here that the absence of warnings of risks that arise from unreasonably expected product use do not render the product defective.

The limit is the result of the Directive itself. Pursuant to Article 6 of the Directive, the reasonably expected use of the product is a factor that needs to be considered when determining defectiveness. As already noted in paragraph 2.3.3.1, the implication of this factor is that a producer prevents a finding of defectiveness by designing a product that not only takes account of the 'intended' use but also other use actions that can be considered to be 'reasonably expected'. This is in line with unwritten producers' obligation under Dutch case law to act as careful producers, which includes their duty to anticipate a certain degree of careless product use or in other words, their duty to protect consumers against their own negligent conduct.[110] It can thus be concluded that the producers' own conceptual model of how the product is intended to be used does not bar liability in the event that a consumer sustained personal injuries resulting from risks posed by other use than intended. Product use within the meaning of the Directive generally involves more than merely the

107 RAPEX Guidelines, p. 42.

108 Grubb & Howells 2007, p. 355. See also in this regard Pape 2009.

109 See *e.g.* for an example the RAPEX Guidelines, p. 35.

110 See *e.g.* HR 22 oktober 1999, *NJ* 2000, 159 (*Koolhaas/Rockwool*); HR 2 februari 1973, *NJ* 1973, 315 (*Warmwaterkruik*); Van Dam 2006, p. 808.

product use that is intended by the producer.[111] This principle is also embodied in public product safety legislation. Furthermore, the sixth recital to the Directive's preamble additionally stipulates that the safety is assessed by excluding any misuse of the product not reasonable under the circumstances.[112] Hence, not all possible uses need to be taken into account in the design of the product. It follows from the Directive that risks resulting from use to which it could not reasonably be expected that the product would be put do not need to be taken into account by producers, because this can be considered 'unreasonable misuse'. This means that product warnings are not required for risks that result from this kind of behavioural action.[113] It can be said that without the warning of these risks, an acceptable or tolerable level of safety is still achieved.

> Consider the superfluous warning statement not to use the hair straightener when taking a bath, or not to use a clothes iron as a hair straightener or the pictogram, illustrating a circular with a red border and the image of a fork and knife, that conveys that the spathiphyllum, a flower plant commonly used as a houseplant, is not suitable to eat. Other examples are those on toasters like 'Do not use this toaster on its side' or 'Do not use this toaster as a source of heating or drying'.[114] A quick search on the internet unfolds the existence of even more real 'stupid' warnings.[115]

Usually risks can be avoided by handling the product with a certain degree of care. Think of following the type of warnings that tell you how to use it safely. It becomes a different story when more and more products contain long lists of warning statements like 'Do not use the product in this way' and 'Do not use it in that way' and so forth. Those warning statements formulate behavioural actions that often do not correspond with behaviour that is likely to happen, but out of fear of liability producers provide them anyway. To battle these types of superfluous warnings and the negative associated costs, it is imperative

---

111 See Grubb & Howells 2007, p. 357, 372.

112 In a similar vein, US Restatement of the Law (Third), Torts: Products Liability, § 2, comment m (American Law Institute 1998, p. 33).

113 See *e.g.* Rb. Maastricht 21 maart 2002, *LJN* 2002, AE0776 (*Versgeperste jus d'orange*); Rb. Zwolle 24 april 2002, *Praktijkgids* 2002, 5921 (*Mini-tampon*); HR 26 september 2003, *NJ* 2003, 660 (*Gekantelde vrachtwagen*), OLG Düsseldorf 20 December 2002, 14 U 99/02, *VersR* 2003, 912 (*Chocolate bar*) and also the German case of 1989 reported in Hodges 1993b, p. 110 where the Federal Appellate Court in Frankfurt held a manufacturer of a pharmaceutical for asthma liable. The court ruled that a warning against the dangers occurring as a result of excessive use was necessary because the drug was intended to be used by the patient himself in dramatic situations.

114 Vanilla Research 2007, p. 24.

115 See *e.g.* the internet site: <www.dumbwarnings.com>, where you can find an extensive collection of warning labels or the entertaining book by Green, Dierckins & Nyberg 1998.

that European product liability law does not interpret the factor of reasonably expected use too widely and that there is guidance in the interpretation of this factor.

Both for producers and courts, it can be difficult to assess what behavioural actions can be regarded as unreasonable misuse and do not need to be described in a warning and what cannot. Ergonomic guidelines and methods can provide a starting point for producers, precisely because this discipline puts the product-user interaction perspective central in the design of products. As discussed in paragraph 3.8.3, a risk assessment begins with predicting accident scenarios on the basis of combining factors that can lead up to an accident. The actions with the product by users form an important factor that contributes to the occurrence of accidents involving consumer products. Generating accident scenarios behind your work desk with your imagination as the only source of information is in all likelihood not sufficient to get a good picture of what accidents can happen. By using various information sources, like epidemiological injury data, consumer complaints, surveys, focus groups and expert opinions *e.g.* from ergonomists, producers can become knowledgeable about the range of likely behavioural user actions and other factors that may lead to an accident.[116] Qualitative observational research or user trials would be particularly helpful in providing a realistic picture of how consumers generally use the product.[117] As a result, this information enables producers to better predict accident scenarios and to thus improve the safety of the product by design. In the aftermath of an accident, this information and the results of their risk assessment can also help producers in explaining why a certain use should fall outside the scope of the Directive.

Guidance might also be sought in how EU product safety legislation deals with the concept of product use. The GPSD speaks of 'normal or reasonably foreseeable conditions of use'.[118] Unfortunately, like the EPLD, it does not further explain what is meant by this.[119] Sector-specific product legislation also takes explicit account of this factor when determining safety and these Directives do contain definitions. The Machinery Directive that was mentioned earlier uses the term 'reasonably foreseeable misuse' next to 'intended use'. 'Reasonably foreseeable misuse' is defined as use of the machinery in a way not intended in the instructions for use, but which may result from readily predictable human behaviour.[120] This definition resembles the general explanation of the ISO/IEC Guide 51, that defines 'reasonably foreseeable misuse' as use of a product, process or service in a way not intended by the supplier, but which may result from readily predictable human behaviour.

116 See also the information sources mentioned in the RAPEX Guidelines, p. 36.

117 See *e.g.* Page 1998 on ergonomics evaluations. See also paragraphs 3.5.3.3 and 3.9.3.3 for more information on information sources.

118 See paragraph 2.4.3.2. The GFL only speaks of normal conditions of use of the food.

119 In all probability, 'normal use' refers to use intended by the product.

120 'Intended use' means the use of machinery in accordance with the information provided in the instructions for use.

Even though the terminology of the definitions differs in some respect, the similarity resides in the requirement that the behavioural actions of users must be 'readily predictable' to regard them as reasonably foreseeable or expected use of the product. Interpretation of this term can again raise questions. Nonetheless, it can be argued that the requirement of 'readily predictable' entails a useful refinement, or rather a useful restriction with regard to the range of behavioural actions that producers need to anticipate through design. Plausibly, 'readily' means that the prediction of a use action is done quickly and easily. Consequently, it can be argued that behavioural actions that cannot be viewed as easily and quickly to predict are regarded 'unreasonable misuse' in terms of the EPLD. A warning related to such a behavioural action is therefore not legally required under the Directive. Equally, if behavioural actions can be readily predicted, than this behaviour should be taken into account when designing the product, for example by means of a warning. If use actions only slightly deviate from intended use, then this indicates that they are easy to predict. Furthermore, if for instance accidents as a result of a certain type of use have been reported in injury registration databases, but this type of use has not been identified and anticipated in the design of the product, this should, in my opinion, point to a finding of defectiveness. In addition, the constraint of 'readily' in respect of predictable behaviour could imply that much research effort on the part of a producer in predicting accidents is not required. This leaves little incentive for producers to collect data on the basis of research, such as observational research, that requires a lot of time and money. Although this effect is undesirable, one must not jump to the conclusion that European product liability does oblige producers to do observational research. The size and nature of the product hazard can also play an influential role in assessing whether a use action can be considered readily predictable. Normally, it can be concluded that if a behavioural action is viewed as unreasonably expected by the producer, the probability that it leads to an accident is usually also assessed to be small, and together with the severity of the injury this combination can lead to a risk of insignificant size. But this conclusion may not hold in the event that the nature of the product hazard can lead to very severe injuries.[121] The combination of the parameter of a low probability of the use action plus the parameter of the severe harm may eventually amount into a risk of significant size. Such a significant risk can require producers to undertake considerable research effort to anticipate several accident scenarios with that product hazard. It follows from this that it is important that producers have an open view when making predictions concerning user behaviour. Expectations about correct behaviour of relevant product users should not be too high. Even though a producer is quite certain that it will not have to deal with this risk later on because the use action seems far-fetched,

121 The RAPEX Guidelines (p. 22) explain that the severity of an injury that a hazard can cause is influenced by several factors. The severity of the injury can depend on the type of hazard, how powerful a hazard is, how long the hazard impinges on the consumer, what body part is injured, what impact the hazard has on one or several body parts and the type and behaviour of consumers.

it remains important that such a use action is identified and taken into consideration during the identification of accident scenarios, since the results of the risk assessment and evaluation can ultimately confirm that it was justified to ignore the risk and to leave any risk-reduction measures aside.

4.4.6 Obvious Risks

Scaling down the number of redundant warnings is also facilitated by the limit that prevents liability for not having warned of an obvious risk that is associated with using a product. A plausible reason for this is that it can be reasonably expected that consumers who are confronted with an obvious risk are – as a result of its obviousness – already warned of the presence of the risk and can take precautions, which makes an additional product warning given by the producer to prevent damage unnecessary.

This view has gained considerable legal support in several ways. The GPSD stipulates that producers must provide consumers with the relevant information to enable consumers to assess the risks inherent in the product where such risks are not *immediately obvious* without adequate warnings and to enable them to take precautions against those risks.[122] Hence, it follows from this public law provision that producers are discharged from their obligation to warn when the inherent risk is immediately obvious. This rule should therefore serve as a guiding principle in European product liability law. This is already common practice. There are courts that have refused to accept defectiveness because of the absence of a warning of a risk that was considered obvious.[123] Also in the academic literature, it has been contended that the absence of warnings of obvious risks should not lead to liability.[124] Furthermore, an overwhelming majority of American courts and authors agree with rejecting failure-to-warn claims when it involves an obvious risk.[125] Comment j of the US Restatement of the Law (Third), Torts expresses this rule. The US Restatement of the Law (Third), Torts explains that warning of an obvious risk in most instances will not provide an effective additional measure of safety. It also considers the possible negative effect of such a warning on other warnings that do matter.[126]

In sum, from different quarters there is impetus to use obviousness as a legal limit against accepting defectiveness due to the absence of a warning of such a risk. Unfortunately,

122 Article 5(1) GPSD.

123 See *e.g.* HR 26 september 2003, *NJ* 2003, 660 (*Gekantelde vrachtwagen*).

124 Howells, Janssen & Schulze 2005, p. 158; Hodges 1993b, p. 105; Miller & Goldberg 2004, p. 436. Cf. Grubb & Howells 2007, p. 370.

125 Owen 2008, p. 654.

126 US Restatement of the Law (Third), Torts: Products Liability, § 2, comment j (American Law Institute 1998, p. 31). The Restatement applies this rule not only to obvious risks, but also to generally known risks. This matter will be discussed further below in the next paragraph.

interpretation difficulties with this category of risks can exist that confuse the application of this limit. These problems are discussed below and recommendations are made with respect to a correct application in view of psychological notions.

A major problem concerns what makes a risk in fact legally 'obvious'. Courts that attach different meanings to this term, may decide in different ways. This is undesirable. It is thus of importance to have clarity on this point, so that Member State courts can rule in a consistent manner that a risk is obvious and that as a result the product is deemed non-defective when it carries no warning.

I consider two elements of significance for the legal test of determining that there is an obvious risk: (1) the product hazard must be considered obvious, which means that the hazardous characteristic of the product is obvious; and (2) the interactions of users with the hazard to avoid or reduce the risk of damage must be considered obvious. Note that as is the case with defectiveness, obviousness is measured objectively.

First there is the obviousness of a hazard. A good indication of obviousness is whether the information about the hazard is already available from the appearance of the design of the product. Warning researchers have expressed the opinion that there is no need for a printed warning in case of obvious dangers, because the product or the environment already plainly conveys its hazardous nature by its appearance.[127] According to Collins Cobuild dictionary, something is obvious 'if it is easy to see or understand'. For the purpose of a legal definition of 'obvious', I would like to extend this above-mentioned description on two points. In view of European product liability law, a product hazard is obvious if it is easy to sense it with one or more of our human abilities and easy to understand.

First of all, 'seeing' should not be considered the only cue that triggers awareness of something. Cognitive psychology has taught us that communicating information happens not only by sight, but also through other sensory channels, such as smell, hearing, touch, and taste. These psychical abilities can play a role in identifying and perceiving a product hazard and should in my opinon therefore have a bearing on determining if a hazard is legally obvious. Clearly, actually seeing the source of the damage when looking at the product should be viewed as the *foremost* indication that the hazard is obvious, since sight triggers awareness easily and contact between the danger and the user is not necessary to become aware of it, but the role of other senses should not be overlooked. So obviousness can be found if it can be said that one of the senses triggers awareness easily, or if more than one sense together trigger awareness easily.

In contrast to sight, it must be noted that there may be a possible disadvantage with awareness through the other senses; to become aware of the hazard, contact with the hazard may be necessary as a result of which the risk can materialise. For example, if a surface is hot, but the design of the product does not tell you this, then awareness of the hazard is

127 Edworthy & Adams 1996, p. 48; Wogalter 2006a, p. 5; Laughery & Smith 2006, p. 421.

only triggered after touching it and then it may be too late. In such a case, I argue that the hazard should not be considered obvious (enough) in a legal way.[128] However, if the design of the product would plainly convey that the surface is hot, *e.g.* using a red glow as a signal or a red light, then I believe that this adequately provides the information that there is an obvious risk, and as a result adding a warning 'Danger, surface can be hot' is not necessary, since a warning, albeit not written, is already integrated into the product's design by means of the glow or the light.

Secondly, I prefer to change the words 'or understand' of the above-mentioned description into 'and understand'. It is important to legally recognise the distinction made in cognitive psychology between understanding (and knowing) on the one hand and seeing on the other hand, because seeing a hazard (or smelling, hearing, touching or tasting it) triggers awareness or consciousness of the hazard more easily than having knowledge of it in our memory. Especially awareness of the hazard at the time of exposure is needed to be able to avoid it. Hence, awareness and not knowledge should not be the decisive requirement for obviousness. Of course, obvious risks are often known too. But known risks do not have to be obvious, they can be hidden to our senses. Furthermore, I believe a hazard can be considered obvious even if people have no or only little previous knowledge of it, in the event that the hazard easily communicates to our eyes or other senses. Because of this communication you learn about it and you acquire the knowledge. Consequently, obviousness should primarily be related to the sensory channels of communication.

> The fact that a cup of tea is hot and that it can lead to serious skin burns if spilled, can be viewed as obvious. Next to the fact that one normally knows that a cup of tea can be hot, it also signals that it is hot, because one can feel the heat when holding the hot cup and you may even see the damp coming off of it. These cues and the knowledge of this make the hazard obvious, and also the consequences of spilling and actions to avoid it.[129]

It has been illuminated above how courts and others can come to the conclusion that a hazard is obvious. The second element of the legal test of obviousness that is proposed here pertains to the interaction with the product. It also is important that the behavioural actions that can lead to damage are obvious or known. If a product hazard is obvious, it

128 This may change in the event that after a user has gained awareness of it, *e.g.* through touching or tasting it, the user continues to come into contact with the danger, whilst this action is not necessary for use of the product. In such cases, liability may be barred as a result of misconduct on the part of the user, or damages may be reduced as a result of contributory negligence.

129 The English *Mc Donald's* case dealt with this risk. The court ruled that persons generally know that if a hot drink is spilled onto someone, a serious scalding injury can result. Whether the risk is an obvious or a generally known risk or both may be difficult to decide. As shown above, I believe that it is rather an obvious risk that is also known.

often means that it is also obvious that some form of personal injury or property damage can take place when exposed to it.[130] For similar reasons, the behavioural actions that can lead to materialisation of the risk are often obvious too. Hence, if the hazard is obvious (the source of damage), the risk of getting hurt as a result of certain use actions is usually obvious as well. For instance, when you see that something is sharp and has moving parts, you know and understand that these product characteristics represent hazards that can lead to damage, like cuts or even amputation. You also know and understand that it is not wise to operate it close to your body without having some form of protection; you need to create some distance between the hazard and yourself to reduce this risk of bodily injury.

Even though my imagination runs short on this point, there might be cases in which the hazard is obvious, but the behavioural actions that can bring the obvious danger into effect and cause injury are not. Normally, if a hazard is obvious, people can produce appropriate behaviour to avoid it and as a result of that, the risk of injury decreases. However, the size of the risk remains high if it can be expected that users are not sufficiently aware of what behavioural actions are safe or risky.[131] It is thus important for courts that – when assessing the obviousness of a risk – they take into account that even if the hazard can be considered obvious, the use actions that increase or reduce the likelihood that the adverse consequences materialise may not. Similarly, it must also be recognised by courts that whilst the danger is obvious, the seriousness of the injury or the likelihood that these can occur may not be. As I have argued in paragraph 4.2.3.4, if this lack of obviousness or knowledge can be reasonably expected, then such information categories should form part of the warning message.

> Warnings on candles can be used as an example to illustrate that risks attached to the use of this product are obvious and do not need a warning. Many candles have a printed sticker on its bottom side or a label draped around the candle that provides product information, including warning information. Usually the hazard information (*e.g.* fire hazard) is left out by the producer, in all probability, because the producer assumes that this information is obvious. The content of the warning messages differ dramatically. They usually contain

130 It seems that European product liability law, at least Dutch law, does not require that all the types of damage attached to a certain hazard need to be mentioned in a warning. For example, as a result of a mechanical hazard, you can suffer cuts, but also amputation. It suffices to describe that serious or permanent injury can happen. In the *Rokersclaim* case, the Dutch District Court (Rb. Amsterdam 17 december 2008, *NJ* 2009, 311 (*Rokersclaim*)) ruled that it is not required to bring to the attention of potential users all forms of diseases, life-threatening or not, that can be caused by smoking cigarettes. The District Court added that if it is clear that smoking can cause the life-threatening disease of lung cancer, the producer is not obliged to bring users' attention to the fact that smoking can also cause lung emphysema and cardiovascular diseases. See also Grubb & Howells 2007, p. 369.

131 This relates to whether the behavioural action can be considered reasonably expected or not.

> information about how to use the candle safely. Some contain pictograms to explain what not to do; others contain only text, which also varies in length. A safety instruction that is frequently present is 'Never leave a burning candle unattended'. Other instruction statements that are regularly spotted on the label are: 'Keep burning candles out of reach of children and pets' and statements of the kind: 'Make sure that the burning candle is not in a draughty place' and 'Keep away from the curtains'.
> Without making a thorough risk assessment, it can be said that the risks attached to a fire hazard can be significant. Notwithstanding that in reality candles are accompanied by warning information, I consider this hazard, the severity of its consequences and the instructions to avoid it to be obvious. They need not be mentioned in my opinion. Users can easily see the actual source of injury, *i.e.* the flame. It is also easy to understand what it can cause and that you should not put it on the floor when there are young children and pets in the environment that can come near the hazard and be exposed to it.

A final comment with respect to problems that judges can face when dealing with these types of risks, is that there are different scales of obviousness. But when can a risk be considered obvious from a legal perspective, so that the absence of a warning does not lead to defectiveness? Unsurprisingly, it has been regularly argued that risks that are *so* obvious need no warning.[132] Furthermore, the Dutch decision of the Supreme Court in the *Gekantelde vrachtwagen* case could imply that, in case of products, liability for the absence of a warning should be denied if the circumstances of the case show that it should have been 'immediately obvious' that it entailed a risky use of the product, even for users who are not constantly cautious and attentive.[133] The GPSD also talks about risks that are 'immediately obvious'. That such risks need no warning seems pointless to say. It becomes a different story with respect to risks that are obvious to a lesser degree. Even though the borderline is thin, there is a difference between a risk being immediately obvious or just obvious. On a scale of obviousness, the former is ranked higher than the latter. Consequently, the bar is set high by the GPSD. The same can be said for the legal test of obviousness proposed above. This definition contains the restriction of 'easily'. It can be argued that something that is easily seen or sensed in other ways is in fact similar to the requirement of being immediately obvious. Having such a high standard is not a bad thing, since the limit presents a bar on accepting product liability for a product that bears no printed warning. It points out to producers that it benefits them to design products with functional

132 Howells, Janssen & Schulze 2005, p. 158; Grubb & Howells 2007, p. 370; Hodges 1993b, p. 105.

133 HR 26 september 2003, *NJ* 2003, 660 (*Gekantelde vrachtwagen*). The original Dutch words were '*aanstonds duidelijk*'. See paragraph 2.2.3.2.

or featural product characteristics that trigger awareness of people so that a printed warning can be left out. It also expresses to producers the importance of investigating whether a certain risk is obvious to the target audience before deciding not to give a printed warning of that risk. Since products become more technologically complex, their hazards have become less obvious. In this regard, I argue to apply a uniform legal standard for obvious risks in European product liability law, and that is the standard proposed above so that there is a coherent and objective way for deciding when a risk is legally 'obvious', with the ensuing legal consequences thereof. This is likely to ease the courts' struggle with regard to what degree of obviousness is sufficient and the way in which to determine whether a risk is legally obvious.

4.4.7 *Generally Known Risks*

The fourth limit relates to generally known product risks. Courts of different European countries as well as academics have expressed that the absence of a warning against a generally known risk does not lead to a finding of defectiveness.[134] As with obvious risks, it can be asserted that warnings for generally known risks are superfluous as the existing general knowledge of the risk itself can be viewed to act as a warning to use the product carefully. As a result, requiring a printed warning for these risks will provide little added safety benefit compared to the situation in which no warning is given. In addition, it can be argued that, as with the other limits, this limit also benefits the effectiveness of warnings, as it contributes to reducing the likelihood that warnings are being overused by producers.[135]

Even though adding a warning is not likely to substantially increase the safety of the product, that does not mean that consumers are not entitled to expect a higher level of safety by means of changing the design of the product. Courts should be open to establish defectiveness if a more effective available design measure could have easily prevented the injury.[136]

Notwithstanding these arguments that support having such a limit in European product liability law, this principle may need to be reviewed on the basis of psychological grounds,

134 See *e.g.* Rb. Maastricht 21 maart 2002, *LJN* 2002, AE0776 (*Versgeperste jus d'orange*); OLG Düsseldorf 20 December 2002, *Anonymous* 1966, 14 U 99/02, *VersR* 2003, 912 (*Chocolate bar*); [2002] *EWHC* 490 (*Bogle v. Mc Donald's Restaurants*); Rb. Amsterdam 17 december 2008, *NJ* 2009, 311 (*Rokersclaim*). See also Miller & Goldberg 2004, p. 436; Grubb & Howells 2007, p. 368; Spier e.a. 2009, p. 135 and comment j of the US Restatement of the Law (Third), Torts that also pertains to generally known risks (American Law Institute 1998, p. 31).

135 See also comment j of the US Restatement (Third) of Torts for such an argument. See *e.g.* Vanilla Research 2007, p. 24 in which research study consumers generally said that they did not read the warnings on everyday products because many of these were considered generally known and common sense and as a result not helpful.

136 See paragraph 4.5 of this chapter on this subject.

since it can be argued that a blanket rule consisting of denying warnings for generally known risks takes insufficient notice of the psychological distinction that can be made between awareness and knowledge. As said before, an important lesson learned from the warning research literature is that having knowledge of a product risk is not the same as being aware, thinking of it at the time needed. Warning researchers have argued that for users to effectively deal with product risks, it is not enough to say that they have knowledge of it. People can have acquired knowledge through the product itself, experience with similar products, training, mass media information campaigns, or other sources of information. Still, a cue can be needed to actually remember this existing knowledge at the critical time and hence to act upon it appropriately. So, providing a warning that contains already known information can in fact have utility, provided that it serves as a reminder.

Although I argue that European product liability law should generally proceed from the viewpoint that the consumers' knowledge will normally lead to awareness as a result of which consumers are able to act carefully, an exception should be made for reminder warnings. For reasons of safety, European product liability law should require reminder warnings, at least under limited circumstances. The warning research literature mentions special circumstances where reminder warnings are particularly useful, such as (1) intense mental work load and involvement in the product task; (2) when foreseeable distractions are expected to occur during product task performance; or (3) where the non-obvious hazard is encountered infrequently and forgetting plays a role.[137] Consequently, in the event that a warning is absent whilst the circumstances of the case indicate that a reminder warning is needed to call existing knowledge into awareness so that users can protect themselves, defectiveness should be established. Note that because reminder warnings pose an exception to the general limit, a cautious approach with respect to applying the exception is recommended. Furthermore, special requirements with regard to the content and form should apply for assessing the adequacy of reminder warnings.[138]

As regards the legal test for determining whether a risk is 'generally known', the basis should be that the risk can be considered known to the 'public at large', which requirement is indicated by the text of the Directive. More specifically, this suggests that to consider it 'generally known' a large majority of the target audience of the product must have knowledge of the risk. This involves the assessment of who are the foreseeable persons that interact with the product and what is their knowledge level, and also the evaluation that a large majority of them possesses this knowledge.

An important indication of whether a risk can be regarded 'generally known' is if it is an inherent risk, *i.e.* if elimination of the risk would result in destroying the functioning or utility of the product. Think of the risks of smoking cigarettes, drinking alcohol, excessive

137 Wogalter & Laughery 2006, p. 901; Laughery & Smith 2006, p. 421; Wogalter 2006b, p. 56.
138 For more information, see paragraph 4.6.2.5.

consumption of sugary and fatty foods.[139] However, most consumer products are technologically complex and the type of risks attached to such products are less likely to be generally known. In addition, whether the information is also delivered through other media and/or transmitted by other sources can be of relevance for determining whether the risk is 'generally known'.[140] This factor is linked to the principle of applying a warning system approach.[141] Additional media channels may involve instruction manuals, package inserts or websites that are used by the producer as a source, but other sources can also communicate the information, such as mass media information campaigns from the EU or the national authorities, or even risk communication by employers. When considering the various channels that have been used as communications, it must be borne in mind that their effectiveness to reach the target audience is not alike.

Two additional comments can be made in respect of generally known risks. It has been submitted by warning researchers that producers should be cautious when making assumptions on the human component, especially with regard to knowledge of risks, since – given their expert knowledge – they may overestimate what people know.[142] This shows how valuable it is for producers to test the warning's design on a group representative of the target audience. The results can point out any wrongly held presumptions on their part and lead to design improvements. Moreover, testing results benefit producers in court, as they can help show that claimant's assertion should be rejected and that in hindsight producers' decision not to warn was indeed justified in the eyes of the law.

Secondly, an issue may be the extent to which a court can impose its own appreciation of whether a risk is generally known, as this appreciation can be decisive for the decision to reject or accept product liability. On the basis of Dutch civil procedural law, Article 149(2) DCCP, generally known facts or circumstances require no proof.[143] Dutch courts can use these facts or circumstances as a basis for their decision, regardless of whether they have been furnished by a party. This rule has consequently an important bearing in civil actions, since courts can omit orders to furnish proof and can quickly deliver the judgement. Even if a generally known fact is asserted as such by a party, a court is not obliged to apply the normal rules concerning the obligation to furnish facts and the burden of adducing proof.[144]

As with producers, the fallacy to overestimate what people know may also thwart courts. It is therefore of importance that Dutch courts, as well as other European courts that have this form of judicial freedom, do not take the conclusion that a risk is generally known too

139 See Howells & Borghetti 2010, p. 469; Grubb & Howells 2007, p. 368.

140 Cf. Howells, Janssen & Schulze 2005, p. 159.

141 See *e.g.* paragraph 4.2.3.8, and paragraph 3.6.2 (source), paragraph 3.6.3 (channel) and paragraph 3.6.10 (warning system).

142 Laughery 1993; Wogalter 2006a, p. 58.

143 This rule also applies to rules relating to general experience.

144 See on this subject Van Boom, Tuil & Van der Zalm 2010.

lightly, as this could mean that an unacceptable part of the target audience is not adequately protected against the product risk.[145] It is inevitable that there is a gap between what people really know and what producers and courts think or expect people to know. This should not be considered problematic as long as European product liability law minimises the gaps between consumer reality and judicial reality as much as possible. In other words, for reasons of product safety it is of essence that the courts' own conclusions do not overly extend this gap. This means that if a court has doubts about whether it concerns a generally known fact, such as the presence of a generally known product risk, and if this fact plays a crucial role in the assessment of liability, it would be wiser to apply the normal evidential rules on this matter, or to provide a party with the possibility to challenge the generally known fact rather than to apply the absolute rule that no proof is required.[146]

4.4.8 Undiscoverable Risks

A special limit relates to the unknowability and undiscoverability of risks. European product liability does not hold producers liable for damage caused by the absence of a warning of a risk unknown at the time of putting the product into circulation. Here, liability depends on the interpretation of the term 'unknown'. Under fault-based liability, 'unknown' refers to the concept of 'unforeseeable', *i.e.* whether a reasonable producer could and should have known of the risk. Whether a risk could and should have been known to a reasonable producer is subsequently determined by the extent of a producer's duty to investigate the harmful characteristics attached to the product's use and the duty to keep abreast of the scientific and technical knowledge in the field. However, if these duties are stringent under fault-based liability, they resemble to a large extent the issue of undiscoverability of risks under the Directive.[147]

This limit differs from the other limits, in that it is construed as a defence by the Directive itself to avoid liability even though defectiveness was established.[148] It concerns the development risk defence of Article 7(e). So, it is not so much a limit against defectiveness, but more a complete limit against liability.[149] According to the Directive's preamble, the defence forms part of creating a fair apportionment of risk between the injured person and the producer. Hence, this limit can be viewed as a shift for safety from producer to

145 This is especially true in situations where the risk is not controlled by other design measures.

146 Cf. Van Boom, Tuil & Van der Zalm 2010, p. 38.

147 This seems to be the case in the Netherlands. See paragraph 2.3.4.1.

148 Note that Member States can derogate from adopting the development risk defence, as this defence is an optional provision of the Directive (Article 15(1)(b)). Most Member States have not done so, including the Netherlands.

149 Provided that this defence has been implemented in the domestic law of the EU Member State. See also paragraph 2.3.4.1.

consumer. The main underlying reason for adopting this optional defence in domestic law is that otherwise, out of fear of increased liability, producers would be discouraged to innovate their products.

Within the liability system of the Directive, the development risk defence protects producers against liability for damage caused by a defect that although present, was not yet discoverable on the basis of the scientific and technological knowledge at the time the product was put into circulation. It follows that producers can escape liability for defectiveness caused by not having warned of an undiscoverable risk when they successfully invoke the defence.[150] The wording of the defence has been explained by the ECJ.[151] This decision stipulates a narrow scope of the defence. The undiscoverability of the defect must be measured objectively. The defence consequently entails that a producer can escape liability for this defect, if he can prove that on the basis of the objective state of scientific and technical knowledge that was accessible, including the most advanced level of such knowledge, it was impossible to detect the defect at the time the product was put into circulation. As regards 'no warning' claims, producers will only escape liability on the basis of proving that even though the absence of a warning against the risk rendered the product defective, there is no liability because the risk had been undiscoverable at the time the producer put the product into circulation.[152] A final note worthy to mention here is that this limit should not be interpreted to mean that if a risk were discoverable, a warning is required. The size of the risk can act as an essential barrier.

4.4.9 *Conclusion*

Paragraph 4.4 dealt with the warning issue as to what risks need not to be warned against and recommended an approach to avoid an increased liability for not having provided a

150 Miller & Goldberg 2004, p. 437.

151 ECJ 29 May 1997 (C-300/95), *ECR* 1997, p. I-2649 (*Commission v. United Kingdom*).

152 The undiscoverability of the defect must be measured at the time the product was put into circulation and not at a later date. Consequently, this factor plays an important role for establishing liability under the Directive. A link can be made to risks that were discoverable only after the marketing of the product, but that needed a warning to prevent harm. After-sale warnings relate to the producers' obligation to provide warnings of risks they know or should have known after the sale of the product in view of the circumstances of the case. The Directive does not require these warnings, since liability depends on when the product is put into circulation. Such claims based on the absence of the inadequacy of after-sale warnings need to be grounded on fault-based liability. See also Article 5(1) GPSD that can be read as to imply a continuing duty to inform consumers of risks, including risks that have been discovered after the products were placed on the market. This warning issue falls outside the scope of this thesis, see chapter 1.
The undiscoverability of a risk can also be linked to the circumstance of uncertainty about whether there is a risk and to the precautionary principle. The relevance of the precautionary principle for product liability and the question whether and how uncertainty about a risk should be viewed as a limit against defectiveness for no-warning claims are interesting issues for future research.

product warning. Since this allegation is a common form of litigation, the approach can be considered of value to European product liability law.

The approach consisted of using limits to the Directive's defectiveness requirement with regard to claims centred on the absence of a warning for a certain risk. Five categories of risks were presented of which I recommended that defectiveness must be denied when a warning was missing. These are risks that have an insignificant size, risks arising from unreasonably expected use, obvious risks, generally known risks[153] and risks that were present yet undiscoverable at the time the product was put into circulation.

To determine whether the product without the warning is defective, the Directive lays down a consumer expectations test. This leads here to answering the question whether the product that bears no warning failed to meet the safety level which a person is entitled to expect, taking all circumstances into account. Case law suggests that many courts in Europe have not stopped using risk-utility factors to establish whether the absence of the warning rendered the product defective, even though these factors originate from fault-based liability. These factors are especially of relevance in cases where the risk is of insignificant size and/or obvious, generally known and/or related to unreasonably misuse of the product. Many legal academics have submitted that the decision whether a producer failed his duty to warn under fault-liability is actually no different from deciding whether the absence of a warning rendered the product defective.[154] Hence, I argued here that it is allowed to conclude that the main part of assessing whether the warning's absence renders the product defective involves consideration of factors relating to the costs of adding a warning and to benefits/utility of adding the warning to the product. Are the costs of adding a warning to the product higher which leads to the conclusion that the product with a warning is not defective? Or does the warning have an expected added value as a result of which the product without the warning can be considered defective?

Some limits, like those with regard to risks arising from unreasonably expected use and undiscoverable risks are laid down by the Directive itself. These limits constitute a complete bar against liability. Albeit implicitly, it also follows from the Directive that producers are absolved from liability when it concerns risks of insignificant size, since the Directive does not demand of producers to design products that are absolutely safe to avoid liability. It must be borne in mind that in case of obvious and generally known risks, the finding of non-defectiveness on the basis of a no-warning claim does not have to automatically preclude a finding of defectiveness on other grounds. Because a warning can often be considered the least reliable measure to protect people from harm, other

153 A warning for generally known risks can be required under the limited circumstances in which a reminder is needed.

154 A difference may lie in the approach to unknown and undiscoverable risks.

design measures that are more effective may as a result be required to render the product non-defective.

Three rationales were given to underpin the adoption of these limits. First of all, it was asserted that these limits to defectiveness do justice to the policy goal of the Directive, as European courts bring about a fairly appropriate apportionment of the risks when giving decisions on the basis of these limits. As a result of not allowing defectiveness due to the absence of a warning of a risk that is of insignificant size, obvious, generally known, undiscoverable, or associated with unreasonably expected use, the burden of safety shifts from producer to consumer. Providing warnings for these risks would not lead to an additional substantial safety effect compared to the situation in which a warning is lacking.[155] As discussed in paragraph 4.3, it was argued here that European product liability law requires producers to provide warnings so that accidents can be prevented or reduced that would otherwise have occurred in the absence of a warning. Accordingly, on the basis of empirical research it can be presumed that the safety of products increases because having warnings present brings an added safety value to the situation. Hence, if it is expected that providing a warning does not substantially raise the safety of a product, a warning should not be required under law in those cases. In other words, if a warning's utility is minimal, because consumers are well able to avoid an accident without being given a printed warning, the warning can be considered unnecessary and should not be required under law. The costs attached to providing a warning can thus be considered of greater importance. Additionally, it could be argued that the limits to defectiveness also encourage consumers to act safely. Thirdly, by accepting these limits European product liability can also contribute to preventing or reducing the negative effects associated with 'overusing' warnings. A review of the warning research literature revealed that there are potential negative effects associated with providing warnings, especially in case of an 'overuse' of warnings. Overusing warnings refers to the circumstance that consumers are being overloaded with too many warnings that accompany the product and/or with too many products that bear a warning. Overusing warnings can adversely affect the cognitive processing of warnings in particular and in general in a number of ways. Concerns with respect to product liability providing incentives to overuse warnings have been noticed in the legal literature. It can be viewed as a common pitfall for courts to assess a no-warning claim in isolation. However, it is imperative that courts are cognisant of the potential negative effects on the effective processing of other warnings that can occur as a result of their decisions that impose

155 The limit pertaining to undiscoverable risks is special since this limit is the result of the implementation of the development risk defence into national law. The main rationale for this limit is to shift the responsibility of these risks to consumers as to encourage product innovation. If a Member State has chosen not to include the development risk defence into domestic law on the basis of Article 15(1)(b) EPLD, then a producer cannot escape liability for defectiveness as a result of the absence of an undiscoverable risk. In those cases, the responsibility for safety does not shift to consumers, but remains with producers.

defectiveness for not having warned against such risks as discussed above. Such decisions can trigger producers to overuse warnings, *i.e.* to provide warnings for every single risk out of fear for liability, whilst these warnings can be considered unnecessary. These negative effects should be viewed as social costs associated with adding an unnecessary warning to a product and these costs should play a role in the balancing process of courts in general, or even at a concrete level as a circumstance in a specific case. In sum, by providing clarity on the matter of what risks need no warning according to liability principles, European product liability can play a role in reducing the unintended effects of overwarning or overusing warnings.

Paragraphs 4.4.4-4.4.8 discussed the limits to defectiveness in more detail. Below, a summary is given of the five risk categories.

Risks of Insignificant Size

A most important bar against accepting liability for the absence of a warning concerns the limit with respect to risks of insignificant size. It was made clear that European product liability must not demand of producers to warn of risks that have an insignificant size, because this would trigger producers to create a warning for every potential risk to minimise their liability. This trend would be detrimental to the effective processing of warnings. Especially with small risks, courts must be cautious not to be trapped into underestimating the costs of providing a warning for such risks and mistakenly rendering the product without the warning defective. The presumption that providing warnings is a cheap and easy measure to protect the interests of potential victims should not be generally held by judges. It was recommended that the social costs associated with overusing warnings need to have bearing too and tip the balance in favour of rejecting defectiveness when courts consider a risk small.

It is therefore of importance that the tool of a risk assessment, which was described in paragraph 4.4.2 and displayed in figure 7, plays a role within the decision making process of courts. A main lesson learned from this tool is that it is important to consider the risk, *i.e.* the combination of the probability and the severity of the injury, instead of putting too much focus on the single factor of the severity of the injury that was caused by the product hazard. A correct focus is of judicial importance, since in the latter situation courts are likely to be more eager to render the product without the warning defective than in the former situation. In addition, the results of a risk assessment can be used as input to select the risks that need to be approached by design, including warnings, and which do not. The risk assessment method laid down in the RAPEX Guidelines of the European Commission may be useful to courts for determining when a risk of a non-food consumer product is of insignificant size, as it provides a table (see figure 8) that determines the level of product risk. It distinguishes four risk levels (serious, high, medium and low). It was asserted that courts can use the risk level of low as a guiding principle for deciding that risks falling under this level generally require no warning.

Risks that Arise from Unreasonably Expected Use

The second limit to defectiveness that was described in paragraph 4.4.5 relates to the manner in which products that have hazardous properties are used. This limit that pertains to denying warnings for product risks that arise from unreasonably expected use can contribute to a reduction of superfluous warnings in the world. It follows from the wording of the Directive that risks resulting from use to which it could not reasonably be expected that the product would be put do not need to be taken into account by producers, because this can be considered 'unreasonable misuse'. This means that product warnings are not required for risks that result from this kind of behavioural actions. Similarly, even without the printed warning of these risks, an acceptable level of safety is still achieved.

It is important that European courts take the interpretation of this factor seriously. If the term is consistently interpreted too widely and producers are being held liable because of the absence of a warning of a risk that was a result of behaviour of which it is considerably questionable that it is reasonably expected use, this trend can trigger producers to provide warnings that are actually unnecessary. Unnecessary warnings should be avoided as this may negatively affect the effective processing of other warnings that are indeed necessary in a number of ways.

To help draw the line between reasonably expected use and those interactions that can be deemed unreasonable misuse, some guidance was found in EU product safety legislation. EU product safety legislation has used the requirement of use that results from 'readily predictable' human behaviour to define the concept of 'reasonably foreseeable misuse', a concept that closely corresponds with the Directive's requirement of use that is reasonably expected. It was argued that the requirement of 'readily predictable' entails a useful restriction with regard to the range of behavioural actions that producers need to anticipate by design. Whether use can be regarded readily predictable can be assessed on the basis of the information sources that have been consulted. Using various information sources is of importance to become knowledgeable about the range of likely behavioural user actions and other factors that may lead to an accident. Generating accident scenarios behind your work desk with your imagination as the only source of information is in all likelihood not sufficient in the eyes of courts to get a good picture of what accidents can happen. Especially ergonomic methods such as qualitative observational research provide a clearer picture of what use actions people do with a product. This enables producers to better anticipate accidents and to improve the safety of the product by design, and in the aftermath of an accident it helps producers explain why this use should fall outside the scope of the Directive. Furthermore the type of product hazard also plays a role in establishing whether a certain use action can be viewed as easy to predict. Some hazards intrinsic to a product have more powerful health effects than others, as a result of which use actions even though not intended still fall under the scope of reasonably expected. This may consequently entail that producers are required to do more research into the occurrence of accidents and to collect more comprehensive data than just injury registration data. Naturally, courts should consider the value of the types of information sources used, since different sources provide different or more comprehensive results. It could even be argued that the use of ergonomics for the design of products is required under the Directive, since ergonomic methods in particular offer ways to gather information on how accidents occur and what product-use interactions are plausible. This would better equip producers to determine the boundaries between reasonably expected use and those interactions that can be deemed unreasonable and outside the scope of liability. Admittedly, the mere application of ergonomic principles does not constitute a barring effect to liability. Ergonomic methods also have their weaknesses. Nonetheless, it can be said that it generally offers producers a reasonably solid escape route.

It was also argued in this paragraph that producers should have an open mind when generating various accident scenarios. Expectations about correct behaviour of relevant product users should not be too high. Even though a producer is quite certain that the use action seems far-fetched, it remains important that such a use action is identified and taken into consideration during the identification of accident scenarios and the assessment of the risks, since the results of the risk assessment and evaluation ultimately confirm whether it was justified to ignore the risk and to leave any risk-reduction measures aside. Risks arising from unreasonably expected use can often be considered of minimal size. Because certain use actions have little to do with the intended use, they are assessed as highly unlikely, which ultimately influences the size of the risk. However, if the severity of the injury can be estimated as very high then its combination with a low probability of the use action may still result into a risk of significant size that needs to be controlled. Showing courts documentation of their risk assessment that explains why they did not take such a risk arising from that particular use action into account will provide a good defence in court to contest the allegations made by the injured party.

Obvious Risks

Another limit to defectiveness that pertains to no-warning claims concerns obvious risks. European product liability law should not encourage producers to provide printed warnings for obvious risks as it was argued that adding these warnings will have little safety utility compared to the situation in which no warning was given. Inspired by psychological notions, a legal test for determining whether a risk can be considered obvious was proposed in paragraph 4.4.6. On the basis of this test, Member State courts can rule in a consistent manner that a risk is obvious and adhere to this limit.

Two elements are of significance for determining whether there is an obvious risk: (1) the product hazard must be considered obvious, which means that the hazardous characteristic of the product is obvious; and (2) the interactions of users with the hazard to avoid or reduce the risk of damage must be considered obvious.

It was suggested here to consider a product hazard 'obvious' within European product liability law, if it is easy to sense it with one or more of our human abilities, and easy to understand. Important indications of an obvious hazard are the cue(s) given by the product's design to communicate to the user that there is a dangerous situation. An essential cue involves whether users can easily see the hazard and in consequence understand what behavioural actions need to be avoided. The reason for this is that sensing the hazard with your eyes triggers your awareness easily. Although sight is the most common and probably the best way of identifying and perceiving product hazards, it is not unimaginable that the danger triggers the user through another sensory modality. If this is likely to be the case, then this form of awareness should contribute to establishing that a risk is obvious. The other senses may not be as powerful as vision, but their role in triggering awareness should not be overlooked. Secondly, something is obvious if it is also easy to understand. I think it is important to make use of the psychological distinction between knowing/understanding and seeing, because seeing a hazard (or smelling, hearing, touching or tasting it) triggers awareness more easily than having knowledge about it. It seems that the label 'obvious' is often used in conjunction with the term 'known'. I believe that the obviousness of a risk should not be put in the same box as generally known risks. Of course, it is not denied here that these concepts are interrelated to each other. If something is obvious, it is usually well-known too. However, something that is known does not have to be obvious.

Generally Known Risks

The third risk category of which it was recommended that the absence of a warning does not render the product defective under the Directive concerns risks that are generally known. European product liability law should generally proceed from the viewpoint that consumers' knowledge will normally lead to awareness as a result of which consumers are well able to act carefully. It is therefore expected that adding a warning will not make a substantial difference to the situation in which the product carries no warning.

Notwithstanding the rationales that support having such a limit in European product liability law, it was argued that a blanket rule consisting of denying warnings for generally known risks takes insufficient notice of the psychological distinction that can be made between awareness and knowledge. I therefore recommended that an exception to this limit should be made for reminder warnings. That consumers have existing knowledge in long-term memory of a certain hazard does not mean that they are always aware of it at the time they are at risk. The purpose of a reminder warning is thus to cue people to remember the needed information at the right time. It was argued in paragraph 4.4.7 that for reasons of product safety, warnings that act as reminders should be required in European product liability law, but only under those special circumstances in which it is expected that awareness is not triggered at the moment users need it to avoid the risk. In these cases, it was contended that a warning of a generally known risk has substantial utility.

This paragraph also provided a framework for deciding whether a risk is 'generally known'. The basis should be that the risk can be considered known to the 'public at large', which requirement is indicated by the text of the Directive. More specifically, this suggests that to consider it 'generally known' a large majority of the target audience of the product must have knowledge of the risk. This involves the assessment of who are the 'foreseeable persons' that interact with the product and what is their knowledge level, and also the evaluation that a *large* majority of them possesses this knowledge. A factor that indicates whether a risk can be considered generally known is the circumstance that without the hazardous product property the product's functionality is severely impaired. Another indication that was mentioned is whether the information is also delivered through other media and/or transmitted by other sources.

Undiscoverable Risks

The final limit, that was discussed in paragraph 4.4.8, entails that there is no liability for the absence of a warning with respect to risks that caused damage albeit undiscoverable at the time the product was put into circulation. This limit should not be interpreted to mean that if a risk were discoverable, a warning is required. The size of the risk can be determining for not requiring a warning for risks that were discoverable.

In accordance with the Directive's preamble, the defence forms part of creating a fair apportionment of risk between the injured person and the producer. The main rationale for adopting this optional defence in domestic law is that otherwise, out of fear of increased liability, producers would be discouraged to innovate.

The limit is embodied in the development risk defence of Article 7(e) of the Directive. It is an optional defence that protects producers against liability for damage caused by a risk which is present, but not yet discoverable on the basis of the most advanced state of scientific and technical knowledge at the time the product was put into circulation. Hence, producers can escape liability for defectiveness for not having warned of such a risk when they successfully invoke the defence. This means that they must prove that even with the objective state of scientific and technical knowledge that was accessible, it was impossible to detect the defect at the time the product was put into circulation. Because of the ECJ's narrow interpretation of this defence, it will be difficult to escape liability.

4.5 When to Warn? Legal Analysis of the Need to Warn in Relation to Other Product Design Solutions

4.5.1 *Introduction*

This paragraph provides a legal analysis of the warning issue of 'when to warn'. As with the warning issues of the previous paragraphs, the legal answer to this question is addressed in the context of the insights from cognitive psychology and ergonomics.

It was concluded in paragraph 4.3 that because warnings have shown to positively influence the behaviour of individuals, empirical evidence warrants the treatment of warnings as precautionary measures in European product liability law. It was further concluded that avoiding accidents should be viewed as the ultimate legal goal of warnings. In view of promoting that ultimate goal, it is vital that European product liability law encourages producers to use warnings as a precautionary measure only when needed and secondly that it requires warnings to be well-designed. The former warning issue is discussed here, the latter issue of warning adequacy will be discussed in paragraph 4.6.

The expression of 'when to warn' can be interpreted in two ways. 'When' can refer to the risk itself: what risks need a warning? This question was the central topic in the previous paragraph, albeit in a negative sense: 'What risks need no warning?'. 'When' can also be interpreted to mean as 'under what conditions are warnings appropriate and preferred risk-reduction measures compared to other design measures?' This question is analysed here from a legal viewpoint. It is particularly of relevance to the situation in which the product that allegedly caused the damage carries a warning, but it can also be of essence in the situation where no warning is provided. It concerns the role of warnings in the design process of safe products. It follows from the warning research findings and ergonomic literature that warnings should be used as a 'last-resort precautionary measure'. This entails an important implication for European product liability and it can be regarded as valuable for the way in which European product liability law should assess the role of warnings as precautionary measures. It is suggested here that this last-resort-measure principle of product warnings should be accepted in European product liability law by designing products on the basis of the hazard control hierarchy model, as it encourages producers to let safety have a prominent role in the design process of products without losing sight of other design goals. This will advance the Directive's implicit goal of promoting consumer safety and secondly contribute to reducing the number of unnecessary warnings, which may in turn help prevent the occurrence of negative effects associated with overusing warnings.[156]

156 See paragraph 4.4 on the subject of overusing warnings and paragraph 3.8.

This paragraph is divided in the following parts. First, the empirical findings of the warning research and theoretical notions that are helpful in comprehending why warnings can generally be viewed as a last-resort protective measure are discussed in paragraph 4.5.2. The idea behind the principle is not new to the law. Various pieces of EU product legislation, mandatory and non-mandatory, contain the viewpoint of the principle and also rulings of European courts indicate that the presence of a warning does not automatically free a producer from liability. Paragraph 4.5.3 describes the legal sources that have included the last-resort-measure principle of product warnings. Next, in paragraph 4.5.4 I recommend that European product liability law should accept the last-resort-measure principle of warnings after discussing the pros and cons of adopting this principle in European product liability law. In the next paragraph, I focus on a special category of warning defects that centre on the 'misuse' of a given warning by the producer and recommend its use by claimants in European product liability law. After that, suggestions are made as to how this principle should fit in with the liability test of the Directive (paragraph 4.5.6). The final paragraph 4.5.7 provides a summary of the recommendations that were made here in paragraph 4.5.

4.5.2 The Last-Resort-Measure Principle of Product Warnings

4.5.2.1 The Degree of Behavioural Effect of Warnings

The conclusion that warnings possess the ability to change behaviour and can lead to a reduction in injuries compared to when the warnings are absent leaves questions pertaining to the degree unanswered.

Two questions with regard to the degree can be distinguished. First, how high is the level of behaviour that corresponds with the warning (relative to the level of risky behaviour that does not correspond with the precautions of the warning). Research studies can measure this behavioural effect, *i.e.* added value, by including a control condition. This enables researchers to assess whether there is a difference between the level of compliant behaviour without a warning (baseline score) and the level of behavioural compliance when a warning is present (compliance score). This type of research measuring whether the presence of a warning improves compliant behaviour compared to the level of behavioural compliance when no warning is given is vital. If it is likely that high levels of self-protective behaviour will occur regardless of the presence of a warning, then warning seems superfluous. On the other hand, if certain safe user actions are not likely to happen when a warning is not present, providing one may be a key instrument to increase safety. The second question concerns how high is the effect of a specific warning design on behavioural compliance (relative to other warning designs)? This is called the effectiveness score and equals the compliance score minus the baseline score.[157] In conclusion, the behavioural effectiveness of warnings can be interpreted in two ways. Normally, it refers to the general observation of the degree of people that produce behaviour that is in accordance with warning compliance in a particular situation. This issue is of relevance for determining whether providing a warning is needed. On a more concrete level, the behavioural effectiveness of warnings can also refer to the effectiveness of a specific warning design. The research findings of studies that measure and compare different warning designs can point out a design format that is most effective and may lead to the highest compliance rate. This issue is of special importance to the adequacy of a warning.

Research indicates that warnings have a moderate role as accident prevention mechanisms. It is difficult to provide a definite answer to how much effect adding warnings generally has on behaviour. Compliance rates of warnings measured in studies often vary in height. Still, it can be concluded from the findings that, even though warnings can modify behaviour, they tend to show a relatively low rate of behavioural compliance by the target audience.[158] As shown by the illustrative studies in paragraph 3.6.9, the degree of behavioural compliance varies among the studies. In a number of cases, compliance rates above 70% were observed, whilst in other studies and under different testing circumstances less subjects produced compliant behaviour and the compliance scores were much lower, around 30%. Moreover, there are also varying levels of effectiveness scores: some are up to 60% or even higher, but scores around 10% and lower were found or no effect at all was also found. For instance, study 4 of Wogalter e.a. 1999b, showed compliance rates of 13%, 43.8%, 68.8% and 81.3%. The lowest compliance score was for the control condition. The effectiveness scores in the warning conditions ranged from 30.8%, 55.8% to 68.3%.

4.5.2.2 Why Warnings Are no Safety Panaceas: Three Interrelated Explanations

Why warnings do not consistently produce high percentages of compliance and effectiveness scores can be explained by the information processing models, such as the C-HIP model, by taking into account the number of factors that influence the warning process, and also by the viewpoint that warnings are active protective measures. As you can notice, these explanations interrelate. They are discussed below.

By viewing warnings from an information processing perspective, it is clear that change in human behaviour does not come about easily. Several stages need to be successful. Moreover, as the model predicts, behavioural compliance will be the least reliably observed effect of a warning.[159] It is thus not surprising that warning studies have confirmed that

157 Edworthy & Adams 1996, p. 9. See also paragraph 3.5.4.

158 Lehto & Miller 1988, p. 261; DeJoy 1989; Hunn & Dingus 1992, p. 497.

159 Lehto & Papastavrou 1993, p. 571.

the percentages of individuals, who notice a warning is higher than or equal to the percentage of persons who read the warning and that fewer people comply with warnings than the number that read it.[160] Take for example the study of Otsubo 1988 again, that measured, on the basis of a questionnaire, whether the subjects saw, read, recalled the content of the warning in addition to observing whether the subject complied with the warning. The results showed that the number who read the warning on the circular saw was two-third the number who noticed the warning, and the number of individuals who complied with the warning was two-third the number who read the warning. For the jigsaw that was perceived as less dangerous, about half of the subjects dropped out at each stage.[161]

The second explanation for why warnings are considered the least effective safety measure corresponds well with the aforementioned information processing perspective. The theory entails that warnings rely on humans to change their behaviour in order to avoid possible harm, whilst there are also product design solutions possible that protect product users from hazards without requiring any behavioural actions of users. The design of the product automatically protects the user. This explanation is based on the distinction between active and passive injury control measures, whereby active measures basically require individuals to engage in self-protective behaviour or to refrain from some behavioural action(s) to avoid possible injury.[162]

A third associated explanation for why warnings generally fail to be safety panaceas lies in the complexity of the warning process that can be evidenced by the variety of variables that interact with each other and that are dependent on the context in which the product is used, the target population and the warning itself.[163] Person variables seem to play an important role in the warning process. As portrayed by the C-HIP model, the flow of information through the C-HIP model is not entirely linear; feedback loops from later stages to earlier stages are possible. Factors related to attitudes and beliefs, such as hazard perception and familiarity, have shown to influence earlier stages. It could even be argued that they broadly determine how individuals approach and respond to warnings.[164] It seems that attitudes and beliefs are at the heart of many bottlenecks in information processing. Addressing incorrect beliefs related to hazard perception and familiarity may overcome not only the bottleneck in the attitudes and beliefs stage, but also the bottlenecks related to noticing and complying.[165] Although the design features of a warning can increase the extent of warning compliance of receivers, the receivers are ultimately the ones that must

160 Edworthy & Adams 1996, p. 51; Lehto 2006, p. 68. See *e.g.* Frantz & Rhoades 1993; Otsubo 1988; Friedmann 1988; Strawbridge 1986; Jaynes & Boles 1990.

161 Otsubo 1988.

162 DeJoy, Cameron & Della 2006, p. 36. See also paragraph 3.9.5.5.

163 Rogers, Lamson & Rousseau 2000, p. 132; Stewart, Folkes & Martin 2001, p. 360; Cox III e.a. 1997, p. 201.

164 DeJoy 1999a, p. 190; Vredenburgh & Zackowitz 2006.

165 Riley 2006, p. 297.

produce self-protective behaviour or not.[166] Notwithstanding that the individuals' expectations may override the best-designed warning message, poorly designed warnings benefit no one.[167]

4.5.2.3 Hazard Control Hierarchy: Warnings as the Third Line of Defence

The failure of the warning research to show the clear superiority of warnings as safety mechanisms confirms the desirability of designing products on the basis of the hazard control hierarchy model that holds that warnings should be considered the third method for controlling safety problems and a last resort. As previously described in paragraph 3.9.5, the hazard control hierarchy model from the ergonomic literature takes account of the limitations of people by prioritising a sequence of hazard control methods from most to least effective in preventing damage to people and property.

The basic sequence is first to eliminate the hazard by designing the hazard out. However, not all hazards can be eliminated or minimised by means of alternative design, for example because of the intended functioning of the product. The second line of defence is to guard against contact with the hazard. The purpose of this strategy is to limit the contact between people and property and the hazard. This can be done in various ways. Using a safety guard that offers a material barrier between the hazard and the user is a form of guarding, but also procedural ways of guarding, such as protective devices that prevent erroneous behavioural actions, can be implemented to reduce the risk.[168] Interlocks, outlocks and inlocks all are design methods that constrain behavioural actions in order to prevent accidents. Think of the so-called dead man switch on lawn mowers that automatically shuts down the power to the blade when the handle is released or an activation mechanism that requires pressing two buttons simultaneously using both hands to activate the hedge trimmer. Thirdly, the model entails that if designing out and guarding are not possible or practical, warnings should be used to deal with safety problems.[169]

The overall finding thus represents an important implication for product design. As argued by human factor specialists such as Lehto, Salvendy, Wogalter, Laughery and many more, warnings should be viewed as an effective supplement, not as a substitute for other hazard-control methods, such as (re)design and guarding. Warnings are a last-resort precautionary measure.[170] Accordingly, it has been argued that providing a warning when it

166 Vredenburgh & Helmick-Rich 2006, p. 381.

167 DeJoy 1989.

168 The Machinery Directive (p. 35) makes a distinction between guards and protective devices. A guard is defined as "a part of the machinery used specifically to provide protection by means of a physical barrier". A protective device is "a device (other than a guard) which reduces the risk, either alone or in conjunction with a guard".

169 McCormick & Sanders 1992, p. 681; Wogalter & Laughery 2006, p. 889; Wogalter 2006a, p. 4; Lehto 2006, p. 65; Statler 2005; Van Aken 1996, p. 27-33.

170 Lehto & Salvendy 1995; Wogalter & Laughery 2006.

is unlikely to be effective as a precautionary measure and without considering other potentially more promising intervention strategies, can be seen as a misuse of warnings.[171]

Because warnings should not be viewed as a cure for bad design, warning researchers have also contended that it is of vital importance that the design of warnings is viewed as an integral part of the design process of the product. Developing a warning after the design of a product is essentially completed is likely to negatively affect the design of effective warnings.[172] The development of a warning is a complex design process and involves more than simply writing a message that warns people of hazards. Considering warnings much earlier allows the warning design process to interact with and overlap the product design process. Such integration can serve to enhance user performance and satisfaction, and help producers to avoid costly design modifications required to eliminate hazards or reduce risks that might otherwise be discovered late in the product development process.[173]

4.5.3 *The Last-Resort-Measure Principle of Warnings in a Legal Context*

4.5.3.1 General

Following the discussion above, it has become evident that warnings are viewed as a last-resort measure because they have a tendency to fall short as a precautionary measure compared to alternative design solutions. This awareness is not new to the legal arena. It has been recognised in various ways. Below, legal sources such as EU product legislation and non-binding, influential documents in product safety policy such as the ISO/IEC Guide 51 show the acceptance of this principle, although some sources do that more explicitly than others. Community sector-specific product legislation applicable to products, such as machinery, consumer products with low voltage, chemicals and toys contain provisions that embody the principle of the hazard control hierarchy. Because many consumer products are covered by these regulations it follows that their designs are predicated on meeting this guiding design principle. The explicit or implicit presence of this perspective in European product safety regulations amplifies that it should be of relevance for the purposes of European product liability. Especially with respect of the products that are expressly covered by product regulations that contain the rule to use warnings as a last resort, it is important that compliance with this rule is tested under the product liability regime of the Directive. Next to public product safety regulations, product liability cases also evidence that courts are not aloof from taking account of this guiding design principle when assessing liability.

---

171 Lehto & Salvendy 1995, p. 2162.

172 Wogalter & Laughery 2006, p. 908; Laughery & Hammond 1999, p. 10; Van Aken 1996, p. 97.

173 Frantz, Rhoades & Lehto 1999, p. 292.

4.5.3.2 EU Public Law Developments

General Product Safety Directive

The wording of the GPSD implies that it agrees with controlling product risks on the basis of the hazard control hierarchy model. Article 5(1) stipulates that the presence of warnings does not exempt any person from compliance with the other requirements laid down in this Directive. This statement could be viewed as to mean that the presence of a product warning does not absolve producers' need to design a product that is intrinsically safe without the warning.

Machinery Directive

Compared to the GPSD, the Machinery Directive clearly recognises the principle of the hazard control hierarchy. As was discussed in paragraph 4.4.3.1, Annex I prescribes the essential health and safety requirements relating to the design and construction of machinery. These include carrying out a risk assessment and adopting measures to eliminate or reduce the risks. This latter procedure is defined as *risk reduction.*[174] As shown below, 1.1.2(b) of Annex I of the Machinery Directive takes account of the hierarchy with respect to preferred design methods by stating that the principles must be applied by the producer in the order given.

> 1.1.2 Principles of safety integration
> (a) Machinery must be designed and constructed so that it is fitted for its function, and can be operated, adjusted and maintained without putting persons at risk when these operations are carried out under the conditions foreseen but also taking into account any reasonably foreseeable misuse thereof. The aim of measures taken must be to eliminate any risk throughout the foreseeable lifetime of the machinery including the phases of transport, assembly, dismantling, disabling and scrapping.
> (b) In selecting the most appropriate methods, the manufacturer or his authorised representative must apply the following principles, in the order given:
> – eliminate or reduce risks as far as possible (inherently safe machinery design and construction),
> – take the necessary protective measures in relation to risks that cannot be eliminated,

174 In a similar way, see the ISO/IEC Guide 5. The EuroSafe Working Group on Risk Assessment speaks of actions for risk reduction, see Rider e.a. 2009, p. 76. The RAPEX Guidelines, also discussed in paragraph 4.4.3, use the term risk management to refer to follow-up action, which is separate form risk assessment and aims to reduce or eliminate the risk, see p. 64. Cf. Article 3(12) GFL.

> – inform users of the residual risks due to any shortcomings of the protective measures adopted, indicate whether any particular training is required and specify any need to provide personal protective equipment.
> (c) When designing and constructing machinery and when drafting the instructions, the manufacturer or his authorised representative must envisage not only the intended use of the machinery but also any reasonably foreseeable misuse thereof. The machinery must be designed and constructed in such a way as to prevent abnormal use if such use would engender a risk. Where appropriate, the instructions must draw the user's attention to ways – which experience has shown might occur – in which the machinery should not be used.
> (d) Machinery must be designed and constructed to take account of the constraints to which the operator is subject as a result of the necessary or foreseeable use of personal protective equipment.
> (e) Machinery must be supplied with all the special equipment and accessories essential to enable it to be adjusted, maintained and used safely.

Toy Safety Directive

Another illustrative example is the new TSD of 2010.[175] Two viewpoints with regard to the misuse of warnings are expressed in this directive. First, the new TSD aims at triggering producers to use toy warnings appropriately. Article 11(1) TSD lays down the obligation that toys shall not bear one or more of the specific warnings where that warning conflicts with the intended use of the toy, as determined by virtue of its function, dimension and characteristics. As is apparent from the 30th recital, this obligation has been introduced by the TSD in order to prevent producers from misusing warnings to circumvent the applicable safety requirements. This has occurred particularly in case of the warning stating that the toy is not suitable for children under 36 months. Apparently, various producers used this warning for toys that were in fact intended for users under the age of 3.

Secondly, according to the preamble, the general safety requirement of the TSD entails that

> Where a hazard cannot be sufficiently minimised by design or safeguards, the residual risk could be addressed by product-related information directed at the supervisors, taking into account their capacity to cope with the residual risk. According to acknowledged methods of risk assessment, it is not appro-

175 See also paragraph 2.4.3.3.

priate for information to supervisors or a lack of history of accidents to be used as a substitute for design improvements.[176]

This means that the Member State authorities that must assess whether a toy, presenting a hazard that is not covered by a specific safety requirement of the TSD, meets the general safety requirement, can determine that the toy fails to meet the requirement, and follow-up actions might be taken when a toy warning had been given whilst a safer design solution was available that would reduce the risk substantially.

ISO/IEC Guide 51

This Guide fully embraces the hazard control hierarchy perspective to reduce risks. This non-binding safety Guide adopts an approach that aims at reducing the risk arising from the use of products to a tolerable level. Figure 4.1 in paragraph 4.4 displays the process. Risk reduction plays an important role in the process as this step is responsible for achieving a tolerable risk. It is noted in point 6f of the Guide that when reducing risks the order of priority should be as follows: 1) inherently safe design; 2) protective devices; 3) information for users.

4.5.3.3 EU Civil Law Developments

It appears that the viewpoint, that a warning will not necessarily provide a defence to liability if the design of the product itself can be considered flawed, has also been held by European courts, although it is difficult to provide an estimation of the extent to which various courts in Europe adhere to this viewpoint, because an extensive and full inquiry into the various product liability laws and cases in Europe was not done.[177]

A good example can be provided by the recent Dutch *Multivac-machine* case that dealt with an employer's liability based on fault for a dangerous industrial machine.[178] Even though the Dutch Supreme Court was assigned to judge the scope of the employer's duty of due care under fault-based liability and not the producer's liability under the EPLD, I believe that the court's reasoning is instructive for the determination of product liability, regardless of the theory, and applicable within the Directive's liability regime. The decision shows that the inclusion of warnings onto machinery will not necessarily lead to the outcome that the machine is reasonably safe. Instead, an employer should rather first focus

176 Recital 29. Note that this part of recital 29 was not yet present in the proposal: Proposal of the Commission of 25 January 2008 for a Directive of the European Parliament and of the Council of the safety of toys 2008, p. 12.

177 See chapter 1.

178 HR 11 november 2005, *NJ* 2008, 460 (*Multivac-machine*). See also paragraph 2.2.3.3. It must be noted that the Dutch liability standard based on fault on the part of employers seems to be interpreted in an overly high way by national courts. Dutch scholars have expressed their concerns about the far-reaching application of liability in this field by the Dutch Supreme Court.

his attention on the safety of the machine itself. The Supreme Court held that a warning of risks that may arise from the use of a hazardous machine will not always be sufficient. This corresponds, according to the Supreme Court, with the circumstance that an employer is obliged to consider the empirical fact that the daily use of a machine or tool may lead to careless behaviour on the part of an employee, even though warnings are present. For this reason, it may be reasonably expected from an employer that he examines whether adequate preventive measures are possible or whether the machine can operate more safely. If this is not the case, an employer must look closely at whether it is possible to warn against the hazard in a sufficiently effective manner.

A case that resembles the one mentioned above, is the German *Floor panel stripper* case that dealt with producer's liability under the German law implementing the Directive with regard to damage suffered by an employee during the use of a machine that falls under the Machinery Directive. The District Court of Düsseldorf held that the producer of the floor panel stripper is liable for a defective design of the product, among others, because the risk of crushing one's fingers could have been reduced by adding a simple protection device. It further adds that the user instructions and warnings against the dangers of misuse did not clear the product of its design defect, since warnings do not replace necessary design features.[179]

Furthermore, in an Austrian case a garden chopper that cut off four fingers of the user's right hand was considered to be defective despite the presence of a warning of that risk, for the reason that the device could easily have been designed with a security guard.[180]

Hence, such cases as described above provide a beacon for the view that warnings should not be treated by law as substitutes for more effective design features and thus that warnings should not be used as a shield against defectiveness for a flawed design. Lastly, even though it seems a bit far-fetched to me, it has also been asserted that it could follow from Article 12 of the Directive[181] that the Directive does not permit warnings to be used to evade liability for a flawed design of the product.[182]

4.5.4 *Recommendation: Legal Acceptance of the Last-Resort-Measure Principle of Warnings*

In this paragraph, it is recommended to accept the last-resort-measure principle of warnings under the liability regime of the Directive. The implications of the hazard control hierarchy

179 Landgericht Düsseldorf 30 November 2005, 10 O 144/04, *NJW-RR* 2006, 1033 ff (*Floor panel stripper*). Reported in *EPLR* with note by S. Lenze, see Lenze 2006.

180 This case was briefly mentioned in Howells & Borghetti 2010, p. 476.

181 Article 12 stipulates that the liability of the producer arising from this Directive may not, in relation to the injured person, be limited or excluded by a provision limiting his liability or exempting him from liability.

182 [2001] 3 *All ER* 289 (*A v. National Blood Authority*), paras. 65, 70; Howells 2005a, p. 146.

model and the consequent last-resort-measure principle for European product liability law are that having provided a warning against a specific product risk does not necessarily mean that producers are absolved from a finding of defectiveness for that risk. The circumstance that it was possible to eliminate or substantially reduce the risk by modifying the design of the product should provide an important indication for finding defectiveness under the Directive.

The reasons for and against accepting this principle are set out below. At the outset of the paragraph it is important to remark that acceptance of the legal implications of this principle does not permit to interpret it as an absolute rule, meaning that warnings can never absolve a producer from a judgement of defectiveness. I recommend, as will become clear below, that the principle should be applied if the circumstances of the case indicate that there is a reasonable design modification feasible but not implemented whilst the circumstances suggest that the use of a warning as a protective measure is an inappropriate and therefore insufficient way to reduce the risk to an acceptable level of product safety. This exercise for determining defectiveness involves an intelligent balancing process of several factors and this will be dealt with in more detail later on in paragraph 4.5.

Various academics have echoed the consideration that warnings should only be used as a last-resort measure and not as a substitute for alternative design methods that are by nature more effective in providing protection and reducing risk. Especially Latin has emphasised that warnings should not be treated as legally acceptable alternatives to safety product designs by referring to research on the human information processing limitations.[183] But also other American academics like Twerski and Henderson have substantiated this principle and they have pleaded for the adoption of a similar type of rule in the new US Restatement of the Law (Third), Torts, which rule has now been set forth in comment l.[184] Comment l of the new US Restatement of the Law (Third), Torts sets forth the relationship between design and instruction or warning and provides that

> 'when a safer design can reasonably be implemented and risks can reasonably be designed out of a product, adoption of the safer design is required over a warning that leaves a significant residuum of such risks. For example, instructions and warnings may be ineffective because users of the product may not be adequately reached, may be likely to be inattentive, or may be insufficiently motivated to follow the instructions or to heed the warnings. However, when an alternative design to avoid risks cannot reasonably be implemented, adequate instructions and warnings will normally be sufficient to render the product reasonably safe.'

183 Latin 1994.

184 Henderson & Twerski 1990. See also Weinstein e.a. 1978.

Lastly, it notes that warnings are not a substitute for the provision of a reasonably safe design.[185] In Europe, legal scholars among whom are Howells, Clark, Miller and Goldberg, and Dutch academics, have contended that warnings should not automatically insulate a producer from liability.[186]

Given the various opinions that have been expressed in this context, the intended underlying reasons for the explicit acceptance of this principle as well as any potential negative ramifications are considered here.

The main reason for accepting such a principle in European product liability law would be that it facilitates designing safe products that have effective design methods to control risks, rather than relying heavily on warnings, that are – by their very nature – less effective.[187] The presumption that consumers should bear the responsibility for safety because they can avoid an accident by simply following the warning that was provided would often turn out to be quite harsh, since cognitive psychology has taught us that following a warning is not seldom a simple task. Conversely, not having such a principle in European product liability could signal to producers that they can absolve themselves from liability by merely giving adequate warnings for the risks attached to the products. Because safety is not the only aspect that is of relevance to the design of products, a consequence of not accepting the principle may well be that eventually safety becomes of minor importance than other design goals, such as product performance, cost, user satisfaction and aesthetics. Hence, it could diminish the incentive for producers to spend optimal attention to safety in the design process. In addition, other circumstances also merit setting high safety requirements to the design of consumer products. Compared to employers, producers are more restricted in their strategies to improve safety. They cannot directly influence user behaviour via rules, procedures, training etc. because they lack contact with product users. Furthermore, it may also be difficult to assess who is likely to suffer injuries from the consumer product because of the variation in product users.

As noted in paragraph 3.9, implementing safety early in the design process is desirable, because making design changes after completion of the design is usually more expensive and can lead to less user-friendly safety add-ons than before completion. Hence, if in a specific case a warning was used instead of a more appropriate safety measure and the producer is held liable, redesigning the finished product to increase safety will be costly. This would in consequence be unfortunate for consumers, since it is likely that they need to pay a higher price for the new redesigned version of the product. This increase in cost can be viewed as a criticism for accepting this principle. However, it would even be more

185 US Restatement of the Law (Third), Torts: Products Liability, § 2, comment l (American Law Institute 1998, p. 33).

186 Miller & Goldberg 2004, p. 432; Howells 2005b; Lenze 2003a, p. 46; Clark 1989, p. 103; Van Boom 2003, p. 32; Giesen 2005; Van Dam 2000.

187 See American Law Institute 1998, p. 16.

costly to change an existing design than to consider safety beforehand and adding safety features when the design is still fluid.

Another important reason for using the principle in European product liability law is that it signals to producers that warnings should not be misused, *i.e.* used as a shield against liability whilst the warning was not the appropriate way to reduce the risk being warned of. Hence, the penalty for having misused a warning is civil liability. Consequently, this principle contributes to having a world full of products that only contain warnings that are truly necessary and that are able to fulfil their legal goal of reducing the risk to an acceptable level so that accidents are prevented or reduced. In addition, the unintended effects that are associated with an overuse of warnings are likely to be minimal as a result of adhering to the principle in law.

A related disadvantage attached to taking consideration of the last-resort-measure principle in European product liability cases is that it can lead to a reduced choice in products on the consumer market. Some producers might decide not to market their products anymore, because they are not able to increase the level of safety without unduly increasing the products' price.[188]

The argument most used against the principle is that it nearly kills the individual responsibility of consumers as it forces producers to design foolproof products. The last-resort-measure principle does not adequately trigger consumers to behave with reasonable care, because liability can still be established even if they have been warned against the risk. In this respect, it has been argued that a finding of liability on the basis of the last-resort-measure principle can be viewed as rewarding consumers for behaving as fools.[189]

Following the discussion of the pros and cons, I believe that the last-resort-measure principle of product warnings should be adopted for its safety incentive and more specifically as it promotes producers not to misuse warnings. It requires producers to be selective in providing warnings. In addition, it does not object to a responsibility of users for safety, it simply embraces human limitations and errors and consequently grounds the perspective of safe design on the notion that it is far more difficult to prevent accidents by influencing behavioural patterns of users than by a product's design that provides automatic protection. As human factor specialists Sanders and McCormick have advocated: 'It is easier to bend metal than to twist arms'. Consumers' responsibility is not ignored as risks arising from unreasonably expected product interactions and insignificant risks do not need to be controlled, not even by warnings.[190] Furthermore, as noted above, acceptance of this principle still makes it possible to consider a warning adequate and to reject defectiveness even though there was a safer design measure available. Its added value is that it forces

188 See Schwartz 1999-2000, p. 56.

189 See Schwartz 1999-2000, p. 56; Ausness 1999, p. 644.

190 See also paragraph 4.4.

producers to make the appropriate trade-offs between the various design goals when designing products.

Empirical research reflects the way consumers handle products. By using ergonomic methods such as user trials and other information sources producers can obtain a reasonably accurate picture of how the target audience perceives the characteristics of the product and how it uses the product. Often, the findings will show that certain consumers are careful, whilst others use it in a dangerous and/or unexpected way because it satisfies their users' goals. Nevertheless, the results will also demonstrate how the majority is likely to use the product. Producers must anticipate the degree of carelessness of the latter group to avoid liability. If the results confirm that many consumers have difficulty with processing the warning information effectively, and that other more effective measures that are available are needed to avoid injury, then we are indeed fools that need 'foolproof' products in certain instances. Likewise, in the event that research, such as a user trial, shows that the majority of the target audience is likely to ignore the product's guard that serves as a protective measure, this circumstance provides an indication that the producers should be allowed to use well-designed warnings instead in order to trigger the user's risk perception and to use the product safely, and hence liability.

4.5.5 Recommendation: Recognition of a Warning Claim Centred on Producer's Misuse of a Warning

4.5.5.1 Warnings Form Part of the Product's Design

My second recommendation follows from what is just suggested above. The previous part of paragraph 4.5 distinguished defects relating to warnings from defects arising from the design of the product. This is standard practice within the field of product liability, given that the traditional classification scheme of defects relating to manufacturing, design or warning is often used by claimants and courts as a source of inspiration for examining why the product caused damage.[191] It must be remembered that the Directive itself does not make a distinction between these categories. It only requires proof of a defect. However, what this means exactly is unclear. The concept of defect remains vague, which underlines the usefulness of having a classification scheme that sets out various forms of possible product defects.[192]

It has been often stated by courts and academics that a product can be defective for the absence of a warning or an inadequate warning even if it is properly designed. This

191 See also paragraph 4.2.4.

192 Cf. Burton J in [2001] 3 *All ER* 289 (*A v. National Blood Authority*), para. 39, who concluded that boxing defects in the traditional way is of no assistance and he proposes a new distinction between 'standard' and 'non-standard' products.

statement shows that they view the categories of warning defect and design defect as distinct defect categories. The categories create independent safety obligations for producers.[193] Warning and design defects are distinct and independent of each other, because fulfilling the obligation arising from one type does not automatically mean that the product cannot be considered defective on another ground. For example, even if there should be no defectiveness due to the absence of a warning because the risk was obvious, this does not mean that the design cannot be held defective for the same risk because a reasonable alternative design of the product that would reduce or avoid the risk was absent.[194]

It has also been held that warning defects and design defects are related.[195] It has even been noted that warnings can be viewed as a special subset of design defects.[196] I agree. The defect distinction between warning and design seems not completely natural to me.[197] An implication that followed from the hazard control hierarchy model and is recommended by the warning research literature is that a safe design is the starting point of a safe product and that warnings should be viewed as part of the design and design process of products. This is obvious since product safety increases along with warning effectiveness. I regard warnings as measures that aim at reducing the product's risk as a result of which accidents can be avoided. Hence, a product is properly designed if the product risks have been reduced with appropriate design methods to an acceptable safety level and this includes the use of warnings. In consequence, if a warning is missing or if the given warning can be considered inadequate then both circumstances negatively affect the safe design of the product.[198] In view of this, it seems peculiar to me that warning defects normally do not fall within the scope of design defectiveness whilst this latter defect category seems to have a broad scope: various design solutions (that aim at eliminating or reducing of the risk), that could have been adopted but were omitted or that were inadequately designed, fall within the meaning of a defect in the design of a product. So, why are warnings not treated as a design defect too, just like other design features that aim at protecting people by reducing the product's risk? Just as with having provided a warning, having implemented

193 Owen 2008, p. 346.

194 See paragraph 10.2 in Owen 2008 about the fall of the patent-danger doctrine in American product liability. This doctrine consisted of a bar from recovery for victims that suffered injuries as a result of an obvious design danger. Nowadays, many courts reject this doctrine in design defect cases, holding that the obviousness of a danger is just one factor in assessing whether the product's design is defective.

195 See Hodges 1993b, p. 106; Van Dam 2006, p. 377; Owen 2008, p. 586.

196 Howells & Borghetti 2010, p. 461.

197 This distinction is most evident in American product liability law and also of significant value, since the US Restatement of the Law (Third), Torts distinguishes between the three defect categories and applies different standards to them. The EPLD does not.

198 Compared to the statement that was given above, I believe that it is better to formulate that a product can be defective for the absence of a warning or an inadequate warning in spite of an otherwise safe design of the product.

a design solution that reduces the risk, does not automatically mean that other available design modifications were not necessary to render the product non-defective.[199]

In conclusion, the design methods (design out, guard or warn) can provide a range of design solutions to control a risk. Because a product risk can usually be approached by more than one design solution, several options are in principle available as an alleged cause for design defectiveness. These design solutions are distinct, because they offer different levels of protection. Hence, establishing non-defectiveness on one ground, *i.e.* fulfilling one duty (*e.g.* providing a warning), does not fulfil another (*e.g.* the need to guard or to design out the hazardous product property). Because warnings are lower in the hierarchical order, it is logical to contend that the use of a warning does not automatically relieve producers of the need to implement a design feature that is in essence more protective. The same can be said vice versa. Even though guarding is of a higher order in the hazard control hierarchy model than warning, that does not mean that protective measures that act as guards automatically absolve the need to provide a product warning. Although the latter usually will be the case, the risk may for various reasons still be considered too high without a warning. Think of dangerous machines that contain additional (reminder) warnings of risks that have already been reduced by other design measures.

By treating warnings as precautionary measures and thus as instruments to achieve safety through the design of products, it follows that it should be recognised by European courts and litigants that the absence of a warning or the presence of an inadequate warning can be viewed as a defect in the design of the product, just as with other design measures, since all products of the product line have this defect (the absence of a design feature or presence of an inadequate design feature).[200] It is the result of the producer's conscious decisions during the design process of products.

An improvement of the traditional categorisation of product defects would be to rearrange the category of design defects by including warning defects and other defects that are inherent in the design process of products and concern all products of the same line. In addition, the category of design defects could be specified in more detail with possible subdefects. This specification of subcategories of design defects should be based on the design methods of the hazard control hierarchy model. The subdesign defects should be ranked by priority as a tool for claimants who want to contest the insufficient level of safety that was provided by the product's design. Just as the hazard control hierarchy model provides a guideline to designers on how to design safe products, claimants should view

199 To increase the safety of a product, think of design solutions such as adding a safety guard to cover a moving part of the product, changing the product's material to reduce the risk, implementing an interlock that prevents accidental use of the product in dangerous situations; improving the adequacy of the guard that was implemented.

200 See also figure 4 in paragraph 4.2. The overlap between the categories of design defects and product information defects is used to depict this link.

the hazard control hierarchy model as a guide in deciding what should be the basis for alleging defectiveness under the Directive (see figure 9).

Moreover, because warnings generally have only a moderate role in increasing the safety level of a product's design, a special type of defect should receive attention in European product liability law and that is *defectiveness related to the producer's misuse of a warning*. This can be done by an explicit recognition and treatment by European courts, litigants and in legal literature of this defect category in relation to warnings next to the common warning claims that centre on the absence of a warning or the presence of an inadequate warning (see figure 4 in paragraph 4.2.4).

This defect type concerns situations in which a user has suffered damage as a result of a risk that has been warned off. The warning itself was used by the producer as a measure to control the risk. Nevertheless, this is viewed as inappropriate, since the warning fails to reduce the risk to an acceptable level. The defect claim presumes that the available safer design solution is the appropriate design measure as it is commensurate with the level of product risk. Recognition and use of this claim is likely to advance the acceptance of the last-resort-measure principle of warnings in European product liability law, since this defect type reflects designing safe products on the basis of the hazard control hierarchy model. As follows from the above, defectiveness related to the producer's misuse of the warning falls within the (wide) scope of design defectiveness, because all the products contain the warning. More specifically, this defect type is viewed as a warning defect because there is a(n) (inappropriately used) warning present. I would like to title this special group of defect claims as claims centred on producer's misuse of the warning. These situations are normally treated as a claim based on the defectiveness of the design of the product, distinct from a warning defect, where the presence of a warning is one of the relevant factors taken into account when determining defectiveness. However, to emphasise the misplaced role/utility that is assigned to the warning as an accident avoiding mechanism by producers, I prefer to use this description.

Note that there are cases where the product hazard is inherent and unavoidable for the functioning of the product. Warning information is the only way to make the product reasonably safe. These cases fall outside the scope of this type of defect, because there are no design options other than warning possible that are more effective in providing protection. These cases can centre on the inadequacy of the given warning by claiming that a better design of the warning was needed to render the product not defective. It may even be argued that the product should not have been marketed at all because of its high risk and low utility.[201]

201 The latter issue concerns products that are inherently dangerous, such as alcohol beverages, other foodstuffs and prescription drugs. This is called 'product category liability' in American product liability law, see paragraph 10.3 in Owen 2008. Such products can be held defective because of a manufacturing defect or because of the absence of a warning or presence of an inadequate warning. However, European courts are

Figure 9 Hierarchical order of possible design defect claims[202]

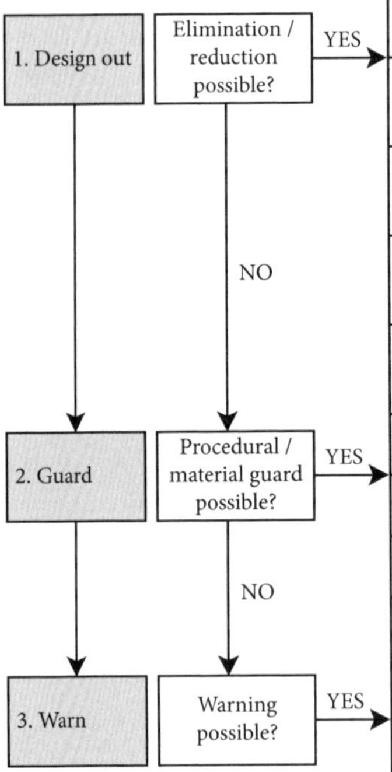

Elimination / reduction used	Procedural / material guard used	Warning used	Priority in claims
Yes	Yes	Yes	1. Reduction inadequacy claim 2. Guard inadequacy claim 3. Warning inadequacy claim
Yes	Yes	No	1. Reduction inadequacy claim 2. Guard inadequacy claim 3. No warning claim
Yes	No	Yes	1. Reduction inadequacy claim 2. No guard claim (misuse of warning) 3. Warning inadequacy claim
Yes	No	No	1. Reduction inadequacy claim 2. No guard claim 3. No warning claim
No	Yes	Yes	1. No elimination / reduction claim (misuse of warning) 2. Guard inadequacy claim 3. Warning inadequacy claim
No	Yes	No	1. No elimination / reduction claim 2. Guard inadequacy claim 3. No warning claim
No	No	Yes	1. No elimination / reduction claim (misuse of warning) 2. No guard claim (misuse of warning) 3. Warning inadequacy claim
No	No	No	1. No elimination / reduction claim 2. No guard claim 3. No warning claim
-	Yes	Yes	1. Guard inadequacy claim 2. Warning inadequacy claim
-	Yes	No	1. Guard inadequacy claim 2. No warning claim
-	No	Yes	1. No guard claim (misuse of warning) 2. Warning inadequacy claim
-	No	No	1. No guard claim 2. No warning claim
-	-	Yes	1. Warning inadequacy claim
-	-	No	1. No warning claim

likely to be very cautious in ruling that an inherently dangerous product is defectively designed even with the presence of adequate warnings because the risks are too high in view of the product's benefits, since such a decision necessitates the product's removal (and probably also similar products) from the market.

4.5.5.2 Potential Obstacles: Popularity of Warning Claims

Having explained the underlying foundations of why warnings have a moderate role in the design of safe products by means of revealing the insights of the warning research literature in this monograph and by recommending the acceptance of this principle in European product liability law, it is strongly hoped that important decision makers in the field of European product liability law have become knowledgeable and have been convinced to act accordingly. Litigants determine for a large part the content of the court's case; they essentially determine the boundaries of the legal dispute. It is therefore of importance that injured litigants investigate whether the above recommended defect claim centred on producer's misuse of the given warning is workable in view of the circumstances of the concrete case. In situations where the product that caused damage to the victim carries a warning that informs of the materialised risk, the ideal action of recovery under the Directive would principally be the warning defect claim centred on the producer's misuse of the warning that accompanied the product. Assuming that claimants can advance a second cause why the product can be considered defective under the Directive, the second claim regarding defectiveness should be centred on the inadequacy of the provided warning. Also in situations where the product bears no warning, this order of priority that respects the principle of warnings as a last-resort measure, can be of importance. In view of the incentive to design safe products and to reduce an overuse of warnings, it is preferred that claimants in such a situation first allege that the product is defective because of the absence of a design feature and secondly assert that the absence of a warning rendered the product defective.

By claimants' use of this category to allege defectiveness, courts are given the ability to take this circumstance seriously into account, whereas this would be impossible if the factor would not have been alleged by the claimant, assuming that courts are bound by what the parties to a suit bring forward and lack the freedom to add such a fact and consider it in their legal assessment. Naturally, if injured claimants make mention of the fact that there was a safer design solution available, courts can further investigate this, for example by appointing experts who can provide insights into the feasibility of a safer design solution including the design trade-offs that have to be made when implementing it into the design.

Unfortunately, there may be problems present that deter litigants to claim defectiveness under the Directive in the preferred order mentioned above. Moreover, courts can also encounter difficulties that would make them more reluctant to deliver opinions about design issues rather than warning issues.

---

202 This figure should be seen as a tool for claimants with respect to a ranking of claims that pertain to design defects that may be applicable to their case. It is not intended to provide a complete overview of all potential claims with regard to design defectiveness.

First, it has been noted in the legal literature that, especially given the experience of American courts, courts find it less problematic to impose liability because there was no warning at all or on the basis of the inadequacy of a warning that was present rather than to impose liability on the basis of a defective intrinsic design of the product. The main reason is that design defect claims challenge courts to second-guess the producers' decisions within the design process. It has been argued that the consequences of judging that the intrinsic design of the product needs to be modified, for example by adding a safety device or guard to render the product safe, can often be far-reaching for the industry. This reason is also valid for the outcome of legal decisions in warning cases, but probably to a lesser degree. Warnings are usually measures that are more practical to implement than other design changes and courts have a tendency to believe that providing warnings is relatively costly and easy to do.[203] Therefore, courts will often by a greater reluctance accept liability on the aforementioned ground in a case in which the problem can be also met by changing the warning's design.[204] A colourful example in this respect is the *Stepladder* case, discussed in paragraph 2.4.5, where, as Lenze points out in his note, the Austrian Supreme Court reverted to the question of the inadequacy of the warning instead of addressing the more prominent question whether the design of the stepladder was defective because it was designed with nylon belts and not metal supports.[205]

Another reason why courts would eagerly reach the conclusion that a warning should have been provided is that they generally put less restriction on the way consumers want to use products. This relates to the autonomy of users to decide which products to use and how to use them. An additional design feature, such as a guard that shields off contact with the hazard, could pose restraints on the ways in which a product can be used. By using a warning instead, that tells consumers not to use the product in that particular fashion, the consumers' freedom of handling the product is not restricted.[206]

Besides that claimants are aware of the circumstance that courts prefer to establish defectiveness based on warning inadequacy or the absence of a warning, instituting warning defect claims may be more appealing to injured victims, because such a claim is not highly technical as in the sense of a design defect claim. It may also be more difficult for producers to persuasively contest the claimant's assertion that changing the warning's design would have prevented the injury which increases the claimant's chances in being successful.[207]

Notwithstanding the advantages that seemingly accompany claims centred on warning issues and that facilitate the judicial decision making process, the advantages should not

203 Howells & Borghetti 2010, p. 476.

204 Miller & Goldberg 2004, p. 576.; Grubb & Howells 2007, p. 14; Owen 2008, p. 584.

205 Lenze 2003a, p. 46.

206 Grubb & Howells 2007, p. 373.

207 Bowbeer, Lumish & Cohen 2000-2001, p. 440; Owen 2008, p. 584.

lead judges to avoid the true matter of the case. Courts should be sensitive to the empirically based viewpoint that warnings are the least reliable and least effective design solution to protect consumers from hazardous product properties. They ought to recognise the circumstances that indicate that a warning is not likely to reduce the risk to an acceptable level and explicitly focus their attention on the question whether another design alternative that is more effective in preventing injuries should have been adopted to render the product non-defective.

4.5.6 *Framework for Determining Defectiveness for Claims Related to Producer's Misuse of the Warning*

4.5.6.1 General

It has been argued above that the last-resort-measure rule pertaining to product warnings should be an accepted principle in European product liability law in a way that it guides litigants in formulating their defect claims and courts to call into question the appropriateness of the design methods that have been used to achieve safety. It was also illustrated that there seems to be enough support for such a rule from various legal sources. This leaves the question unanswered that pertains to how the principle can find its way within the liability test of Article 6, *i.e.* the cornerstone of the Directive's liability regime. The answer is of specific interest for products that have warned against the risk. Therefore, the remainder of this paragraph deals with how the defectiveness test should be applied to products that have a warning of a risk that injured the claimant, *i.e.* in respect of claimants that use the above recommended allegation that the product warning has been misused by the producer.

A fundamental starting principle is that acceptance of the last-resort-measure rule of warnings in European product liability law should not be interpreted to mean that producers can never rely on warnings to circumvent the decision of defectiveness. Warnings *can* be suitable means to reduce risk and to avoid a finding of defectiveness under the Directive. There are instances in which producers can escape liability for defectiveness by having provided a warning; it can then be asserted by defendants that under those circumstances the warning can be considered an appropriate way to avoid accidents. Furthermore, acceptance of the principle should not lead to the conclusion that if there was an alternative design option available that would have controlled the risk more effectively but was not included in the product's design, the product should be deemed defective on the sole basis of that, as safety is not the only goal that needs to be taken into account when designing products. Designing products involves making trade-offs between various design goals.

The wording of the Directive shows that it does not require producers to design the safest version of products.[208] Member State courts should thus apply this principle with care.

The defectiveness test of the EPLD requires a determination of the level of safety a person is entitled to expect of a product. Hence, the formal test of defectiveness concerns judging the (legitimate) safety expectations of persons. As with products that have an obvious or generally known risk that caused the damage suffered, the test of the Directive can be problematic to utilise in cases of products that have a warning against the risk that materialised. Because expectations about a product and its safety are often formed on the basis of a product's appearance, consumers can run a high risk of being unsuccessful in such cases. Hence, if there is a warning present against the risk that materialised, courts could quite convincingly argue that the test permits rejecting defectiveness, because the warning enables consumers to form a good idea about the dangerousness of the product and to know what to expect. Indeed, warning studies have shown that the presence of a warning increases perceptions of product hazard and compliance behaviour compared to its absence. Equally, in the event that a risk can be considered obvious or generally known, courts could contend that consumers cannot expect to get more protection against this risk through the product's design.

As was noted earlier in chapter 2, many American and European academics recognise that this ambiguity is a serious flaw of a test based on consumer expectations.[209] As Owen points out, the consequence of this flaw is that it effectively rewards producers for failing to adopt cost-effective measures to remedy obviously unnecessary dangers.[210] The consequences of applying the test in such a way are too harsh and contrary to the test's intention I believe. It involves a narrow focus on only one circumstance, in other words that there is a warning present,[211] whilst the test requires taking all the circumstances into account. Furthermore, there is little consideration of the moments of carelessness and inattentiveness that can lead to certain user harmful actions that all humans are susceptible to. People can accidentally fall, make a wrong move, forget, be distracted, be unaware of the dangerous use action or highly misperceive its dangerous nature etc. In addition, such a reading of the test seems to concentrate on 'actual' expectations whilst it must be borne in mind that the Directive's test imposes a standard of expectations that consumers are 'entitled' to have.[212] In my opinion, this latter interpretation of the test allows the judgement that even in the presence of a product warning, persons can be entitled to expect a higher level of safety of the product.

---

208 Article 6(2) lays down that "A product shall not be considered defective for the sole reason that a better product is subsequently put into circulation".

209 See *e.g.* Stoppa 1992, p. 213.

210 Owen 2008, p. 506

211 Or that the risk is obvious or generally known.

212 See also paragraph 2.3.3.2.

As we have learned from the hazard control hierarchy design perspective, this higher level of safety can be offered through methods of designing away (guarding) and designing out. This directs the assessment under the defectiveness test towards analysing whether an available design solution that is more effective in reducing risk than a warning should have been present in the product's design to render it non-defective. Risk-utility or cost-benefit factors[213] should then play a role in the assessment of defectiveness of the product with the warning, as these factors make it possible to assess whether optimal trade-offs between the various design goals have been made. In this respect, it is argued here that if – on balance – the available design solution that is more effective in avoiding the risk can be considered a better design measure than the warning, the public at large is entitled to expect that such a design feature is used as a precautionary measure against the risk instead of or in addition to the warning. Accordingly, the application of such a notion of defectiveness with regard to products that have a warning truly reflects the recognition of the last-resort-measure principle of warnings under the Directive's liability regime.

In conclusion, a narrow application of the consumer expectation test should be avoided with regard to products that have a warning, an obvious and/or known risk. Factors such as the increase in size of the risk as a result of expected inadvertent and inattentive moments of users and the feasibility of implementing a design solution that would substantially increase the product's safety level should also be considered when determining the safety level persons generally are entitled to expect with regard to such products.

Because of this flaw, as well as other flaws, with regard to employing a consumer expectation test many American courts, scholars as well as the US Restatement of the Law (Third), Torts have favoured the risk-utility test as the sole, principal standard for adjudicating defectiveness in relation to a product's design and warnings.[214] According to some, the test entails a narrow 'micro-balance' of the costs and benefits of an untaken design feature that would have prevented the claimant's harm.[215] European academics have not so much contended that the defect test of the EPLD should be revised and solely based on risk-utility. Rather, it has been regularly argued that it is a matter of interpretation: although the test of the Directive is expressed in terms of consumer expectations and not risk-utility, that does not mean that is not allowed to use risk-utility considerations when determining what consumers are entitled to expect, especially in case of complex design defects, since all circumstances must be taken into account according to the text of the Directive. It has

213 These include the availability of a safer alternative design feature, the cost and feasibility of adopting the alternative design solution.

214 Owen 2008, p. 508; US Restatement of the Law (Third), Torts: Products Liability, § 2(b), comments d, f (American Law Institute 1998, p. 14).

215 Owen 1996-1997. In consequence, the risk-utility test should not be interpreted as a 'macro-balance', which involves a global weighing of all the risks of the product against the product's aggregate utility.

been said that the Directive's test can be seen as a composite or combined test.[216] But what is precisely meant by this? In what manner do the risk-utility factors play a role? More specifically, how should a composite test be applied to products that have a warning?

Risk-utility factors are especially well-suited and unavoidably needed to determine the defectiveness of products that have warned against the risk, since by means of these factors a comparison can be made between the costs and benefits of a warning as a risk-reduction measure and those of a proposed safer design solution, that derives from the method of guarding or designing out. Furthermore, because costs and benefits are the central components, it makes it possible to consider various factors by translating them either in terms of a cost (disadvantage of the design solution) or a benefit (advantage of the design solution). I believe that the product with the warning can be considered defective under the Directive, if the comparison shows that the balance of costs and benefits of the alternative design solution is better than the balance of the costs and benefits of the warning.[217] In other words, a product that has a warning to reduce the risk does not provide the safety consumers are entitled to expect, whenever the warning can be considered less cost-effective than the absent alternative design solution. The interpretation of the test in such a manner entails that the test is couched in terms of consumer expectations although the actual assessment to answer the question of defectiveness boils down to employing a risk-utility analysis. Hence, although the Directive does not require the claimant to prove that a better alternative design solution was available, the absence of proof of such circumstances is likely to seriously hamper the success of claiming that the product that already has a warning should be considered defective.

Below, a framework is provided that further explains which factors should play a role and how they should be applied/balanced in case of adjudicating defectiveness under the Directive's test of a product that has a warning against the risk that caused the injury. Note that, as is the case with any other test, it is admitted that by including risk-utility factors the difficulty remains of how much weight should be attributed to a single factor within the balancing process; what is more important: an increase in safety or more user satisfaction? Even though the approach of assessing the defectiveness of products that have warnings described here can be considered consistent, the outcome of such decisions may still be uncertain, given that courts may vary in their assignment of weight to the individual factors.

216 Miller & Goldberg 2004, p. 416; Howells & Borghetti 2010, p. 463; Stoppa 1992, p. 217. A similar type of approach(es) has also been adopted by various American courts, see Owen 2008, p. 532 ff. It follows that the American experience in finding a combination of the two tests of consumer expectations and risk-utility has been a source of inspiration to European product liability law for interpreting the Directive's defect requirement.

217 Cf. Miller & Goldberg 2004, p. 418.

4.5.6.2 Factors Related to the Size of the Risk

The first component of the test involves the basic factors that determine the risk level. They form the starting point of the balancing process, since the size of the risk itself basically indicates the type of design measure(s) that is appropriate to reduce the risk. Evidently, the severity of the possible damage by the product hazard and the probability that this damage occurs are the parameters that determine the risk.[218] The user action to which the product was put influences the size of the risk. It can contribute to the conclusion that the given warning was an inappropriate measure. The more foreseeable the product use is that can lead to materialisation of the risk, the higher the risk generally is and the more likely it is that only a warning against the risk is insufficient to reduce the risk to an acceptable level. So, in accordance with the last-resort-measure principle of warnings, it is preferred to design out risks arising from intended use or guard against them than to use a warning instead, compared to risks that are related to user actions that are less likely to expect. Such risks are of significant size so these need to be dealt with first by design. If it is already highly questionable whether the user action can be considered reasonably expected, then the (given) warning will suffice or may even not be required.

4.5.6.3 Chosen Design Solution: Factors that Indicate the Inappropriateness of Using a Warning

The second component of the balancing test relates to the costs and benefits of having provided a warning as a precautionary measure to reduce the risk. Obviously, this balance was believed to be positive for the producer, as he has used this specific measure to control the risk. The claim based on the producer's misuse of the warning indeed questions the producer's assessment on this point. It presumes that the safety benefits, the warning's utility, arising from the use of the warning have been overestimated.

A key point for a successful claim based on this allegation is to show that the circumstances of the concrete case point to the direction that providing a warning is an inappropriate measure as it does not sufficiently reduce the level of risk because it is not effective enough, and that more safety through design was needed to render the product non-defective. It is therefore of importance that claimants include factors that indicate that providing a warning was an inappropriate way to control the risk at the time the product was put into circulation. In other words, that there is a *mismatch* between the size of the risk and the warning's safety benefits that were assessed by the producer. Hence, factors indicating that the warning can be expected to be an inappropriate precautionary measure are those that suggest that the likelihood that the warning will be followed by users generally is low. Likewise, they indicate that the behavioural effectiveness of the warning is not what it ought to be in order to prevent a sufficient number of accidents.

218 See also the RAPEX Risk Assessment Guidelines that explain how a risk should be assessed.

As described in paragraph 3.6.6, a non-exhaustive list of potentially influential factors pertaining to the effectiveness of the stages of the C-HIP model, including behaviour, were identified by the warning research studies as important. Precisely because these factors pertain to the effective processing of the substages and the end stage of the behavioural effectiveness of product warnings, they ought to be viewed by European product liability law as valuable indications of whether the product should have been made safer by a design modification or by changing the design of the warning itself. Factors that have empirically proven to significantly influence actual compliant behaviour are location, interactivity, colour, channel, warning symbols, hazard perception, costs of non-compliance, explicit information, the familiarity belief, costs of compliance, social influence, stress and mental work load.[219]

Moreover, the warning research findings have also indicated that some of these factors particularly emphasise the need to use warnings as a last resort. *High familiarity* with the product can be a vital cause why most users do not follow the warning, thereby increasing the likelihood of injury. As was pointed out in the previous chapter, Wagenaar has contended that risk communication, such as product warnings, will be of little value for people who are familiar with the product, because their behaviour occurs on a routine basis and these people rarely evaluate the risks and consider the accidents in advance. Observational studies have confirmed the finding that consumers are not always aware of the risks when using consumer products. This shows that not all individual human behaviour is based on consciously analysing the risks and benefits prior to performing their behavioural actions, as is presumed by decision making theories such as the value-expectancy theory.[220] The legal implication arising from this is that in the event that familiarity is high and product use is likely to occur on a routine basis, a (reminder) warning is less appropriate than a more protective design measure, and that a warning is less appropriate as the only measure to control the product risk. In a similar vein, the existence of a *low hazard perception* of the majority of product users also points in the direction that other more effective design solutions are needed than merely warnings. A *high cost associated with compliance* has also shown to be a potential hindrance. If the costs of complying with a warning are likely to be high in the foreseeable situation of product use, this circumstance can also be a hint for adopting other safety solutions.

In sum, the presence of one or more of these circumstances in a concrete case can be helpful for claimants to argue that the warning has been 'misused' by the producer. Because the safety benefits of the warning are expected to be low in light of these circumstances, it is expected that the warning is not able to avoid a sufficient number of accidents that

219 The factors of adding information components, adding information about the warning's source, perceived control and people's prior injury experience have shown to influence the *behavioural intentions* of complying with a warning.

220 Wagenaar 1992, p. 279; Weegels & Kanis 2000; Van Duijne 2005. See also paragraph 3.9.5.7.

happen as a result of the risk. Note that these circumstances not only indicate that a more effective design method is needed to render the product safe, they may also point towards producers' need to undertake more research effort to produce a well-designed warning. Hence, these factors can also be valuable when alleging defectiveness on the basis of the inadequacy of the given warning.

4.5.6.4 Alternative Design Solution: Relevant Factors

The third component necessary for the balancing process concerns the costs and benefits of the other design solution. Clearly, the above-mentioned claim is grounded on the circumstance that, on the basis of the hazard control hierarchy model, there is a design solution available that is more effective in controlling the risk. If not, then what is normally left is the allegation of the inadequacy of the given warning.

The benefit of the alternative design solution entails the safety benefits that are expected to result from the use of such a design measure. In other words, it concerns avoidance of the damage (personal and property) that results from materialisation of the risk in question. Because the proposed design solution is ranked higher in the hazard control hierarchy model, the effectiveness of the measure, that is, the number of accidents that can be avoided is in consequence also likely to be higher than with a warning. Such a statement is safe to make, since design solutions that guard or design out the hazard are by nature more reliable in their task of preventing accidents. As a result, the chance of having a 'mismatch' when using such a design solution to reduce the risk is, in contrast with warnings, limited.

The costs of the alternative design feature should be explained broadly. It relates to the sacrifices that need to be made for any design goal. It includes factors such as the financial costs of implementing the safer design solution for such a product, any loss of usefulness in the product that the design change may cause, and any new dangers that the design solution may introduce.[221] Usefulness can also be interpreted broadly by including various design goals, for example a loss of productivity or performance of the product, reduced user satisfaction, which in turn may relate to a comfortable use of the product, or an aesthetic design.[222]

In conclusion, the approach discussed here to determine whether a product that contains a warning of the risk provides the safety consumers are entitled to have is straightforward, as it analyses whether the warning given by the producer or another safer design solution suggested by the claimant, as a measure to avoid accidents that result from the specific risk, is better in terms of safety benefits and costs (in a broad sense). However, because next to safety there is a variety of intangible factors that are of relevance for determining defectiveness, the assessment remains difficult.

221 Owen 2008, p. 524.

222 See also paragraphs 3.2.3 and 3.9.2.

The example of portable music players illustrates the recommendations that have been made in this paragraph. Assume that a consumer sues a producer of music players for damages with regard to hearing impairment he allegedly suffered as a result of the use of the producer's music player.[223] The instruction manual provided a printed warning that informed the user of the risks to hearing. The victim argues that the music player is defective as listening to music with use of the player has caused permanent hearing damage.

In 2008, the independent Scientific Committee SCENIHR,[224] that provides the European Commission advice on emerging health problems, made an assessment on the threat of music players to hearing.[225] It follows from the report that the risk for hearing damage depends on sound level and exposure time. One of the conclusions entails that listening to music at 80 decibels or less is considered safe, no matter for how long or how often personal music players are used. This is the sound limit that has been set to protect workers. As regards sound levels above 80 decibels, hearing damage may occur if personal music players are regularly used for excessive periods of time over several years. Listening to loud sounds over many hours per day entails a similar risk as listening to an even louder sound for a shorter period per day.[226] The Scientific Committee has noted in its report that users of personal music players risk permanent hearing loss if they listen for only 5 hours per week at high volume control settings (exceeding 89 decibels) for at least 5 years. This would exceed the current limits in place for noise allowed in the work place (80 decibels for an 8 hours working day).

In response to the report, the European Commission established the decision 2009/490/EC that lays down a safety requirement for personal music players: 'Personal music players shall be designed and manufactured in a manner that ensures that, under reasonably foreseeable conditions of use, they are inherently safe and do not cause hearing damage'.[227] On the basis of this requirement standards set by standardisation bodies need to be established. More specifically, the requirement entails that exposure to sound levels shall be time limited to

223 See SCENIHR2008.

224 The Scientific Committee on Emerging and Newly Identified Health Risks.

225 SCENIHR 2008.

226 In order not to increase overall exposure, each 3 decibels increase in sound levels must thus be compensated by halving the listening time. For instance, listening to a personal music player at 95 decibels during 15 minutes per day is equivalent to being exposed at work to 80 decibels during 8 hours per day (technical norm) or to an even louder sound of 107 decibels during 1 minute per day, under the assumption that these exposures are repeated over a long period of time.

227 Commission decision 2009/490/EC of 23 June 2009 on the safety requirements to be met by European standards for personal music players pursuant to Directive 2001/95/EC of the European Parliament and of the Council (*OJ* 2009, L 161/38)

avoid hearing damage and warnings must also be added to products in order to inform consumers about existing risks. Although there are developments in this area, there is at this moment no technical standard that defines a maximum sound limit for personal music players. Some personal music players currently available can even generate up to about 120 decibels, which is equivalent to an aircraft taking off nearby.

As I have recommended, it would be preferable in view of promoting safety by design, that claimant lawyers in such a situation institute a legal action against the producer of the music player on the basis of the Directive's liability regime and that they claim that the product is defective because the risks of hearing impairment should have been reduced to an adequate safety level by adopting a design that has an acceptable sound limit. Consumers are entitled to expect that music players that carry a risk of permanent hearing damage during reasonably expected use have a design that adequately protects consumers.

General suggestions that can be made in respect of such a claim under the European product liability law is that the risks of injury can be quite severe because of their irreversible nature. They concern temporary and permanent hearing loss, ring in the ears, difficulties understanding speech in noisy conditions, acoustic isolation from the environment, learning and memory impairment, and increased blood pressure and heart diseases. Although further research is needed to adequately assess these risks, the scientific report indicates that certain consumers, such as adolescents, are vulnerable to hearing damage, since they often set the volume high and/or listen to music for many hours per day. This should thus be viewed as reasonably expected use that producers should take into account in the design of the product.[228]

As is clear from above, the risk is avoidable. Several design solutions are available to control these risks. On the basis of the hazard control hierarchy model, the first design method within the order that relates to eliminating the hazardous property is obviously not possible. The second design method of guarding does apply in the form of the use of a design feature that limits the maximum volume of the music to an acceptable sound level. This could be optional or not. Naturally, it is difficult to determine the acceptability of the sound limit. Levels of music players of 110 decibels (which is equivalent to a chain saw) or higher should be considered unacceptable. Below these levels, it seems to be more difficult to asses the appropriate level. Perhaps the limit of 89 decibels (men-

228 See also Commission decision 2009/490/EC of 23 June 2009 on the safety requirements to be met by European standards for personal music players pursuant to Directive 2001/95/EC of the European Parliament and of the Council (*OJ* 2009, L 161/38).

tioned above) or even 80 decibels (limit in work environment) should be considered as acceptable. Claimants should focus on the absence or inadequacy of this design feature, as a warning alone should not be regarded sufficient in view of the risk of permanent hearing impairment. It would be wise for claimants to introduce circumstances that indicate that the warning should not be used as a substitute for a safer design and thus that the presence of a warning against the risk does not provide an adequate level of safety that persons are entitled to expect of music players. Because the product is a product that is regularly used on a frequent basis for a certain amount of time, users are likely to become familiar to the risk and the associated warning and may underestimate it. People may also underestimate the risks, because it concerns a risk that often materialises only after long-term exposure. In addition, people may have difficulty protecting themselves against the risks, as it can be difficult to assess when the situation is risky. In addition, because of noise clutter in the environment, it can be reasonably expected that users increase the volume to a risky level, such as maximum level, in order to hear their music well and hence, not follow a warning's message. This can be seen as a cost of warning compliance.

The benefits of adding an acceptable sound limit relates to safety as the risks of developing hearing impairment reduce. It provides automatic protection. On the other hand, the costs of adding a sound limit include the financial cost of adding such a feature. Cost also involves a potential loss of utility of the product, such as the decrease in enjoyment people may have of listening to music at high volume, or a decrease in the ability to actually hear the music well when one uses the products in a noisy environment. Hence, this latter cost aspect may have some bearing and should thus be taken into account by courts. In addition, the inclusion of a protection limit seems to be a relatively easy feature to implement and in consequence should not pose a bottleneck for accepting defectiveness.

A final note with respect to claims centred on producer's misuse of the warning is that a finding of defectiveness does not prevent taking erroneous behaviour on the part of the user into account. In accordance with Article 8(2) of the Directive, the users' contribution to the accident can be translated into bearing some responsibility for the losses they partly incurred. Hence, because a warning related to avoiding such accidents was present, a user's failure to follow that warning (or to act upon the obvious or generally known risk) can play a role in determining the amount of damages, especially in the event that the warning itself was not inadequate. In contrast, if the warning itself can be criticised on several points, a reduction seems less obvious.

4.5.7 Conclusion

Paragraph 4.5 called for attention to treating warnings in European product liability law as a last-resort measure to reduce a product's risk and avoid accidents. This expression stems from the ergonomic design perspective of the basis hazard control hierarchy model that provides prevention methods in order of preference: design out, guard, warn. An appeal was made to European courts and injured litigants to recognise and adhere to this principle, which means that having provided a warning against a specific product risk does not necessarily absolve producers from finding the defectiveness for that risk. Even though paragraph 4.5.3 showed that EU product safety policy and European product liability law already more or less recognise the principle of the last-resort measure, the contribution of this paragraph lies in providing the underlying explanations why warnings should be viewed in such a manner.

There is firm empirical justification for this principle. Although the warning research findings generally show that the presence of warnings leads to more safe behaviour than in the absence of warnings, the degree to which warnings have an added value is often not huge. Research indicates that warnings have a moderate role as accident prevention mechanisms. Three interrelated explanations were offered in paragraph 4.5.2 for why warnings are no safety panacea and why they form the foundations of the last-resort-measure principle that governs the need for warnings. The first explanation can be offered by the information processing perspective. Before a warning can end in safe behaviour, several cognitive stages of information processing need to be successfully passed. Hence, behavioural compliance will be the least reliably observed effect of a warning. Furthermore, this corresponds with the viewpoint that warnings can be considered active protective measures that require users to perform or refrain from certain user actions each time the product is used and are not passive measures that require no behavioural input from people. Lastly, a consideration of the various factors (relating to the design, the person and the environment of use) that influence the warning process also elucidates why following warnings is generally a complex task.

The first recommendation with regard to this warning issue in a legal context was presented in paragraph 4.5.4. It was suggested that European product liability law must adopt the last-resort-measure principle of warnings. The last-resort-measure principle should not be interpreted as an automatic way of finding defectiveness in situations where a warning against the risk was provided whilst a safer design solution was available. In addition, the principle does not entail that a warning can never free a producer from a judgement of defectiveness. The main rationale underlying the adoption of the principle in law is that it provides an incentive to producers to improve the safety of a product by design. If there is no explicit acceptance of such a rule in product liability law, it is expected that producers will be less driven to increase the level of safety of their products through implementing design solutions that are more effective for protecting consumers against risks than warnings. A second important argument entails that it contributes to discouraging producers from providing unnecessary warnings that lead to an overuse of warnings and the occurrence of the associated negative consequences. A major argument likely to be used against the principle is that it places too little responsibility on consumers. Although this point may seem valid at first sight, the principle of warnings as a last-resort-measure does not object to the responsibility of users for safety, it simply embraces human limitations and errors and consequently grounds the perspective of safe design on the notion that it is far more difficult to prevent accidents by influencing behavioural patterns of users than by a product's design that provides automatic protection. As human factor specialists Sanders and McCormick have advocated: 'It is easier to bend metal than to twist arms'.

I recommended in paragraph 4.5.5 that warnings should be viewed as part of the design process of products. To emphasise that warnings are design measures that have their own contribution in achieving safety by design, it should be recognised by European courts and litigants that the absence of a warning or the presence of an inadequate warning can be viewed as a defect in the design of the product, just as with other design measures, since all products of the line have this defect (*i.e.* the absence of a design feature or presence of an inadequate design feature). It is the result of the producers' conscious decisions during the design process of products. So, placing a warning on a product that tells you of the risk and how to avoid it should not allow the conclusion that a material or procedural guarding mechanism that blocks any contact between the hazardous product property and people is not required to render the product non-defective. Similarly, a producer that has used a guard to provide protection ought not to be automatically insulated from defectiveness because defectiveness may still be established for the absence of a design solution that designed the hazard itself out of the product. To put it in another way, even if the hazard has been designed away, that does not preclude the conclusion that the hazard should actually have been designed out of the product to render the product reasonably safe. Of course, for a finding of defectiveness other circumstances also play a role, *e.g.* that this latter change must not unduly restrict the functioning of the product. Hence, I agree that these are distinct obligations that pertain to designing safe products. These are distinct because the design solutions based on these controlling methods offer different levels of protection. They are all independent which means that one design solution that reduces the risk does not preclude the necessity of another design solution. Defectiveness can always be imposed on another design defect ground, although some are more likely than others; it is more likely to establish defectiveness because only a warning is present and not a safety guard, than finding defectiveness because there is a guard and no warning. But these claims about design are also related because they all address how safety can be improved by the design of the product.

Special attention was called for the defect claim that centres on producer's misuse of a warning, *i.e.* that even though a warning has been provided, the product can still be found defective because, in view of the circumstances of the case, a more effective design measure, which was adopted from the hazard control hierarchy model, was required to render the product non-defective. In consequence, lawyers of injured victims involved in a product liability case carry the important task of instituting legal actions under the Directive that are inspired by this preferred way of approaching product risks. Because warnings are related to design defects and can be viewed as part of the design of a product, it is often possible for claimants to assert that the product is defective because of its flawed design with regard to the risk in question and because of the presence of an inadequate warning or a lack of a warning against the same risk. If the circumstances of the case permit it to allege both deficiencies, it is of importance that claimants take account of the last-resort-measure principle of warnings and first claim that the alternative design solution that is more effective should have been implemented. Only after having pointed out to the court that there was a safer version of the product possible and feasible, claimants should focus on the warning defect and contend that this rendered the product defective. In contrast, if they allege defectiveness because of the inadequate design of the warning that was provided, whilst the heart of the problem concentrates on the absence of an alternative design solution, the outcome of the decision (inadequate warning) would not advance the incentive to design products that have an optimal safety level.

The final recommendation was made in paragraph 4.5.6 and concerned the approach of applying the defectiveness standard of the Directive to products that warned against the risk that allegedly caused the damage. It suggested to claimants who use the claim centred on a producer's misuse of the warning which factors should be presented to the court, and it recommended to European courts a consistent manner of applying these circumstances in order to reach a satisfying decision about a product that already contains a warning against the materialised risk.

The defectiveness test of the EPLD requires a determination of the level of safety persons generally are entitled to expect of a product. A serious flaw of this test is that it can be problematic to utilise in cases of products that have a warning against the risk that materialised. Because expectations about a product and its safety are often formed on the basis of a product's appearance, courts could be eager to conclude that the test permits rejecting defectiveness, because the warning enables consumers to form a good idea about the dangerousness of the product and to know what to expect. Although this flaw of the test is often cited in relation to its effect on obvious and generally known risks, this criticism is also of special concern for those cases in which it can be alleged that a producer has misused the warning in order to reduce his chances of being held liable, because obvious and generally known risks as well as warnings influence expectations.

Unlike manufacturing flaws that can be compared with the design standard intended by the producer and in consequence failed to meet consumer expectations, defects in the design of the product (including warnings) have no defined product standard against which the defective design of the product can be judged. The integrity of the design of the product itself is being challenged in court. Therefore, as regards the assessment of the defectiveness of a product with a warning, concentrating on factors that indicate the desirability of a safer design solution over a warning is hardly inescapable.

I believe that the product with the warning can be considered defective under the Directive, if the comparison shows that the balance of costs and benefits of the alternative design solution is better than the balance of the costs and benefits of the warning. In other words, a product that has a warning to reduce the risk does not provide the safety consumers are entitled to expect, because the warning can be considered less cost-effective than the absent design solution. Non-consideration of risk-utility (or cost-benefit) factors when determining what level of safety consumers are entitled to expect of products with risks that have been warned of may lead to wrong judgements, because the test would be too narrowly focused on the fact that the safety expectations are lowered because of the presence of the warning. Instead, the test should focus on assessing the level of safety consumers are 'entitled' to have. This directs the defect assessment towards analysing whether an available design solution that is more effective in reducing risk than a warning should have been present in the product's design to render it non-defective. For this reason, risk-utility factors should be allowed within the liability test of the Directive to determine the defectiveness of a product that warned against the materialised risk as they enable a comparison between the given warning and a proposed safer design solution.

More specifically, when asserting that the product with a warning is defective, claimants should include factors that indicate that providing a warning was an inappropriate way to control the risk. In other words, that there is a mismatch between the size of the risk (*i.e.* combination of severity of injury and probability) and the warning's safety benefits that were assessed by the producer. Factors indicating that the warning is an inappropriate precautionary measure are those that suggest that the likelihood that the warning will be followed is low and that therefore the safety benefits of the warning are insufficient to reduce a sufficient number of accidents. The warning research literature is instructive in this respect as it has offered a list of factors that have empirically proven to influence the effective processing of warnings. Other relevant factors relate to the cost-benefit analysis of proposing another design solution to control the risk that is more effective, based on the hazard control hierarchy model. These include the availability of a design solution/feature, the effectiveness of it in controlling the risk (*i.e.* the safety benefits that arise from avoiding those type of accidents), the feasibility of implementing this design solution in terms of the cost of implementing this design solution. It was argued that cost should be interpreted widely; it entails the financial costs of adopting the design solution, but also the decreased usefulness of the product as a result of the adoption, and also the creation of new hazards.

4.6 How Should Consumers Be Warned? Legal Analysis of the Adequacy of Product Warnings

4.6.1 *Introduction*

In paragraph 4.6 the spotlights are on the legal analysis of the warning issue of how to warn. An overwhelming majority of the warning research studies have been dedicated to this issue. The empirical insights that have emerged with regard to how to design warnings so that they are more likely to be effective in producing safe behaviour of warning receivers are instructive for the assessment of the adequacy of warnings under the Directive. They have been a source of inspiration for the recommendations that are made here on this issue of how to find a warning legally adequate. As noted in the introduction of this book, the primary legal sources offer little guidance with respect to when the presence of a warning renders a product defective. How does the test need to be applied with respect to product warnings? And what are the relevant criteria and factors that can guide courts, defendants, claimants and academics in making their judgements about a specific warning?

The above-mentioned questions are dealt with here. In paragraph 4.6.2, the recommendation is made with regard to the standard for determining whether the warning that was provided by the producer is legally 'adequate'. The framework of this test proposes a distinction between warning adequacy criteria and warning adequacy factors. Warning adequacy is a concept that is comprised of a number of warning adequacy criteria (or requirements). These warning adequacy criteria can, as was described earlier in paragraph 4.3, simultaneously be viewed as legal subgoals of a product warning. This paragraph explains the warning adequacy criteria in greater detail. Subsequently, it demonstrates an organised approach towards the basic factors that should be viewed as relevant for the legal assessment of the warning adequacy criteria. In addition, a toolkit filled with guiding principles for courts and other interested parties is provided (paragraph 4.6.3), that is helpful for the evaluation of whether the criteria for warning adequacy have been met. The final paragraph (4.6.4) provides an outline of the aforementioned recommendations.

4.6.2 Recommendation: Framework for Defectiveness for Claims Related to an Inadequate Warning Design

4.6.2.1 General

The allegation that the given product warning is inadequate and renders the product defective is popular in product liability litigation.[229] Both under fault-based and strict product liability theories, it has been an accepted legal viewpoint that good warnings can render an otherwise unsafe product safe, whilst inadequate warnings may be the reason why a product is deemed unsafe.[230] This, however, leaves unanswered the question regarding what is in fact a good or an adequate warning according to European product liability law. To gain understanding of what constitutes a legally adequate warning, the wording of the Directive is not exceptionally helpful. Obviously, the Directive encompasses a defectiveness test which reads that the product must provide the level of safety the general public is entitled to expect. It follows that defectiveness can be established in the event that the product that contains a warning fails to provide the level of safety people are entitled to expect, because the associated product warning can be considered inadequate. Unfortunately, although the text of the Directive implicitly notes that warnings can be of relevance for defectiveness, it does not spell out when a warning can be considered inadequate and what factors are particularly of importance. Direct guidance from the law ends here and a lack of clarity creeps in.

Given that the text of the Directive and the decisions of the ECJ provide too little guidance on this point, it would be a wise exercise to consult other legal sources. The GPSD sets forth the general informational obligation for non-food products, stating that producers must provide consumers with the relevant information to enable them to asses the risks inherent in a product and to take precautions against those risks.[231] Tests suggested by academics to determine a warning's adequacy correspond considerably with the GPSD's provision. For example, Hodges suggested that it is relevant to consider whether the warning has allowed the user to adequately identify, assess and, if so wished, avoid or minimise the hazard.[232] Howells argued that warnings should be specific and enable consumers to avoid the risk.[233] What is striking about these tests for warning adequacy is that only relevant warning information should be given and that not only a good assessment of the risk is important, but also avoidance of the risk. Does this mean that warnings should

229 As formulated in paragraph 4.2.4 of this chapter, this claim can be considered a warning defect, that consists of the situation in which the warning that has been provided was inadequately designed and seemingly caused the defectiveness of the product.

230 Grubb & Howells 2007, p. 357; Hodges 1993b, p. 103.

231 Article 5(1) GPSD. See paragraph 2.4.3.2. See also the specific informational requirements as laid down in the TSD and the Machinery Directive.

232 Hodges 1993b, p. 100.

233 Howells 2005b.

guide consumers in their assessment of the risk by communicating the relevant information in a particular way? And when it is likely that consumers' assessment of the risk is inaccurate, must the design of the warning help steer users in a correct assessment? These questions are I believe of essence and they are addressed below in the following paragraphs.

Although it remains unclear to me whether there is an overall accepted concept or standard of warning adequacy in European product liability law, it seems that a range of terms have been used to give meaning to the term adequate. Warning requirements that are commonly mentioned in case law, academic literature and EU product safety legislation to assess a warning's adequacy is that the information about the risks must be relevant, easily understood, in an appropriate Community language(s), legible and clearly visible. Likewise, it was shown in chapter 2 that on the basis of a number of decisions of Member State courts, a non-limitative catalogue of factors can be given. Although arbitrary, this list provides an indication of the factors that courts regard as important for the test of warning adequacy. They include: the probability that a product danger emerges; the degree of harm arising from that; nature of the hazard; noticeability of information; legibility of information; unambiguous, understandable language; complete information; location; language; knowledge/expertise of the user; and public law legislation or standards.[234]

All together, a clear picture of what the legal test is for warning adequacy, what the warning requirements are and what relevant factors are has not been presented clearly in European product liability law. The next paragraph will provide a framework for determining the defectiveness on the basis of an inadequate warning and it deals with the questions raised above.

4.6.2.2 The Test: Four Warning Adequacy Requirements

The lack of clarity that surrounds warning adequacy can be explained by the uncertainty with regard to the warning issues of what is a product warning and what should it accomplish according to European product liability law. It was already recommended in paragraph 4.3 that European product liability law should be explicit about the legal rationale and legal goal(s) of product warnings, because this exercise is helpful for getting grip on the legal building blocks of an adequate warning. I remarked there that although legislation, legal literature and case law scantly openly discuss the legal goal of a product warning, there is considerable support to assert that the main underlying rationale for requiring warning duties in product liability concerns preventing or reducing the number of accidents that are caused by the absence of an adequate product warning. The reason for the argument is that warnings have traditionally been treated as precautionary measures to protect the interests of potential victims from damage. Notwithstanding this viewpoint on warnings, it seems that the warning adequacy test currently used by courts mainly revolves around

234 See paragraphs 2.4.5 and 2.4.3.

providing hazard-related information that is sufficiently comprehensible, and also noticeable and legible. This treatment presumes that by giving the comprehensible information to warning receivers product safety is sufficiently achieved. Nevertheless, as shown before an essential implication arising from the C-HIP model is that for a warning to lead to safe behaviour providing a comprehensible warning is not sufficient. The warning process consists of a number of cognitive stages and each single stage needs to be successfully completed before information flows through the next stage and before it finally ends in behavioural compliance. The warning research findings are consistent with this prediction of the model. Hence, it appears that there is an incongruity between accident prevention as the underlying rationale for requiring adequate warnings and the concrete requirements of noticeability, legibility and comprehensibility to assess the adequacy of warnings.

In view of my recommendations made in the previous paragraphs of how a product warning should legally be defined and what the general legal goal of a product warning is, it should thus hardly come as a surprise that the general test of warning adequacy should be linked to the warnings' ultimate goal of accident prevention. I therefore recommend determining the adequacy of a product warning on the basis of whether the design of the warning has the potential to sufficiently influence behaviour in the way prescribed by the warning. If this can be answered in the affirmative, then the responsibility for safe use fully shifts from producer to consumer and any losses should be borne by the latter party.

To further determine whether a warning meets the general test, the primary model within cognitive psychology's history, that is, the information processing model, can again aid European product liability law. As was already briefly discussed in paragraph 4.3, the receiver stages of information processing can be viewed as *subgoals*, since they need to be reached in order to reach the ultimate goal of accident prevention. They should concurrently be viewed as legal requirements or criteria to assess the adequacy of warnings. Accordingly, the legal test of the adequacy of warnings is comprised of these requirements.

It follows that the receiver stages of the C-HIP model can be used to draw up 5 corresponding legal requirements. Nonetheless, I suggest a modification. For the legal test, the requirements related to the stages of attitudes and beliefs and of motivation are combined into one legal requirement. Whilst the C-HIP model treats the stages of attitudes and motivation as separate before behavioural compliance occurs, other models of the warning process lack such a division. This distinction has been made to theoretically shape the model as explicit as possible. The distinction may be less clear though in the real world. This could be evidenced by the circumstance that some warning researchers describe a certain factor as influential at one stage whilst others apply it at the other stage.[235] It has

235 For instance, in Wogalter 2006b the factor of personal relevancy is discussed at the attitudes and beliefs stage, whilst in DeJoy 1999b and Riley 2006, it is discussed at the motivation stage. In addition, cost of compliance is often described at the stage of motivation (Wogalter 2006b; Riley 2006), but also at the attitudes and beliefs stage (DeJoy 1999a).

been noted in the warning research literature that beliefs, attitudes and motivation can serve as an explanatory link between comprehending warnings and the behavioural response.[236] Hence, both stages play an essential role in altering behaviour and many of the influential research factors found at both stages correspond. Likewise, the factors that have empirically shown to influence motivation also have shown to influence actual behavioural compliance. The important beliefs with respect to warnings that have been studied in the warning research literature, such as hazard perception familiarity belief and beliefs about the cost of compliance, provide or contribute in an essential way to the receivers' motivation to produce compliance behaviour.[237] A persuasive warning does this; it concentrates on the motivational aspect of a warning by persuading users to use precautionary behaviour when encountering potentially hazardous products.[238] Consequently, given the large overlap between the stages and the similarity of the influential factors, I have chosen to combine these stages for the purpose of creating the legal warning adequacy requirement that lays down that the warning should be sufficiently persuasive. In all, the test for determining whether a product warning is legally adequate depends on the evaluation of the following warning adequacy requirements:

1. the product warning is sufficiently salient;
2. the product warning is sufficiently legible;
3. the product warning is sufficiently comprehensible and memorable;
4. the product warning is sufficiently persuasive.

It is important to note that my starting point for determining the legal adequacy of a product warning is that there is a difference between requirements or criteria that pertain to the adequacy of warnings on the one hand, and factors that influence these requirements on the other hand. As shown just now, the warning adequacy requirements have been derived from the information processing stages and they form the main criteria that need to be met in order to find a warning adequate. Warning adequacy factors provide substance to whether a requirement has been complied with. They relate to the factors that have shown to influence the stages of the warning process.[239] These are presented in paragraph

236 Riley 2006, p. 296.

237 DeJoy 1999b, p. 221.

238 Cameron & DeJoy 2006, p. 302. See also paragraph 3.4.

239 Note that besides the receiver stages, the stage of delivery can also be a cause of a warning's failure. Delivery is a key function of the source (*e.g.* the producer) of the warning. A warning must reach the target audience before it can have an impact on one or more sensory modalities. So, in this respect it can be viewed as a basic warning adequacy requirement. Whether a warning is delivered depends *e.g.* on the warning's location. If it is not affixed to the product, it can get lost. The type of channel can also influence this stage, as a warning may not be delivered to people with sensory deficiencies. An illustrative case is the English *Klunk klip* case briefly described in chapter 2, where the user instructions were not available because of a second-hand purchase. In view of what is said above, the court could also have ruled that even if the instruction manual had not been lost or delivered with the product, the warning was not sufficiently located in view of the size

4.6.2.4. For instance, legible text is a requirement rather than a factor. The size of the warning is viewed as a factor to determine the adequacy of a warning rather than a requirement.[240] The relation of size as a factor for warning adequacy might at first glance be mixed up with it being a requirement, because size is a design feature that can be of relevance for determining the adequacy of a warning in several ways (*e.g.* salience, legibility, compliance). Clearly, I believe that European product liability law and its decision makers are knowledgeable of this difference. This distinction has not, at least not yet, come satisfactorily to the fore I believe. The recommended framework for determining warning adequacy intends to change that.

4.6.2.3 Rationales for Proposing this Warning Adequacy Test

After explaining the foundation of the legal test for warning adequacy, the reasons why claims centred on the defectiveness of a given product warning under European product liability law should be applied in this way is discussed next.

First and foremost, recommending this test stems from the fact that it synchronises the law with the psychological model of the warning process; it reflects the insights on how individuals process warning information and make choices on the basis of that processing. Having legal requirements for product warnings that facilitate the occurrence of behavioural compliance by consumers is likely to reduce the number of accidents resulting from warnings that do not sufficiently attract and hold people's attention, or that are not understandable and persuasive enough. Merely giving comprehensible hazard-related information to consumers to educate them about unknown risks or to remind them of familiar risks is not sufficient for safe behaviour. For reasons of product safety, this can be viewed an ill-considered approach towards adjudicating warning adequacy under European product liability law, since important factors such as the cost of compliance and the influence of hazard perception and other associated beliefs are in consequence hereof legally set aside under the traditional standard. Warning researchers have argued that in relation to the other processing stages, the stage of beliefs and attitudes is more powerful,

---

of the risk of misuse. A tag could have been attached to the device to advise and remind users of the appropriate amount of slack. This type of argument was however rejected by the judge in the English *Tampon* case. Furthermore, the stages of source and channel form part of the C-HIP model. Research studies have shown that source and channel characteristics can influence the warning process. For this reason, I consider the role of these stages to be that of a warning adequacy factor rather than a warning adequacy requirement.

240 Note that sector-specific product safety legislation can lay down more specific warning requirements. These may be generally treated as factors under European product liability law, but because of their importance it has been translated into a mandatory requirement under EU public law. For example, § 1.2.1 of Annex I of the CLP Regulation determines the dimensions of the labels of products that contain hazardous chemicals. It states that each hazard pictogram must cover at least one-fifteenth of the surface area of the harmonised label but the minimum area shall not be less than 1 cm^2. Failing to meet such a requirement should therefore be viewed as a significant indication for rendering the warning inadequate under the Directive. See also paragraph 2.4.3.

since attitudes and beliefs broadly determine how individuals will react to warnings.[241] They often are motivators themselves and can affect processing at earlier stages. Producers are not encouraged to take the information processing stages that succeed comprehension and their corresponding factors into account when designing warnings whilst these are a prerequisite for the process. Hence, it entails an incomplete view of the warning process, which increases the likelihood that these stages form bottlenecks and cause warnings to fail.

Certainly, it is admitted that this may impose a more stringent outlook on warnings in European product liability law than is currently the case. In this respect, the Dutch Supreme Court has adopted a new yardstick in the recent *Jetblast* decision with regard to the assessment of the adequacy of warnings under Dutch civil liability law.[242] This test is quite similar to the one proposed above as it links the test to behavioural response that is intended by the warning. The former does however not explicitly refer to the requirement that the warning should be persuasive.

I believe that it is not too high of a criterion to additionally require from producers that the warning should also be designed in a sufficiently persuasive way. When it boils down to marketing the product, the use of persuasive techniques is very popular. Why can and should the employment of these insights from persuasion theory and literature not be applied to product warnings? Various academic disciplines, such as social psychology, marketing and communication deal with theory and research on persuasion. The warning research literature also provides warning design implications from the persuasion literature. Producers may be reluctant to provide persuasive warnings because they perceive them to negatively affect sales rates, since such warnings emphasise the negative aspects of the product and in consequence reduce the attractiveness of the product compared to others products that have less prominent and explicit warnings. As wisely remarked by Cox, if there is a potential conflict present between the selling efforts of producers and the requirement to provide adequate warnings as a result of which sales reduce, this loss should be attributed primarily to the hazardousness of the product itself and not so much to the warning.[243] This perceived effect of warnings on consumers' purchase intentions/decisions has also been a topic of interest with warning researchers. Studies indicate that there is no strong finding that confirms the producers' concern.[244]

The second argument for making this recommendation follows from what is just mentioned above. The test elaborates further on the traditional view of warnings as pre-

241 DeJoy 1999a, p. 190; Young & Lovvoll 1999, p. 45; Riley 2006, p. 297.

242 The judgement of the Dutch Supreme Court stipulates that a warning can be regarded as an adequate measure to protect against a certain danger if it is to be expected that the warning will lead to an act or omission as a result of which the danger is avoided, see paragraph 2.2.3.3. For a critical view, see Giesen 2005; Haak 2006. I have advocated in Pape 2006 that this test is also applicable to liability cases that deal with product liability.

243 Cox III 2006.

244 See paragraph 3.6.6.7.

cautionary measures in civil liability law. A warning is by its very nature a design measure, in the form of a safety communication that aims at protecting the safety interests of potential victims. Because of this protective trait, it can be argued that consumers are entitled to expect that warnings do in fact have a design that invites consumer protection. Another characteristic of product warnings is that they, unlike most other product design measures, require a behavioural response of recipients because they need to engage in or refrain from a behavioural action to avoid possible harm every time they use the product. The measure does not provide automatic protection because it does not remove the hazard by itself. Given this particularity, it can be argued that the adoption of this type of measure to control the risk by producers enforces them to design a warning that takes account of the factors that are likely to influence whether or not the warning produces the intended effect of safe behaviour and accident prevention.

An argument raised against the test might concern that it leaves little scope for the user's responsibility for safety. Admittedly, because of the requirement that warnings should be sufficiently persuasive, producers are obliged to design a warning that has the potential to correct erroneous beliefs that are expected to be present and guide users' expectations in the right way. The adverse consequence of fully shifting the burden of complying with the warning from producer to consumer when the warning has reached the stage of comprehension is that the effectiveness of the subsequent stages is borne fully by consumers. But, should this responsibility fully rest with users if the costs of compliance are so high that only a few warning receivers will engage in self-protective behaviour and comply with the warning whilst the producer has knowledge of this circumstance and has reasonable means to reduce the cost? Similarly, should the consumer's claim fail when the warning could have been designed with features that would have communicated the hazard level more accurately and intensely to influence a false perception of the hazard so that the injury would have likely been avoided? My answer is no. I prefer to view the warning process as a shared responsibility. Consequently, the effectiveness of the later stages is also a shared responsibility between producer and consumer. Naturally, the responsibility in the later stages rests for a large part with the users, since it is the receiver characteristics (*i.e.* their expectations about the hazard, its severity and the behavioural actions of the warning) that are generally prevailing in these stages and secondly because they need to take the final step themselves of following the precautionary statements expressed in the warning. They are often the most suitable party that can reduce or avoid the harm and they thus carry a responsibility to control the risk. However, this does not automatically mean that there should not be a responsibility for producers in these final stages. In contrast, as far as it concerns the earlier stages, the responsibility to avoid accidents is also a shared one, but it considerably rests with producers as several warning design features can be employed that dominate whether a warning is noticeable, legible and comprehensible.

Another pitfall, or argument that is likely to be raised against the application of this yardstick, concerns the interpretation of the test in such a way that the inadequacy of a warning in a concrete case is established by the mere fact that this warning did not have a positive effect on the behaviour of the injured claimant. Hence, the occurrence of an accident as a result of a risk that was warned against implies the inadequacy of the warning. This interpretation is incorrect. The test of warning adequacy should not be interpreted in black and white; a non-effective warning is not tantamount to a finding of defectiveness. Evidently, this is not a realistic legal goal or test for warning adequacy. No single warning, how well-designed it may be, can guarantee 100% compliant behaviour. It embodies an overly simplistic view towards warnings and it imposes an absolute form of warning liability.

4.6.2.4 A Framework of Warning Adequacy Factors

The list of factors distilled from European case law and legal literature in chapter 2 to determine whether a warning is adequate is far from perfect. Factors and requirements seem to have mingled. There is no structure and it seems to be made up in an arbitrary way. This is a natural consequence of the fact that assistance from court decisions with respect to identifying any relevant factors depends on the number of claims brought before courts that deal with warning adequacy. Furthermore, decisions are sensitive to the circumstances of the concrete case. Moreover, the use of a certain factor may be more based on a subjective judgement of litigants and courts than on empirical evidence.

The warning research studies have identified variables – here referred to as factors – that can influence the way in which people process warnings.[245] It is argued below that building an organised framework of such factors benefits European product liability law, since it provides a structured list of relevant factors that should be guiding in the assessment of whether the requirements pertaining to warning adequacy have been met. Figure 10 provides this framework.

Classification of the Factors by Information Processing Stage and Type

Several classifications have been suggested in the warning research literature to organise the variety of factors that mediate the warning process. A first categorisation concerns linking the factors to the information processing stages they can influence. In addition, the factors can be subdivided in a more detailed way, by organising them by their type. A popular distinction concerns the one between personal and warning variables.[246] Another frequently used categorisation concerns factors relating to the characteristics of warning receivers, design characteristics of the warning itself, and situational factors.[247] Lastly, they

245 See also paragraph 3.6.

246 Rogers, Lamson & Rousseau 2000.

247 See *e.g.* Lesch 2006; Laughery 2006 and Kalsher & Williams 2006; DeTurck 2002.

can also be categorised into intrinsic (related to the warning design) or extrinsic (not related to the warning's physical design) factors.[248] This latter distinction is broader than the former two, because the category of extrinsic factors covers a range of factors more than just person variables (*e.g.* hazard perception, personality). It also includes factors that are more related to the environmental context or situation of product use, which can relate to the social context (*e.g.* social influence) or the environmental context (weather conditions, visual clutter and auditory noise). There are thus subtle differences in how the research literature categorises the factors. For example, some have classified cost of compliance as a person variable, whereas others call it an extrinsic or situational variable. Similarly, social influence can be regarded as extrinsic, but defining it as a situational factor would be more explicit.

Regardless of the precise distinction at hand, the key implication for European product liability law that follows from cataloguing the factors is that it is not sufficient to only take characteristics of the warning design into account when developing a warning. Also factors that relate to the target audience and to the context in which the warning occurs, need to be taken into account when determining the adequacy of a warning.

This view should not be considered as new. As regards person factors, the language of the Directive indicates that it takes the viewpoint of product users into account, as it centres on the safety expectations that the general public are entitled to have. More specifically, product liability acknowledges that the user's knowledge of the product hazards is relevant for the adequacy of a warning. As regards situational factors, the GPSD for example stipulates that producers should provide information that enables consumers to assess the risks throughout the normal and reasonably foreseeable period of product use. This indicates that during the use of the product, the associated warnings can be affected by environmental conditions and that the durability of a warning's design is of importance. Despite these clues, it has not yet been explicitly accepted that the adequacy of a warning (design) depends on factors that can be divided into characteristics that relate to the environment, the target audience and the warning design itself.

248 Vredenburgh & Helmick-Rich 2006; Wogalter e.a. 1998, p. 144.

Figure 10 Warning adequacy factors

Salient	Legible	Comprehensible & memorable	Persuasive	Compliance
- Size - Colour - Borders - Signal words - Symbols - Location - Interaction - Hazard perception - Familiarity belief - Information seeking mode - Environmental conditions - Channel	- Colour - Symbols - Location - Hazard perception - Familiarity belief - Environmental conditions - Letter case - Font style - Alternative label design (size) - Layout	- Receiver abilities (language skills, reading abilities, knowledge level) - Signal words - Colour - Symbols - Explicitness	- Hazard perception - Message framing - Control perception - Familiarity belief - Prior injury experience - Risk-taking attidude - Personal relevancy - Costs of compliance - Severity of consequences - Response efficacy - Self-efficacy - Social influence - Time stress - Mental workload - Source	- Location - Alternative label designs (size) - Interaction - Colour - Channel - Symbols - Hazard perception - Severity of consequences - Explicitness - Familiarity belief - Costs of compliance - Social influence - Time stress - Mental workload

Basic List of Research-Based Factors

To aid European product liability law, important factors identified by the warning research literature as potentially influential have been organised and listed above in figure 10. The factors that have shown to influence actual behavioural compliance have been included as well. The list provides a basis for courts, injured parties and defendants to act upon when dealing with the adequacy of a warning. The fact that this list is based on empirical research should justify and promote the use of the factors in European product liability law. It has been submitted by warning researchers that these factors need to be considered first for the design of initial warnings. Hence, the list gives producers, claimants and courts a head start in the probable factors that govern the legal adequacy of a warning. Furthermore, the use of this list should enhance a more consistent approach amongst European courts in assessing the adequacy of a warning. This list of factors does, however, not intend to be complete. There may be other factors that are influential too, but that have not yet been researched or only to some degree.[249] Future research may identify additional factors, reaffirm the value of a factor, or lead to a changed viewpoint. It is thus of importance that producers keep abreast of the latest developments in this area.

The Balancing Process of the Research-Based Factors

Courts have the task to weigh the relevant circumstances of the case. Because there are so many factors that can influence the adequacy of a warning, this is not straightforward. As

249 See paragraphs 3.6.9.1 and 3.6.12.

with claims centred on the producer's misuse of the warning, the starting point of the process of balancing the relevant circumstances of the case should be to distinguish in general three important components of factors within the test. I believe that these factors form the building blocks of claims that are based on the assertion that the product is defective because the concrete warning is inadequate. Presenting these factors to the court provides claimants with a comprehensive, well-composed allegation, which in turn should increase their chances in succeeding:

1. factors that relate to the level of risk involved;
2. factors that relate to the given warning;
3. factors that relate to the availability of a better alternative design of the warning.

The starting point for the balancing process is determination of the level of risk.[250] This concerns the nature of the hazard, the severity of the possible negative consequences and the probability that the actions of users lead to materialisation of the type of risk that has been warned against.

Notwithstanding that the Directive is based on a no-fault liability regime, the use of the defectiveness test to answer whether the warning can be considered adequate is dominated by notions of reasonableness, because the nature and size of the risk ought to determine the design of the warning.[251] Different hazards need different warnings as a result of which the content and form of product warnings vary. Various warning design features can be used to appropriately connote the risky situation. This tool is called hazard matching and is described further as a guiding principle in more detail below. For instance, the explicitness of information, especially about the consequences, has been found to be an important factor for several warning adequacy requirements as shown in figure 10. Even though it is an influential factor, it should not be permitted under European product liability law that it is used to present each piece of information for each hazard in an explicit way. One implication of the warning research results is that where products pose significant risks and severe consequences, explicit warnings should be used to communicate the information. Risks of smaller sizes, with less severe consequences should not have explicit warnings.[252] Hence, this should be handled skilfully by producers. Moreover, courts should be cautious when assessing whether the absence of explicit information was justified in view of the risk that is being warned of, and also in view of the other warnings attached to the product, before rendering the warning inadequate on this basis.

---

250 See paragraph 4.5.6.

251 See also paragraphs 2.4.2 and 2.4.5. The US Restatement of the Law (Third), Torts adopts the view that the adequacy of a warning must be measured on the basis of reasonableness. It requires a risk utility balancing: US Restatement of the Law (Third), Torts: Products Liability, § 2(c), comment i (American Law Institute 1998, p. 30).

252 Laughery & Smith 2006; Edworthy & Adams 1996, p. 68.

The second component relates to the costs and benefits of the given warning. It is obviously of importance that claimants clearly argue that there are factors indicating the inadequacy of the given warning to provide a sufficient level of safety. To pinpoint a probable cause for the inadequacy of the concrete warning, the factors that have shown to influence actual compliance behaviour provide a first impression.[253] A more detailed approach would be to analyse which warning adequacy requirement was likely to be a bottleneck in the warning process. Having done that, one can focus on whether the research-based factors for this particular warning adequacy requirement have been sufficiently taken into account by the defendant. For instance, if the design of the warning suggests that it is not very salient, claimants and courts should zoom in on the influential factors associated with this warning adequacy requirement and assess whether they are present or not, and if so, whether they are present in an adequate way in view of the risk. Figure 10 can be used as an investigative tool to European product liability law.[254]

If a factor has not been taken into account, this can be indicative for claimants to argue that the warning is defective, since there is empirical evidence that this factor can influence the adequacy of a warning. The absence of a particular factor, such as a warning design feature, relating to a warning adequacy requirement should not be interpreted too stringent. As noted earlier, defectiveness (of a warning) does not depend on the assessment of one single factor. Moreover, it has been repeatedly argued that the Directive does not require the best possible version of the warning that has included all possible design features and is designed in the best way possible taking all relevant factors into account. Furthermore, there are differences in the relative weights of the factors of the warning process. For example, the absence of a border around important warning information is likely to have less influence on the effective processing of the warning than the absence of a warning symbol. Unfortunately, although the research provides indications, it is still too soon to make robust, general statements about the relative effects of the factors on an intermediate stage or behavioural compliance.[255]

Similarly, if the specified factors have been taken into account in the design of a warning, this should not yet lead to the conclusion that the warning message is adequate. The adequacy should depend on the way in which these factors have been considered by the producer. For example, the presence of a warning symbol does not automatically mean that the warning adequately communicates the information. This depends on its design

253 See paragraph 3.6.9.

254 Figure 10 is based on the information processing perspective and the corresponding warning research findings. For a more detailed analysis, courts and injured parties can use the C-HIP model, which is of course the designated tool.

255 See paragraph 3.6.12.

and the criteria that have been used by the producer to reach the decision that the symbol is adequate.[256]

The third component concerns the factors associated with an alternative version of the warning. These factors should relate to the costs and safety benefits of a changed design of the warning. Claimants often argue that the warning was inadequate because there was a better design of the warning available that would have avoided such accidents as the one of the claimant, *e.g.* that the warning could and should have been made more specific, fuller, more prominent, with more impact etc.[257] It must be borne in mind that although the Directive does not require claimants to show that there is a better warning available, I believe that this component is closely linked with answering the question why a warning is considered inadequate. By arguing that the warning is inadequate, it presumes that there is a better version possible that would have prevented such accidents. For example, the assertion that the warning is inadequate because of the absence of a symbol that would alert, inform and motivate users to comply with the warning, implies that the new and improved version of the warning is one that contains a symbol. Likewise, the argument that the specific risk should have been expressed more impressively, predicts that the alternative warning provides this information in a much more explicit way.[258] Hence, addressing these factors within the assessment of warning adequacy therefore benefits the case. In addition, a judgement that takes account of how the warning could and should have been designed may even have a preventive effect, as producers have been given directions as to how their warning should be designed according to product liability standards.

Evidently, the alternative design of the warning is proposed by claimants because of its expected safety benefits. Given that many factors can influence the adequacy of a warning's design, it is almost always possible for claimants to propose in the aftermath of the accident changes that are likely to increase the warning's utility as a precautionary measure compared to the warning that was given. This ease of proposing a warning that seems to be safer can hinder the balancing process of courts since it may erroneously encourage them to rule in favour of claimants. Because of this, courts should carefully consider the costs associated with changing the design in such a way. As noted before, assessing the costs of a warning can provide obstacles for courts. Courts should not routinely presume that the costs are minimal and thus lower than the expected safety benefits of the proposed warning. The costs related to an alternative warning design should not only relate to the financial costs of designing the warning, but also to the research costs of investigating how to create a well-designed warning on the basis of the state-of-the-art of

256 Note that standards that specifically deal with the design and evaluation of symbols may provide direction on this matter.

257 Miller & Goldberg 2004, p. 576.

258 See *e.g.* the *Stepladder* case, discussed in paragraph 2.4.5.

the warning research literature. Moreover, costs also relate to the trade-offs that need to be made in the design of the warning and/or the product. As with the intrinsic design of the product, the design process of a warning also involves making trade-offs between various design goals. The trade-offs that producers are faced with during the design process may be more apparent when it comes down to the assessment of whether a safer design solution, like a material guard, should have been implemented instead of providing the warning against the risk.[259] Nonetheless, such trade-offs also occur during the design process of warnings and need to be given recognition in European product liability law. Adding or changing one factor of the design generally affects other warning adequacy factors and as a result possibly the adequacy of a warning too. This can plague courts when applying the test. An often used example is the conflicting situation of adding more hazard-related information to the warning because of its relevance, whilst adding this amount can negatively affect the brevity of the warning and thus pose constraints on the warning process as well as the processing of other warnings attached to the product. Another trade-off can relate to the placement of the warning. A prominent location is often on the product itself, but this may affect the aesthetic appearance of the product. Lastly, a likely trade-off consists of the need to tackle an incorrect belief about the hazard by using various design features that influence hazard perception and accentuate the risk without losing sight of the principle of hazard-matching.

> An example, which was already used in the previous paragraph, concerns the warnings associated with portable music players. As advocated there, claimants should first base their legal action on the notion that in spite of the presence of a warning, the absence of a safer design solution, that is, an acceptable sound limit, renders the product defective. It concerns a mismatch between the risk and the chosen precautionary measure to control it.[260]
> Secondly, claimants can attack the warning itself. Suppose the instruction manual contains the following warning on page 6: 'High volumes with an in-ear headphone can increase hearing impairment and can distract your attention from for example traffic. Always use an acceptable volume for you and the persons around you.' The information is preceded by a symbol of a triangle containing an exclamation mark.[261]
> As was noted earlier, long-term exposure to excessive sound can cause hearing damage. The consequences of listening to high sound levels can in the long

259 For example, the trade-off can entail a decrease in the functioning of the product or users' satisfaction of operating the product whilst the safety level of the product increases.

260 See paragraph 4.5.6.4.

261 This warning is derived from a music player's instruction manual found on the internet (translated in English).

term cause severe injuries, such as permanent hearing loss, ringing in the ears and difficulties in understanding speech in noisy environments. The risk depends both on the level of the sound and on the exposure time, but more research is needed on the users' exposure time and sound levels to provide comprehensive risk assessments. The Scientific Committee noted in its report that users of personal music players – if they listen for only 5 hours per week at high volume control settings (exceeding 89 decibels) would exceed the current limits in place for noise allowed in the work place (80 decibels for an 8 hours working day). Users listening for longer periods risk permanent hearing loss after 5 years. This approximates to 5-10% of the listeners, which may be between 2.5 and 10 million people in the EU.[262] Nonetheless, because of the increasingly widespread use of personal music players on a regular basis, many consumers in Europe, especially vulnerable groups such as adolescents, put themselves at risk.[263] Hence, it should be considered reasonably expected use that a number of people listen often to music at high or even maximum volume for longer periods. In reply to the findings of the report the European Commission established a decision that contains a safety requirement. It stipulates, among others, that 'personal music players shall provide adequate warnings on the risks involved in using the device and to the ways of avoiding them and information to users in cases where exposure poses a risk of hearing damage'.[264]

A general suggestion that can be made in respect of the adequacy of such warnings under the European product liability law, is that placement should be considered a key factor. I believe that a warning that is (only) placed in a music player's manual should not be considered sufficient. It follows from the warning research literature that it is important to place a warning close to the hazard in time and/or space. So, the absence of the warning that is not close to the risky use situation in time and/or in space provides an important indication for its legal inadequacy. Given that many users frequently use the product and also for longer periods of time, it appears that people tend to listen to the music without much conscious thought of the risks that are involved. In addition, the risk of hearing damage often occurs gradually and in consequence consumers may not be aware of or perceive the situations that they repeatedly expose themselves to as risky. It looks like the familiarity belief may play an important role. An interactive or dynamic warning may provide a good solution as it has

262 SCENIHR 2008.

263 SCENIHR 2008.

264 Commission decision 2009/490/EC of 23 June 2009 on the safety requirements to be met by European standards for personal music players pursuant to Directive 2001/95/EC of the European Parliament and of the Council (*OJ* 2009, L 161/38).

been suggested to reduce the effect of familiarity. Important warning information can appear (close in time to the hazard) when the user exceeds the safety limit, such as warning text on the player's display, and/or an auditory warning signal, and/or flashing light. A benefit of an auditory warning signal is that it provides an additional warning that is transmitted through the auditory channel and it alerts the user who may be distracted by the environment. On the other hand, there are potential costs associated with a dynamic warning, such as a sound or a flashing light, as users may find the interference by the warning during listening intrusive or annoying. A static warning (close in place to the hazard) can also be chosen, for example, a tag or a sticker attached to the player to make the warning noticeable. Nevertheless, an interactive, dynamic warning is more preferred because it only warns when the risk appears during use and when it is thus relevant to them, rather than a static warning that always provides the warning and leaves it up to the user to decide when they are at risk. In addition, it has been suggested that people become less habituated to a warning when it is an interactive, dynamic warning rather than a static warning.[265]

All in all, even though it may be difficult to assess the risks properly and even though it may be difficult to determine whether one of the warnings mentioned above is sufficient or that more of these warnings should have been included, it is argued here that the mere printed warning in a manual should not be considered sufficient in respect of the risks associated with the reasonably expected use of a music player.

Furthermore, it is important that the warning provides relevant information about the risks. It could be argued that although people are generally aware of the fact that loud sound can damage your hearing, there is a lack of specific knowledge within users of what sound levels are acceptable for certain periods of time. Hence, the instructions that users should listen to 'acceptable' sound levels for 'limited' periods or that 'high' volumes should be avoided should be considered vague and inadequate with respect to communicating it in an understandable way. In addition to the other warning, more detailed information about the negative consequences and what use conditions (volume and exposure) are risky can be provided in the manual. This requirement corresponds with the Commission decision's requirement that determines that information about the risks, when the risky situations can occur and how they can be avoided are necessary.

265 See *e.g.* Laughery 2006, p. 475 on sensory technologies for warnings.

Wish balloons have been on sale in Dutch shops since 2007 and they have become increasingly popular to use for special occasions. They are originally from Thailand. These wish balloons are sky lanterns that operate in the same way as hot air balloons. They are made out of thin rice paper. The air in the balloon is being heated by an open flame that hangs below the balloon. Before releasing the balloon into the sky, consumers can make a wish (and write it on the balloon). The balloon stays in the air as long as the flame stays alight. When the flame self-extinguishes, the wish balloon floats back to the ground, where the rice paper will biodegrade. The wish balloons are accompanied by instructions for use. They usually contain the information that they should be used when there is little wind (force 2 or less) and no rain or fog. They should not be used in close vicinity of high buildings, trees, brush or dry hay, or near an air field. Obstacles should be at least 30 metres away from the lanterns. Persons younger than 18 years old should not release the balloon.

The Dutch Food and Consumer Product Safety Authority assessed the risks associated with wish balloons in a report to the Dutch government to determine if measures need to be taken to reduce the risks.[266] The findings of the report showed that on the basis of the risk assessment procedure as described by the RAPEX Guidelines, various accident scenarios can be generated with serious consequences.[267] The calculated risks were determined as medium and low risks, and thus not negligible. The report suggested several measures to control the risks. The Authority advised the Dutch minister to adopt the following measures: (1) an explicit warning including a list of user instructions; (2) to make an industry agreement to develop a design of a wish balloon that reduces the risk, *i.e.* by changing the design of the burner; and (3) to ban selling the old wish balloons when the new designs are available on the Dutch market. The Authority has, subsequently, set requirements with regard to the design of the product's burner and minimal requirements for the content and form of the accompanying warning and announced to enforce these agreements and to take action against wish balloons that do not meet these new requirements.[268]

Parties who allege to have suffered damage as a result of the use of the wish balloon (by themselves or another party) can start proceedings on the basis of

266 VWA 2009; VWA november 2009.

267 *E.g.* the scenario that the balloon is released in wind conditions (force 2 or higher) that make it unstable. The swinging motion causes it to catch fire. It falls on flammable dry material, *e.g.* grass or heath land in summer. A bush fire occurs or a house fire (with persons in the house). Another type of scenario is that the balloon is released in wind conditions that make it lose hot air. The balloon lands with still active burner on flammable dry material, *e.g.* grass or heath land in summer. A bush fire occurs or a house fire (with persons in the house).

268 VWA december 2009.

the Directive's liability regime. It followed from the reports described above that injured parties may succeed when claiming that the design of the product, namely the burner, does not provide sufficient protection if it does not meet the new requirements. Such a claim should be given priority as was discussed in paragraph 4.5. Secondly, the injured party should allege that the warning, *e.g.* the one given above, is inadequate and renders the product defective.

To succeed, claimants must focus on whether the warning is designed in such a way that it is reasonably expected that it will sufficiently influence users to use the lantern under safe conditions. It followed from the above-mentioned evaluation of the risk that it concerns a fire hazard with possible severe consequences and a small probability. It is a medium or low risk, which justifies the presence of a warning.

What factors may distort the utility of a warning that accompanies such products? It is likely that the warning is not sufficiently persuasive for users to follow relevant use restrictions. It seems that the cost of compliance is the factor that influences this warning adequacy requirement the most. The costs of following the warning can be expected to be high. People buy these balloons to honour special occasions, such as weddings, birthdays or funerals. The happening of such an occasion is likely to induce people to use the balloon, irrespective of the warning's use restrictions. Furthermore, consumers can experience difficulties in estimating whether the weather conditions, especially the wind, are suitable for using the balloon. In this regard, ignoring the use restriction with regard to the wind can be viewed as reasonably expected use.

Because it is likely that the costs are perceived to be high by users, the effort to follow the warning's precautions should be made as low as is reasonably feasible by the producer. However, it seems that the use conditions cannot be changed in order to lower the perceived costs. Another implication from the warning research literature that remains to lower costs is to have a warning present that emphasises the costs of non-compliance. It is thus important for this type of product to clearly communicate the fire hazard with its severe consequences. In consequence, a claimant should argue that the design does not sufficiently influence users to behave safely. In view of the size of the risk, the given warning design did not adequately express the risk because of a lack of relevant information and informative warning design features. On the other hand, if the warning does have such features, like an appropriate signal word, the use of an appropriate colour, and explicit information about the consequences, this should indicate that users carry the responsibility to motivate themselves to follow relevant instructions. Having suggested changes that would only minimally improve the warning's design should not lead to a finding of defectiveness.

4.6.2.5 Special Assessment of the Adequacy of Reminder Warnings

This paragraph focuses its attention on the legal adequacy of a special type of warnings, namely a *reminder warning*. Reminder warnings are warnings that serve as a cue to call the risk into awareness, which information would in the absence of a reminder warning be dormant in long-term memory, so that warning recipients can produce safe behaviour at the right time.

As shown above, one of the requirements relates to providing a comprehensible and memorable warning. The adequacy of the subrequirement with regard to memory will often be of minor importance, as the emphasis will be on warning comprehension. Nonetheless, this subrequirement should receive special attention in specific circumstances. In paragraph 4.4.7, it was already argued that even though a risk can be regarded known to the foreseeable target population, specific circumstances of the case can give rise to the conclusion that the absence of a reminder warning renders the product defective. Moreover, the content and form of reminder warnings should be adjusted to their function.

When warnings are primarily intended to function as reminders of known risks, the emphasis should be on attention.[269] Because knowledge already exists, the amount of information can be reduced. In addition, symbols are likely to be more effective than textual information. Text is unlikely to be read because people who need a reminder already have the knowledge and their behaviour occurs at a skill-based or rule-based level as implied by the levels of performance model.[270] In consequence, within the legal framework for evaluating the defectiveness of a reminder warning, courts should in particular take the legal subgoals that represent the beginning stages of the warning process into account, such as attracting and maintaining attention to the warning. Naturally, the other requirements may also be of relevance in the design of a reminder warning. However, the adequacy of a reminder warning should mainly turn on whether the warning is salient and legible enough. The factors that influence the warning adequacy requirements of sufficient salience and legibility play a role in the assessment of whether a reminder warning is adequate. To name a few: factors such as symbols, signal words, colour, location and size.

4.6.2.6 Special Assessment of the Adequacy of Purchase Warnings

I have previously argued in paragraph 4.2.3.3 that it is possible to distinguish between product warning types. A distinction was made between on the one hand safe use warnings that allow users to use the product safely, because complying with the warning ought to lead to avoiding damage resulting from materialisation of the risk. Most warnings, including reminder warnings, are of this type. On the other hand, there are purchase warnings that allow consumers to decide on the basis of the warning information whether

269 Wogalter & Laughery 2006, p. 902; Lehto & Papastavrou 1993, p. 590.

270 Lehto 1992, p. 131. See also paragraph 3.4.4 for more information about this model.

to use/consume the product or not. The risk is present during product use by consumers in general or by a specific consumer category (such as allergic consumers) and cannot be avoided by using it carefully, only by not using the product at all. I recommended making this distinction between these two types of warnings and to consider this differential characteristic when evaluating their adequacy. It was made to aid European courts in understanding that the treatment of warnings with regard to adequacy should not always be similar.

As the definition suggests, purchase warnings primarily aim at guiding the purchase decision of potential users and not the decision of how to use the product safely, since avoiding the risk of harm is tantamount to not using the product at all. It is therefore more desirable to draw people's attention to the warning before purchase than after purchasing and before actual use. As with reminder warnings, the primary requirements of the adequacy of these warnings are linked to the early stages of the warning process. It is vital that purchase warnings catch the attention of potential buyers, and if so that the information is easy to read and comprehensible enough with regard to the risk so that they can make an informed decision about whether to buy and accept the risk during use or not. Hence, a number of warning adequacy factors that can be of relevance to assess a purchase warning's format and content are the presence of a signal word (that also conveys information about the hazard), a border around the warning, a prominent location of the warning (*e.g.* on the package or the product contrary to the manual) highlighting the warning text, and a legible letter size. However, the adequacy test of these warnings – in contrast to 'safe use' warnings – should not be concerned with questions related to the later stages of the warning process, such as whether the information is sufficiently persuasive to motivate recipients to follow the warning. It can thus be said that having bought a product that contains a legally adequate purchase warning essentially means that the buyer has accepted the risk as well as the possible damage that results thereof.

The circumstance that a purchase warning warns a specific category of people that the product is not suitable for them because they will be unable to avoid the risk during use, does not mean that the risk is unavoidable for the producer by means of a product design change. It was described in paragraph 4.4.4 of this chapter that product warnings are generally considered a suitable and sufficient measure to address the needs of a specific foreseeable subgroup within the target audience that can be at risk, like the elderly, people who have certain diseases that limit the use of products or children who may unwillingly come into contact with the product's hazards. However, precisely in the event that a purchase warning advises a specific subgroup not to use the product, whilst the risk is not unavoidable for producers, the question arises whether the purchase warning is misused by the producer to circumvent other design options. As was discussed in great detail in paragraph 4.5, warnings should not be misused by producers to avoid implementing design changes that provide more protection. On the other hand, even if the risk can in principle be avoided by implementing a safer design change, consumers may not be entitled to expect this change because of its associated high costs of implementing. In such cases, the risk can thus be considered avoidable in theory, but unavoidable with reasonable means.

> Examples of purchase warnings are the toy warnings on the basis of the TSD that specify user limitations, such as the warning 'Warning. Not suitable for children under 3 years'.[271] The TSD requires that these warnings are preceded by the word Warning(s). Furthermore, it explicitly stipulates that warnings which determine the decision to purchase the toy shall appear on the consumer packaging or be otherwise clearly visible to the consumer before purchase, including in cases where the purchase is made on-line. In consequence, the adequacy of this warning depends on whether the warning meets the TSD requirements. Moreover, European product liability law can provide additional requirements, *e.g.* with regard to the presence of other warning adequacy factors. Another purchase warning is one that points out that the product is not suitable for users of certain weight categories. A good example of such a warning is one inspired on an American case.[272] It concerns the warning that informs consumers that the product, *i.e.* an exercise bike, is not suitable for obese users. The adequacy of this warning depends on whether the warning was salient enough for potential buyers to notice, whether it was legible and whether it was sufficiently clear to understand the risk that was involved. It is submitted that the mere sentence: 'This product is not suitable for obese users' will often prove insufficient to render the warning non-defective. For example, the presence of a signal word, the use of a border, placement of the warning on the topside of the product's packaging, specifying what is meant by obese users by describing the maximum weight are some of the factors that are of relevance here.

---

271 See for examples of bad purchase warnings on toys that are sold throughout Europe: ANEC & BEUC 2008.

272 The case discussed in Dutcher 2006, p. 644, concerned a woman who used an exercise bike that was purchased from Wal-Mart. When the victim mounted the exercise bike and pedalled for only three to four revolutions the bike collapsed. The victim who weighed nearly 500 pounds fell off the bike backwards and sustained serious injuries. The woman filed a complaint against the producer of the bike and claimed that the bike was unreasonably dangerous because it did not provide a warning to obese users. The trial court ruled in favour of the defendant, but the Court of Appeal reversed. In short, the Court of Appeal ruled that the damages resulted from a foreseeable use of the exercise bike. The circumstance that the bike was manufacturered in accordance with relevant standards did not lead to an automatic escape from liability. The standards required that the bike was designed to withstand use by adults weighing op to 250 pounds, which encompasses greater than 98.5% of the US adult population. The court also took into account that the producer designed and marketed the bike specifically for the obese. It could have easily and reasonably included a short statement in the owner's manual describing the maximum weight for the bike.

The warning on a tampon box relating to the risk of TSS can also be considered a purchase warning, as all women are exposed to this inherent risk during intended tampon use. For this reason, I believe that it is of importance that the warning is placed on the package itself and not solely in the accompanying leaflet, because potential buyers should be given the opportunity to see and read it before making purchasing decisions.

4.6.3 *Recommendation: A Toolkit for Assessing the Warning Adequacy Requirements*

4.6.3.1 General

The warning research literature offers a richness of design guidelines or guiding principles implied by the results of studies and these – unsurprisingly – form a basic toolkit for European courts when confronted with a claim centred on the inadequacy of the product warning. A number of design principles that are essential to the warning design process are discussed here. It can be said that their significance during the development of warning designs compels European courts to engage in an evaluation of whether one or more of these guiding principles have been satisfactorily applied. The relevance of the guiding principles to European product liability law is explained in greater detail below.

4.6.3.2 Guiding Principle: Hazard Matching

The design guideline of hazard matching is instructive for European courts.[273] Hazard matching has been defined as matching the strength of the hazard implied by the warning to the specific level of hazard of the situation being warned about. Its primary advantage is that it improves the informativeness of warnings, because hazard-matched warnings not only inform that a threat is present, but also indicate the severity of that threat so that warning recipients have more information on which to base their behavioural decision making. Another benefit is that it helps prevent habituation, since not all warnings look alike.[274]

Various warning design features can be adopted to facilitate hazard matching. Although signal words and colour have attracted the most research attention, other design features such as symbols, border width, font and letter size could also be manipulated to convey different hazard levels.[275] In general, warning studies have supported the finding that different signal words produce different levels of perceived hazard. Words such as DEADLY and DANGER are consistently interpreted by the participants as implying high levels of

273 See also paragraph 3.6.6.3.

274 Hellier e.a. 2000, p. 579. See also paragraph 3.6.5.7.

275 See *e.g.* Adams & Edworthy 1995; Silver & Braun 1993; Edworthy & Adams 1996, p. 37.

hazard perception, whereas terms like NOTICE and IMPORTANT are consistently interpreted as implying less hazard. Nevertheless, studies have also shown that certain signal words are less helpful as hazard indicators.[276] Research has selected a broad list of signal words that represent a wide range of arousal strength. These signal words produced fairly consistent results among varying groups from the US, such as the elderly, non-native English speakers, children and the UK population.[277] Warning researchers have noted that this list provides a useful source of information for producers to assist in the determination of the appropriate signal word for the given product hazard to connote the severity of injury and the likelihood of injury. As regards colour, the research has consistently shown that people in western cultures understand that red connotes hazard. Red generally connotes the highest hazard level, followed by orange and yellow. However, the distinction in hazard connotation of orange and yellow is less clear. Previous research also suggests that black can communicate a high hazard level. According to the literature, it must be borne in mind that warning design features such as colour and signal words can interact and lead to misinterpretations of the intended hazard connotation.[278] Combinations of colour and signal words that convey different levels of hazards must thus be avoided.[279]

European courts should adjudicate warning adequacy claims on the basis of this design principle. Because the comprehensibility of a warning is one of the most important legal requirements (*i.e.* subgoals) of a warning and because hazard perception of users is one of the determinant factors of behavioural compliance, it is a vital finding for European courts to have affirmed that a warning's design features, such as signal words, colour and symbols can be employed to communicate the level of hazard or risk associated with product use. In consequence, the presence and adequacy of such design features to connote the appropriate amount of hazard or risk provides an important indication of a warning's adequacy. The content of textual information is commonly viewed as a means to express the level of risk. However, because of these robust research findings, courts can also take other characteristics of the warning into account when determining whether the warning sufficiently communicates the risk.

Furthermore, this principle provides a way to distinguish the product's hazards from each other and to prioritise them when a product is accompanied by more than one

276 Hellier & Edworthy 2006, p. 411; Rogers, Lamson & Rousseau 2000, p. 125.

277 Edworthy & Adams 1996, p. 31; Silver & Wogalter 1989; Wogalter & Silver 1990; Wogalter & Silver 1995; Braun & Silver 1995; Hellier e.a. 2000. The ANSI Z535.4 standard only uses the signal words DANGER, WARNING and CAUTION to convey high to low degrees of hazard in a warning message. Warning researchers have recommended expanding the list of signal words for the reason that this limited number of signal words for many warnings may lead to an overuse as a result of which people become habituated to them. Another reason is that research indicates that the difference between WARNING and CAUTION is small or not present.

278 Leonard, Otani & Wogalter 1999, p. 172; Edworthy & Adams 1996, p. 30; Lesch e.a. 2009.

279 Braun & Silver 1995, p. 2219.

warning. Warning researchers Edworthy and Adams have advocated that the most effective warnings should be reserved for the most serious situations and other less serious situations should be graded correspondingly.[280] This hazard-matching aspect of warnings is desirable, as people are entitled to expect that different risks have different warnings. The principle suits well with the legal principle that the adequacy of a warning is based on reasonableness; its design must be tantamount to the size of the risk.[281] This is what consumers are entitled to expect of a warning as a safety communication. In this regard, an injured party can claim that the warning is inadequate because the design of the warning connotes a lower level of risk than is actually present, as a result of which the claimant was misled with regard to the seriousness of the risk. Although it is less likely that a claimant would argue that the warning is inadequate because the warning design feature(s) is/are too intrusive compared to the risk, it is important that European product liability law also signals the message to producers that warnings should not denote a higher level of risk than the risk that is actually present during product use. A low level risk should not be accompanied by a warning design that implies a high level of risk. This can dilute the strength of signal words and warnings in general. So, the signal word DEADLY should not be used for products that frequently cause only mild injuries.

> The Dutch Court of Appeal in the *Nagelstyling* case explicitly holds that the nature and size of the risk of developing a serious allergic reaction when using the defendant's products are of relevance for the nature of the warning that can be required of the producer.[282] The Court of Appeal confirms the defendant's statement that the risk of developing an allergic reaction when using the defendant's products is not such that it necessitates a specific warning on the label of the product. It concludes that there is insufficient reason to hold that the simple warning statement with regard to the irritating nature of the product and the precautions 'Avoid skin contact' and 'Discontinue use immediately if redness or other allergic symptoms occur' are inadequate. Hence, this decision with regard to adequacy is understandable in light of the small risk of contracting an allergic reaction. Because the risk is small, it should not be required – conform the hazard matching principle – that the warning is specific. Specific should be interpreted here as to mean that the information is expressed in a more explicit way. Explicit information, which is an important factor for the warning process, should be reserved for the more serious risks.

280 Edworthy & Adams 1996, p. 12.
281 See also paragraph 2.4.5.
282 See also chapter 2.

4.6.3.3 Guiding Principle: Prioritisation of Warnings

A principle that European courts should consider when evaluating the adequacy of a specific warning is how the producer has prioritised the hazards or risks that have been warned of. Warning researchers have declared prioritisation as a principle that should be kept in mind when designing warnings.[283] The principle of prioritisation concerns the ordering of hazards when multiple hazards exist that accompany a product. It often relates to deciding which hazards and their associated warnings to emphasise or de-emphasise. This can be done on the basis of warning design features that have shown to influence people's hazard perception, taking thereby into account the principle of hazard matching. Important considerations that should be taken into account for the prioritisation process are the probability of injury, the severity of injury, known/unknown by the target population, important information (*e.g.* safe use) and practicality. According to warning researchers, the general rule is that warnings that convey the most important information and can result in the most likely and most severe injuries if not complied with have a high priority in the rank ordering of warnings.[284]

A most important lesson following from this principle is that the legal adequacy of a warning against a specific risk should be viewed in the context of the other risks that carry a warning. Being labelled as higher priority may in consequence entail that these risks should be listed first in an aggregate warning message and with more emphasis (*e.g.* more salient warning design features) than the others mentioned. It can also include attaching the warnings of higher-priority risks directly to the product and the others in a supplementary component of the warning system. Likewise, a lower-priority warning is likely to be less prominent or intrusively designed.

Clearly, the process of prioritisation and the decisions made on the basis thereof should be seen in light of other design considerations. Circumstances that also play a role in determining the number of warnings that should be mentioned on the product or on the product's label or not, are the amount of space available, the amount of given information and the associated potential problem of information overload and warning overload, or the issue of aesthetics with regard to placing warning information on the product itself. Constraints such as the size of the label can provide a good explanation regarding why a warning of a specific risk is placed in a separate supplement. It must be borne in mind that moving extra warning information to a product's label may reduce the noticeability and legibility of the available information in an unacceptable way. When this threshold is reached is difficult to predict. There are alternative warning design labels though that can reduce such problems and the absence of such a label may not justify the producer's decision

283 See paragraph 3.6.5.6.

284 Vigilante & Wogalter 1997, p. 284; Wogalter 2006a, p. 7; Wogalter & Laughery 2006, p. 906; Wogalter, Conzola & Vigilante 2006, p. 490.

to place a warning elsewhere. Furthermore, in the event that there is a considerable amount of space available on a product's label, but only a small amount is used to convey the warnings of the different risks, this circumstance may put producer's decision not to mention the warning of the materialised risk on the product or label into a different light. It may lead to the conclusion that even though the content of the risk information is accurate, the warning is inadequate, as I believe that it is important to consider the weight of warning information relative to other product information, especially marketing information. The reasonableness of this imbalance ought to be taken into account by European courts.

It follows that the principle can be employed favourably by both legal parties. Suppose that a consumer has suffered injuries as a result of a risk that was conveyed in the accompanying manual, but not with the warning information presented on the product itself. The injured party should allege that the product is defective because the warning against this risk was not sufficiently noticeable and legible. Because the warning was embedded in the separate manual, the discussion can centre on this warning system component and whether the warning was salient and legible enough. Moreover, the claimant should additionally argue that the warning should have been provided with the other warning information on the product instead of in the manual, because the risk should have been viewed as high(er) priority. The producer may also be able to put forward a strong defence against claimant's argument that the warning is inadequate on the basis of this principle. As with the principle of hazard matching, prioritisation provides a substantiated explanation why certain warnings of risks have received a preferential treatment in design and why others have not. So, in the event that there were several risks attached to the product that were of higher priority, the producer can argue that because this risk was of lower priority and because of design trade-offs that needed to be made (such as available space), there was sufficient reason to place this warning elsewhere or to design it not in a similar way as other warnings.

4.6.3.4 Guiding Principle: Brevity of a Warning

Another guiding principle that should receive legal attention concerns the design principle of brevity of a warning.[285] This principle should be taken into account by producers as failing to do so increases the likelihood that warning recipients experience an information overload.[286] As was previously noted, an information overload relates to the inability to adequately process information at a given time, *e.g.* because the warning contains a great deal of information, or because the warning is surrounded by and embedded in other warnings or other product information.

285 See also paragraph 3.6.6.6.
286 See also paragraph 3.6.5.6.

European product liability law has a traditional focus on assessing whether complete, *i.e.* enough information has been provided so that consumers are able to make safety related decisions. However, it appears that European courts pay scarce attention to the issue of whether there is an extensive amount of information that is detrimental for making safety-related decisions. Howells noted that if too many warnings are given, this can obscure the crucial messages, but he thinks that courts will be slow to impose liability on that basis.[287] Such reasoning is sensible, since empirical research has not yet clearly affirmed when overloading occurs. Nonetheless, this lack of certainty should not be a free pass for European product liability law to avoid consideration of the potential problem of overloading warning recipients all together. Of course, the ideal situation would be that courts can properly identify whether there is too much information present to process at once and as a result find the warning and the product defective. However, often there will be border-line cases where the decision does not come easily.

Because overloading first affects the information stages of attracting and maintaining attention to a warning, the problem of an information overload can implicitly be addressed by alleging that the warning is inadequate because it was not salient and legible enough to stand out and hence, that it should have been designed more salient and legible. This can be done by specifying the relevant factors that influence these warning adequacy requirements (*e.g.* colour, highlighting, an outline format, location). Helpful design solutions that have been proposed by the warning research literature to reduce the likelihood of overloading are prioritisation and employing a warning system.

Furthermore, information overload also influences the succeeding stage of warning comprehension and memory. Complex information is more effortful to process than simple wording that is easy to read and recipients are less likely to expend substantial amount of time to encode and understand a difficult warning. The same can be said for the amount of information that is given. An advantageous allegation that can be made in this respect by claimants is that the warning itself or the warning label in general is not brief enough because it contains irrelevant information as a result of which the essential information could not have been adequately processed. Claimants should point out that there is an information category in the warning message itself that is obvious or generally known and thus irrelevant and unnecessary. Or, that there are other warnings present of risks that are obvious, generally known, minor, related to unreasonably expected use that should have been omitted because they negatively affect the processing of this warning.[288]

In consequence, if the absence of an important warning adequacy factor(s) or the presence of a factor(s) that is/are poorly implemented into the warning's design indicates

287 Howells 2005a, p. 145; Grubb & Howells 2007, p. 376. His argument seems to imply that it refers to the issue of preventing an overload of warnings in the world in general as well as the issue of an information overload at a given time as was discussed in paragraph 4.4.2. See also Taylor 2007, p. 5.

288 See paragraph 4.4.

that the warning is not salient, legible and/or comprehensible enough, courts should be wary of the possibility of an information overload and take the claimant's allegation(s) into serious consideration. For future decisions, it would even be more constructive if European courts that are being confronted with the issue rule in a transparent way why and how they find that the warning should have been designed in another way to render the product non-defective.

> In the study of Vanilla Research discussed in chapter 3, the case study involved a toaster that was accompanied by a manual that contained a list of 38 warnings. This example illustrates the potential problem of an information overload because of the amount of warning information that needs to be processed, but also of a potential warning overload, given the many warnings that were present of which a number seem to be irrelevant and unnecessary. The finding was that many consumers do not read such warnings. Consumers provided solutions to making the information more effective, which solutions boiled down to making the information more salient and legible. The solutions were simplifying and shortening the information, making it more visible by highlighting, using stickers on the product with the key warnings, and providing a short list of dos and don'ts.[289]

4.6.3.5 Guiding Principle: The Warning System

The Design of a Warning System with Several Warning Components

A design principle that has gained prominence in the warning research and that is educational for the assessment of warnings under European product liability law is the approach to view warnings as warning systems.[290]

It has been argued by ergonomists that it is a too narrow view to think that a warning is only a sign or a portion of a product label. A warning system refers to the different means and formats (warning components) to communicate a warning message. The components of the warning system are not necessarily identical in terms of content and purpose. Different components can be necessary to communicate the information effectively to different groups within the target population.[291]

The value of this principle for European product liability law is the insight that a warning should not be treated in isolation, but in the context of the warning system that

289 Vanilla Research 2007, p. 23-25.
290 See paragraph 3.6.10 and paragraph 4.2.3.8.
291 Wogalter 2006a, p. 7; Wogalter & Laughery 2006, p. 905; Rousseau & Wogalter 2006, p. 154.

is employed by the producer. Hence, the system as a whole should be under evaluation when assessing the adequacy of the warning.

There are multiple channels and modalities to transmit the messages and choosing several components enhances the likelihood that the warnings are effectively processed by people. Hence, one single warning message may not be enough for producers to get away with liability, for example in view of the considerable amount of important information that needs to be communicated and a good placement of the warning. The circumstances of the case determine the appropriate component(s) to communicate the hazard-related information effectively. Courts should recognise, on the facts of any particular case, what the components of the warning system are and how they interact and complement each other to communicate the warning information effectively. If there is only one warning, this raises the question whether this is sufficient in view of the risk, which includes the issue of whether the warning should have been communicated by a different component or together with another component. This also entails that courts should be sensitive to the variety of channels and modalities to transmit the message. Evidently, only using the internet as a warning component should generally not absolve producers from finding the defectiveness. Furthermore, under certain conditions, it can be argued that it should have been required to incorporate an auditory warning (tone or voice) in the product's design to communicate the potential risk instead of, or in addition to the traditional printed warning. Similarly, the use of dynamic warnings (that *e.g.* flash or change over time) may be preferred to static warnings.[292] Because the technological developments in this field enable the use of alarm tones, voice, and dynamic warnings, judges should be given the discretion to make such considerations when assessing the adequacy of a warning system under the Directive.

> There is evidence in case law indicating that there has been support for this view.[293] In the English *Tampon* case, the warning on the tampon box against the risk of toxic shock syndrome as well as the enclosed leaflet was taken into consideration when determining the adequacy of the warnings as a whole.[294] The victim had argued that it was foreseeable for the tampon manufacturer that the personal leaflet might not be kept and/or read and that in consequence, the health warning on the tampon package should have included full, *i.e.* more information than was the case. According to the judge, the small amount of information in the warning on the box did not render the warning inadequate

292 Laughery 2006, p. 475; Wogalter, Conzola & Smith-Jackson 2002, p. 223; Kalsher & Williams 2006, p. 316. See also paragraphs 3.6.3 and 3.6.12.

293 See [2000] *PIQR* 95 (*Worsely v. Tambrands Ltd*); OLG Celle 29 January 2003, 9 U 176/02, *VersR* 2004, 964 (*German mixed concrete*).

294 [2000] *PIQR* 95 (*Worsely v. Tambrands Ltd*). See chapter 2.

> because it was clearly visible and because the enclosed leaflet contained additional information that described the hazard-related information about toxic shock syndrome in a sufficiently adequate way. She contended that the use of this 'dual system' was justified in view of the small likelihood of the risk materialising and the life-threatening consequences.

Warning System Components Can Differ in Function and Weight

An important aspect of a warning system is that the components within a specific warning system can have different functions. This thus asks for different designs. In addition, given these different functions, the warning components are not necessarily of equal importance. For example, warnings about the risks attached to the product mentioned on the producer's website should generally receive less weight than warnings placed in an accompanying package insert.

The implication for European product liability law is that it is of importance that courts view the design of warning components in light of their functions within the system. They must be cognisant of the differential weights of the warning components when assessing the adequacy of the warning system.[295]

> Taking the above-mentioned English judgement as an example again, it is clear that the internal leaflet aims at providing detailed information, including hazard-related information, whilst the warning on the package is mainly designed for attracting attention so that it can direct the readers to the other component for detailed information. From the perspective of a warning system, the judge took the right approach by assessing the adequacy of the warning on the basis of whether it was clearly visible to tampon users, and by judging the warning message in the leaflet on legibility, completeness and the absence of ambiguity. Nonetheless, the judge may have been too quick in deciding that the warning on the box sufficiently attracted the attention of tampon users. It is unclear to me how she arrived at the opinion that the warning was visible enough. Did the judge take factors into account such as the location of the information on the package (backside, bottom), warning design features that enhance the salience of the warning (*e.g.* signal word, bold type, colour, a border); the embeddedness of the warning in other information; the layout of the available information. In my opinion, the mere sentence on a box that there is a hazard cannot be considered adequate enough to make this warning noticeable even though the risk is small. Especially in view of the fact that the warning on the box contained little information, its ability to alert users to the hazard and to

295 Laughery & Wogalter 2005, p. 30-39.

the availability of the enclosed leaflet for detailed information became significant. Even though the claimant here was an experienced user who had knowledge of this risk, inexperienced young females, such as the one in the Dutch case,[296] also form part of the foreseeable target audience that buy and use the product. In view of considering these circumstances, the warning on the box may not have been sufficiently effective to attract attention so that it will be read and comprehended. This latter aspect is also connected to the next topic that is discussed below.

Warning System Components Can Reach Different Target Users

The warning systems approach also appreciates the variety between warning recipients within the target audience. As was shown in paragraph 3.7 of the previous chapter, warning research studies have indicated that the effectiveness of warnings is related to the extent to which a warning is compatible with the needs and capabilities of the target audience.[297] Several warning components within the warning system can be designed to reach the target audience when the target audience consists of foreseeable subcategories whose characteristics diverge on relevant dimensions. The warnings can each be attuned to the important needs of the subgroups within the target audience. On the other hand, producers may opt for providing one warning, because of all the variability in warning receiver characteristics.[298] The warning is then directed to the general target audience. In such cases, the warning research literature recommends to design a warning that addresses the needs of people at the lower end of sensory capabilities (*e.g.* reduced visual acuity in older adults), cognitive competence (*e.g.* understanding technical information and language, reading abilities) as much as possible because this increases the likelihood that those people with such limitations also adequately process the warning.[299]

This aspect of the warning system as well as the warning research literature on person variables provides European product liability law the insight that different users require different warnings. Therefore, when determining the adequacy of a warning, a decisive question can be whether the warning's design is sufficiently matched with the warning receiver characteristics that are of relevance in a specific case. Hence, it follows that a warning specifically designed for professional users should in principle not be considered legally adequate to consumers, *i.e.* sufficiently comprehensible, as they differ in receiver

296 Rb. Zwolle 24 april 2002, *Praktijkgids* 2002, 5921 (*Mini-tampon*). See chapter 2.

297 Important receiver characteristics (*i.e.* factors) that affect how people approach and respond to warnings are hazard perception, product familiarity, demographic variables (such as age and gender), competence (with regard to technical knowledge, language and reading ability), culture, literacy and personality factors.

298 Take for example the aforementioned case, where it is advised that the warning's design of tampons also takes account of young inexperienced users who have no knowledge of the risk and for whom the warning needs to be salient.

299 Wogalter 2006a, p. 7; Wogalter & Laughery 2006, p. 905.

characteristics especially with regard to knowledge of the technical jargon that is used.[300] On the other hand, when producers choose to create only one warning design for the general target population of the product, it can be worth challenging whether this was the appropriate approach to sufficiently reach the different categories in the target audience or whether it renders the product defective. In such a case, it is important that courts consider whether the warning's design can also be sufficiently processed by people – like the injured party – with *e.g.* lower levels of cognitive competence, experience, knowledge and sensory capabilities.[301]

> It follows that to escape liability, a producer should not use the warning he specifically designed for professional users, such as workers, if the product is also supplied to consumers.[302] This insight was evidenced in the Dutch *Betonmortel* case.[303] The District Court ruled that the warning rendered the concrete fluid defective, mainly in view of the fact that the end-user of the product was not a professional user of concrete fluid; the warning contained too little information for unskilled users to know that specific work clothing was needed. The High Regional Court in the *German mixed concrete* case took the other direction.[304] It also assessed the design of the warning in light of the knowledge level of its users. However, it held that because the product was intended for professional, skilled users and because it was not a standard consumer product the warning was sufficiently adequate. Consequently, it shows that the court did not really take into account the fact that private costumers (*i.e.* consumers) could buy and use the product. I would rather have seen the Higher Court consider whether the warning was sufficiently adequate even for consumers, instead of the court's viewpoint that it was sufficient enough to shape this warning to the needs of professional users, since they are the intended audience of the product. Similar to the difference between intended and reasonably foreseeable use, producers should also anticipate that others than the intended

300 Recital 10 of the GPSD indicates a similar line of thought. It states that products which are designed exclusively for professional use but have subsequently migrated to the consumer market should be subject to the requirements of this Directive because they can pose risks to consumer health and safety when used under reasonably foreseeable conditions. This consideration implies that there is a difference between product use by professionals and consumers and that this can subsequently affect the level of product safety that needs to be provided. Hence, this may necessitate changes in design of the product, including the warning. Similarly, see recital 15 and 1.7.4.1 of Annex I of the Machinery Directive.

301 Wogalter & Laughery 2006, p. 907.

302 Warnings for prescription drugs that are directed to physicians may provide information that is more technical than the language on the label of the pill containers that get to consumers, see Laughery 2006, p. 468.

303 Rb. Middelburg 13 juli 2005, *JA* 2005, 104 (*Betonmortel*). See chapter 2.

304 OLG Celle 29 January 2003, 9 U 176/02, *VersR* 2004, 964 (*German mixed concrete*). See chapter 2.

public can use the product. Since the product was delivered to commercial and private customers, the producer ought to have taken the characteristics, in particular the lower knowledge level, of private customers into account when designing this warning, or he should have used two warning components within the warning system.[305] I admit that there is evidence to contend that the warning was adequate enough for professional users. In this regard, I find the warning's format sufficiently salient and the warning's content sufficiently adequate to deny liability. Less information is needed, because professional users can be expected to have a higher knowledge level and they can infer additional information from the information that was provided. However, when consumers are also likely receivers of this warning, I find it questionable whether this warning's content can still be considered adequate. Even though there was more information available than in the Dutch case, it is uncertain whether the producer was entitled to assume that the nature of the hazard was known to consumers and that they were capable to infer from the printed instructions that merely covering your skin does not provide enough protection: special clothing is required. Hence, whether the assumptions made by producer were valid should be the core point of discussion in such a case. In this respect, having based the content of the warning on an expert's judgement or even better, the results of a test on a representative sample of the target audience, should be regarded as strong evidence of the adequacy of the warning's content that in all probability will free producers from liability. This leads me to the discussion of the next principle of warning evaluation that can offer guidance in product warning cases.

4.6.3.6 Guiding Principle: Testing a Warning's Design

Another tool that can be of use when evaluating the adequacy of a warning message that belongs to the product is the principle of testing. As discussed in paragraph 3.6.11, many warning researchers recommend including testing or evaluation of the effectiveness of a warning as an integral part of the warning design process.[306] Testing can be done during the process of designing the product and the warning, but the warning can also be tested after all design activities have been completed. There are different dimensions that can be measured (the intermediate information processing stages and behaviour) and with different

305 Note that this decision deals with producer's obligation to warn under fault liability. My guess is that the outcome would have been in favour of the injured party had the case been brought under the regime of the Directive, since it follows from the Directive that the characteristics of users of which it is reasonably expected that they use the product are relevant for determining defectiveness.

306 See also paragraph 4.2.3.7.

types of measures (objective or subjective). It should be done on a sample that is representative of the target audience of the warning.[307]

Testing is for several reasons important. Even though design guidelines have been followed, the warning may not be adequately processed because of the unique characteristics of products, environments and warning receivers.[308] Applying design guidelines does not guarantee that the warning will be adequately processed in the real world. This latter process is likely to increase when the design guidelines have been based on the empirical findings of the warning research. Note however that warning design guidelines that appear in voluntary standards do not have to be supported by empirical evidence.[309] Furthermore, guidelines can conflict. Think of the possible dilemma of providing explicit information about the hazard, the instructions and the potential negative consequences versus the guideline of brevity or limited surface area to print the warning. Testing the warning can offer a means to gather feedback on whether the right design trade-offs have been made and it can give ideas for design improvements. For example, it can show whether the information provided is explicit enough, whether the assumption that the hazard is obvious and needs no information is accurate or if the size of the warning affects legibility excessively. Regrettably, conducting research can be costly, time consuming, difficult and even dangerous for participants in instances of examining actual behaviour.[310]

Because testing a warning forms such a fundamental part of creating a well-designed warning, European product liability law ought not to disregard this principle when assessing whether the warning can legally be regarded adequate. In view of the abovementioned disadvantages that can be associated with testing, requiring producers that they need to have tested the effectiveness of their warnings on all dimensions of information processing including behaviour prior to putting the product into circulation is too extreme. The costs would for many producers, especially small companies, in all probability outweigh the benefits of providing the warning. But, carrying out some sort of testing to evaluate the adequacy of their warning can, in my view, actually be required from producers, especially with regard to the assessment of whether the warning is understood as this is a significant subgoal of many warnings and whether the warning motivates consumer to comply, as motivation is the closest link to actual behavioural compliance.

It has been noted in the warning research that compared to behavioural compliance, comprehension can be assessed easily, quickly and at low costs. There are well-established methodologies involving memory tests, open-ended response tests, interview techniques and so forth that can be used. The positive and negative results of such a study provide feedback with respect to what information was understood, what information category is

307 See also paragraph 3.5.

308 Wogalter 2006a, p. 9.

309 Lehto 1992.

310 Wogalter, Conzola & Smith-Jackson 2002, p. 226; Laughery & Smith 2006, p. 427.

critically needed and what can be omitted to increase the level of comprehension.[311] This information source is valuable since the understandability of a warning is a function of the users' knowledge and experience. The circumstances of the case can also warrant an evaluation on other dimensions such as whether the warning's design is salient, persuasive enough to motivate users to behave safely. When objective measurement of behavioural compliance is not feasible, behavioural intentions can be measured as a proxy. In such a study, participants are asked to answer questions about whether they would comply with a warning for a particular product or hazard. Warning researchers have noted that even though such studies will generally reflect higher levels than actual compliance, they can be useful for determining whether a warning is likely to be effective as well as for comparing experimentally manipulated warning designs to determine which would probably be most effective.[312] Such measurement is far less expensive and the difficulties of designing an experiment that observes actual behaviour can be reduced.

A major benefit of having carried out these evaluations is that producers can use the documentation of the test results to contest plaintiff's claim that the warning was not sufficiently adequate. Hence, to escape defectiveness such statements have a more powerful effect than having a warning that is merely designed on the basis of assumptions that have been made about the information processing capabilities and limitations of the target audience. Courts should thus recognise this effect. Warning designs that are based on research results from evaluations that indicated that the majority of the target audience understood the warning adequately in fact prove that the warning is sufficiently comprehensible to the target audience.[313] The same goes for warnings that have been designed using test results that measured the effectiveness of other stages, including intentions to comply and actual behaviour. Consequently, this should in principle be considered sufficient evidence to deny a claim. The circumstance that the warning was inadequately processed by the injured party should in such a case be borne by the injured party. Nevertheless, attention should also be paid to the way in which the study was designed and how the data were collected. For example, the validity of the producers' statement is weakened when doubt can be raised with regard to whether the sample that was included in the study was representative of the target audience. Because regard should be taken to such information sources, courts should use experts to help understand the statistical information that pro-

311 Wogalter & Laughery 2006, p. 907.

312 Wogalter & Laughery 2006, p. 907.

313 The threshold of a 85% comprehension score by a representative user group can be used as a criterion. Both the American standard on symbols (ANSI 2002b) and the ISO standard (ISO 2004) provide information on testing procedures and acceptance criteria to evaluate the comprehensibility of symbols. They suggest at least 85% comprehension by a sample of 50 individuals representative of the target audience for a symbol that is used without accompanying text before an acceptable level of comprehension is reached. The symbol must also produce less than 5% critical confusions (*e.g.* the symbol elicits an opposite meaning/action). See Deppa 2006.

ducers have submitted and to assess the quality of the research method that was employed to obtain the results.

4.6.4 Conclusion

Paragraph 4.6 addressed the major warning issue of 'How to warn?' in light of the findings from the warning research literature. It provided recommendations pertaining to the adjudication of the liability yardstick of the adequacy of a product warning according to European product liability law. More specifically, it discussed what the legal test is and should be for warning adequacy under the Directive, and in view of this proposed test what the requirements with respect to an adequate warning are and which relevant factors should be used to assess these requirements.

The paragraph started with a discussion of how courts commonly apply the test of the Directive to evaluate the content and form of a given product warning. The Directive's test requires that the product must provide the level of safety the general public is entitled to expect. The allegation that the product is defective, because the product warning provided by the producer should be considered inadequate is a popular allegation in product liability litigation. Although 'adequacy' appears to be the magic word in European product liability law for the assessment of a warning, the Directive provides little guidance on what governs the adequacy of a warning. Because the safety expectations that people have of a product are influenced by the warnings that accompany the product, it is important to have clarity about how to legally measure the adequacy of warnings.

It seems that there is no real, coherent test present in European product liability law to determine the adequacy of a product warning. Key terms that have been used by European courts and academic commentators to interpret the concept of warning adequacy are whether the warning is sufficiently visible, legible, or comprehensible. In addition, it was shown in chapter 2 that from a number of decisions of Member State courts, a (non-limitative) catalogue could be formed of relevant considerations with regard to a warning's adequacy. Hence, these appear to be of importance and include: the probability that a product danger emerges; the degree of harm arising from that; visibility; legibility; unambiguous language; complete information; nature of the hazard; location; language; and knowledge/expertise of the user and public law legislation or standards.

I argued that the ambiguity that governs the standard of warning adequacy could be explained by the uncertainty with regard to the warning issues in respect of what is a product warning and what should it accomplish according to European product liability law. A link was made with the analyses of these warning issues in paragraphs 4.2 and 4.3. Paragraph 4.3 already noted that European product liability law should be explicit about the rationale and legal goal of product warnings, because this exercise is helpful for getting grip on the legal building blocks of an adequate warning.

It was asserted that it appears to be standard practice in European product liability law at this moment to consider warnings legally adequate insofar as they provide relevant information to product users in a noticeable, legible and understandable way. This test does, however, not entirely correspond with the viewpoint held in European product liability law that warnings are precautionary measures. As advocated in paragraph 4.3.4, the warning research literature has taught us that for a warning to lead to safe behaviour providing a comprehensible warning is not sufficient. In consequence, it looks like the current test for warning adequacy presumes that warning receivers will be persuaded by the warning and motivated by it to act safely. Hence, the effectiveness of these stages is borne fully by consumers.

In view of these observations, it was subsequently recommended in paragraph 4.6.2.2 to determine the adequacy of a product warning on the basis of whether the design of the warning has the potential to sufficiently influence behaviour in the way prescribed by the warning. If this can be answered in the affirmative, then the responsibility for safe use fully shifts from producer to consumer and any losses should be borne by the latter party.

To further determine whether a warning meets the general test, the stages of information processing should act as requirements to interpret the legal concept of warning adequacy. It was concluded that the test for determining whether a product warning is legally adequate depends on the evaluation of the following warning adequacy requirements: (1) the product warning is sufficiently salient; (2) the product warning is sufficiently legible; (3) the product warning is sufficiently comprehensible and memorable; (4) the product warning is sufficiently persuasive. Because of the great overlap between the stages of beliefs and attitudes and motivation of the C-HIP model and their corresponding factors, the stages were combined to generate the legal requirement of a persuasive warning.

Several considerations were given why claims of defectiveness based on the inadequacy of a given product warning should be applied in this way. First, the test synchronises the law with the psychological model of the warning process; it reflects the insights on how individuals process warning information and make choices on the basis of that processing. Having legal requirements for product warnings that facilitate the occurrence of behavioural compliance by consumers is likely to reduce the number of accidents involving products with inadequate warnings. Another reason for advocating this test was that it elaborates further on the traditional view of warnings as precautionary measures in civil liability law. A warning is by its very nature a design measure that aims at protecting the safety interests of potential victims. It can be argued that consumers are entitled to expect that warnings do in fact have a design that invites consumer protection. Merely giving comprehensible hazard-related information to consumers to educate them on unknown risks or reminding them of familiar risks is not sufficient for safe behaviour. The adverse consequence of shifting the burden of complying with the warning from producer to consumer when the warning has reached the stage of comprehension is that important factors that can affect the subsequent stages, including behaviour, do not have to be taken into account by the producer when designing the warning. These factors that are largely dominant in the final stages can also influence the effectiveness of earlier stages, such as attention. As a result, the effectiveness of these stages is borne fully by consumers, and this should not be viewed as a proper allocation of the responsibility for safety.

It was submitted that the test probably imposes a more stringent outlook on warnings in European product liability law. Nevertheless, the use of persuasive techniques is very popular in the field of consumer products. Why can and should the employment of these insights from persuasion theory and literature not be applied to product warnings? In addition, another plausible argument raised against the test can be that it leaves little scope for the users' responsibility for safety. The producers' responsibility is to assist this decision making process by providing a warning that is designed in such a way that it sufficiently persuades people to decide to comply with the warning. There still rests a responsibility on users to use the product safely in accordance with the warning, as it is ultimately up to the consumer to produce the compliant behaviour. I prefer to view the warning process as a shared responsibility. There should be a remaining responsibility for producers to take the factors into account that affect the later stages in the warning process that succeed comprehension, because these are essential to the warning process. Consequently, the effectiveness of the later stages is also a shared responsibility between producer and consumer. Nevertheless, the responsibility in the later stages rests for a large part with the users, since it is the receiver characteristics (*i.e.* their expectations about the hazard, its severity and the behavioural actions of the warning) that are generally prevailing in these stages and secondly because they need to take the final step themselves of following the precautionary statements expressed in the warning.

Another argument that is likely to be raised against the application of this yardstick, concerns the interpretation of the test in such a way that the inadequacy of a warning in a concrete case is established by the mere fact that this warning did not have a positive effect on the behaviour of the injured claimant. This interpretation is incorrect and undesirable. The test of warning adequacy should not be interpreted in black and white; a non-effective warning is not tantamount to a finding of defectiveness as no single warning, how well-designed it may be, can guarantee 100% compliant behaviour. Such an application of the test would stand in the way of the incentive effect of European product liability law to create well-designed warnings that can protect the health and safety of consumers. The Directive does not demand all warnings to be 100% effective in protecting consumers.

Following the discussion on the relevant warning adequacy requirements, paragraph 4.6.2.4 addressed the warning adequacy factors that should be considered when evaluating whether a warning adequacy requirement has been met.

It followed from chapter 2 that the adequacy of a warning has often been determined on the basis of factors that have previously been identified in case law or factors that intuitively are considered to be of importance by legal parties or courts. In view of this arbitrary process of selecting factors, it is possible that the weight of a factor is misjudged or that an influential factor has inadvertently been excluded from the assessment. It was argued in this paragraph that guidance can be found in the warning research literature, as it provides influential factors that can be used as a basis. The use of this list should enhance the use of a consistent approach among European courts when applying the defectiveness test to warning adequacy. Moreover, because this list is supported by empirical evidence with respect to how people process warnings, it is argued here that the use of these factors in the assessment of warning adequacy is expected to improve the safety level of products. As a result, figure 10 was drawn up which provides a list of the factors identified by the warning research literature as potentially influential and organised on the basis of the warning adequacy requirements. The factors pertaining to actual behavioural compliance are included as well.

The paragraph also discussed how all the relevant research-based factors should be balanced. It was recommended that the starting point of the process of balancing the relevant circumstances of the case is to generally distinguish three important components of factors within the test. These factors should form the building blocks of claims that are based on the assertion that the product is defective because the concrete warning is inadequate. They entail (1) factors that relate to the level of risk involved; (2) factors that relate to the given warning; and (3) factors that relate to the availability of a better alternative design of the warning.

With regard to the first group of factors, the fault-based notion that the nature and size of the risk of injury determine the design of the warning should be maintained under the liability regime of the Directive. Various warning design features are available to producers to realise a warning that appropriately connotes the risky situation.

Naturally, the factors indicating the inadequacy of the given warning are of paramount importance. Analysing the warning adequacy requirements can pinpoint a probable cause for the inadequacy of the concrete warning. This helps to determine whether the research-based factors for this particular warning adequacy requirement have been adequately taken into account by the defendant. Figure 6.1 can be used as an investigative tool for claimants and courts in this respect. If a factor is absent or poorly taken into account, claimants can argue that the warning is defective since empirical evidence supports the influence of this factor on the adequacy of a warning. As regards the adequacy of a reminder warning, one can primarily zoom in on the factors relating to the warning adequacy requirements of salience and legibility. As far as a warning can be classified as a purchase warning, of a risk occurrence of which is not dependent on how the user uses the product, the adequacy should mainly be determined by whether the warning is sufficiently salient, as the warning normally affects the purchase decision. Naturally, the comprehensibility of the information should also be considered vital as consumers need to decide on the basis of this information whether or not they want to expose themselves to this risk. In light of these functions, the requirement of persuasiveness should not be considered of relevance for both types of warnings.

Finally, the factors that relate to a better, alternative version of the warning also deserve attention. These factors should relate to the costs and safety benefits of a changed warning design. Evidently, the alternative design of the warning is proposed by claimants because of its expected safety benefits. By having pointed out the flaws of the given warning that consists of missing factors or poorly addressed factors, it normally follows what the improvements should be for the alternative warning design. However, because it can be asserted that in the aftermath of an accident it is a rather easy task for claimants to make suggestions for a warning that is likely to be better in avoiding injuries associated with the materialised risk, claimants and courts ought to be alert in considering the costs of changing the warning too. These costs may entail the design trade-offs that are involved when adding extra information to a warning or making the warning more impressive.

Having explained the basis for determining the defectiveness of a warning by specifying the warning adequacy requirements and the warning adequacy factors, the final recommendation in paragraph 4.6.4 offered a toolkit filled with a number of special design principles. Courts can use these to facilitate the evaluation of the adequacy of a warning. The tools are derived from the warning research literature and involve design principles that facilitate the effectiveness of warning designs. These include matching the warning's design to the risk, prioritisation of the available warnings, brevity of a warning, the warning as a warning system containing several warning components, and the judicial value of having designed warnings on the basis of testing results rather than only design guidelines.

4.7 Summary

Aim

This chapter is the heart of the book, as it analyses the five key warning issues from a European product liability perspective by undertaking a behavioural approach. The legal analysis entailed using the insights from cognitive psychology and ergonomics, specifically the warning studies and research literature, to examine whether the presumptions that are made about behaviour and products and warnings in European product liability law (*e.g.* do warnings really prevent damage?) are correct. Furthermore, the analysis involved providing explanations supporting why European product liability law deals with a warning issue in a specific way (*e.g.* why are warnings viewed as a last resort measure?). Moreover, the behavioural insights were obtained to better understand the warning issues in European product liability law and to provide support for suggested improvements (recommendations) on how courts and litigants should deal with warnings under the defectiveness test of the Directive.

Structure

Before this legal analysis took place, chapters 2 and 3 discussed the topic of product warnings from two different perspectives.

Chapter 2 discussed warnings in the context of European product liability law. Chapter 2 contained a concise discussion of the current state of the two product liability theories in tort (fault-based liability and strict liability) that are popular in the Netherlands (paragraphs 2.2 and 2.3). The central part of the chapter (paragraph 2.4) explored the consensus on product warning requirements for determining defectiveness under the Directive's liability system on the basis of studying case law of a number of Member States and legal academic literature. Its focus was on interpreting the defectiveness standard, especially in relation to warnings, and not on other requirements of the Directive.

Chapter 3 gave a comprehensive analysis of important warning topics with regard to the interaction between human behaviour and products and product warnings, such as theoretical perspectives of the warning process, the type of factors that influence the effective processing of warnings, the associated design guidelines to facilitate the design of effective warnings, and the role of warnings in the design process of safe products. Various empirical warning studies were described in detail to illustrate the analysis. The main source of information has been the warning research literature that has evolved over the last 30 years. Many warning researchers are ergonomists or psychologists and they apply the knowledge and the methods to carry out studies that investigate the effectiveness of a warning design. The research activity has been substantial and various articles, book sections and books have been published providing qualitative and quantitative reviews of the warning research covering a certain period. As a result of this growing body of empirical warning research and reviews thereof, basic principles and guidelines pertaining to the design of warnings and safe products are at hand. Not only producers, public law policy-makers, researchers and safety specialists, but also European product liability law can learn from these insights. Note that in chapter 1 consideration has been given to the potential limitations attached to the value and use of the findings of the warning research literature in a legal setting.

Recommendations

On the basis of the legal analysis, the aim of the monograph was to provide a toolbox filled with the recommendations for courts and litigants on how to deal best with the relevant warning issues. The conclusions of the previous paragraphs of chapter 4 are grouped together below in order to summarise them.

The Legal Meaning of the Term 'Product Warning'

After the introduction, chapter 4 started with the warning issue of 'What is the legal meaning of product warnings?' so that subsequently the other warning issues could be addressed.

Three recommendations were made. First, I proposed in paragraph 4.2.3 to legally define visual warnings as safety communications: they are intended to communicate relevant information about the hazard so that undesirable consequences can be avoided or minimised. 'Relevant' refers to conveying important different hazard-related information categories. Warning messages should generally contain the basic information regarding the type of hazard, the consequences of the hazard and the safety instructions, unless one or two information categories can be omitted. This shows that the term 'warning' is interpreted widely here (see figure 3). The exceptions with regard to omitting an information component are that the information component is obvious, generally known or can be inferred from another category.

Four main reasons were given to explain why European product liability law should interpret visual product warnings widely. A key reason is that the research findings and literature on the effectiveness of warnings support the notion that the content of each warning should contain these information categories. These are in principle prerequisite for enabling users to protect themselves against the hazard. Secondly, it provides clarity to injured parties with respect to which starting point to choose when they want to claim compensation. In the event that one of the categories is missing, the adequacy of the warning message has been possibly undermined and the claim should consequently be based on product defectiveness due to the presence of an inadequate warning. When no warning information at all is given, it should be argued that the absence of a warning message renders the product defective. Thirdly, this definition guides producers with respect to the informational content of a warning. Lastly, it may trigger producers to consider testing the content of the warning on comprehensibility to ascertain whether one or two information categories need to be omitted and whether other means to communicate the information need to be included. Producers may also be encouraged to employ a warning system to ensure that the warning information that is relevant and considerable in amount can be communicated effectively. These activities are important because they can secure that the principle of brevity of a warning message is considered sufficiently.

Secondly, it was recommended in paragraph 4.2.4 to use the categories of product information defects, design defects and manufacturing defects under the Directive, since warning information is not the only form of product information that has the ability to influence safety expectations, which embodies the defectiveness test under the Directive. Product information defects subsequently consist of warning defects, marketing defects and other potential product information defects that can influence safety expectations. I also recommended using the term 'warning defects' to refer to defects in relation to product warnings. Warning defects can then be subdivided into product defectiveness as a result of (1) the absence of a warning; (2) the presence of an inadequate warning; and (3) the inappropriate use of a warning as a precautionary measure whilst a higher level of safety was needed (see figure 4).

The final recommendation in paragraph 4.2.5 advocated for maintaining the linguistic distinction between 'warning defect' and 'duty to warn', since these terms refer to distinct legal theories.

Underlying Rationale for Requiring Warnings in European Product Liability Law

The second warning issue analysed in a legal context concerned the short but complex question of 'Why warn?'.

I argue that for a good understanding with respect to which requirements European product liability law should lay down for warnings, it is important to have a clear picture of why warnings are required by law in the first place. Unfortunately, there is a dearth on legislation, case law and legal literature that extensively discusses the underlying rationale and goal of imposing warning duties on producers. Nonetheless, it was shown in paragraph 4.3.2 that there is considerable support for the argument that the main underlying rationale for requiring warning duties in product liability concerns preventing or reducing the number of accidents that are caused by the absence of an adequate product warning. The reason for this argument is that warnings have traditionally been treated as precautionary measures to protect the interests of potential victims from damage. As concluded in paragraph 4.3.3, an important lesson that can be learned from the warning research studies in this respect is that this treatment is justified. Warning researchers have done a number of studies that have led up to the general conclusion that warnings can indeed influence behaviour in a positive way. Empirical evidence thus warrants their use as precautionary measures in European product liability law.

Paragraph 3.4 showed that it is not difficult to see the link between the questions why warn and what is the goal of a warning according to law. In accordance with the consensus amongst warning researchers, it was recommended here to consider accident prevention or reduction as the ultimate goal of product warnings in European product liability law. Requiring adequate warnings should be viewed as means or instruments intended to accomplish prevention or reduction of unnecessary accidents that happen as a result of the presence of inadequate warnings. Equally, the information embedded in a warning is the 'vehicle' to reach the destination of accident prevention (see figure 5). For warnings to be adequately processed and end in safe behaviour, this information must meet a sequence of conditions. This can be explained by the C-HIP model. The C-HIP model describes behavioural compliance with a warning as the culmination of a sequence of several information processing stages within warning receivers. The information processing stages must all be successfully completed for the warning to end in safe behaviour. Hence, the information processing stages can be considered requirements for effective warnings. For this reason, I recommended in this paragraph to consider each information processing stage as a *subgoal* that needs to be achieved in order to reach the goal of accident prevention.

It was also argued in paragraph 4.3.4 that it is helpful for European product liability law to focus on what are the main goal and subgoals of product warnings according to law, since this exercise can uncover insights that can be useful for the legal test of the adequacy of warnings (see figure 6). Because the information processing stages can be viewed as subgoals, they can also be considered to give meaning to the determination of when warnings should be considered 'legally adequate' as a result of which the legal responsibility for safety shifts from producer to consumer. This raises the question regarding at what point on the legal line of responsibility for safety there is a shift from producer to consumer. Even though European product liability law commonly treats warnings as protective measures, it appears that European product liability places much emphasis on reaching the subgoal of giving a comprehensible warning. Such a legal treatment presumes that by giving comprehensible information to warning receivers, product safety is sufficiently achieved. However, as shown by the C-HIP model and figure 6, more cognitive steps within the receiver or subgoals need to be completed to reach safe behaviour. It can thus be argued that these final two subgoals are currently not or not sufficiently considered by law. This deficiency may call for a re-evaluation of the legal test of warning adequacy. This is done in more detail in paragraph 4.6.

Legal Analysis of the Risks that Need Not Be Warned Against

The third warning issue concerned the warning issue as to "What risks need not be warned against?". Paragraph 4.4 recommended an approach to avoid an increased liability for not having provided a product warning. Since this allegation is a common form of litigation, the approach can be considered of value to European product liability law.

The approach consisted of using limits to defectiveness that pertain to claims regarding the absence of a warning for a certain risk. Five categories of risks were presented of which I recommended that defectiveness must be denied when a warning was missing. These are risks that have an insignificant size, risks arising from unreasonably expected use, obvious risks, generally known risks[314] and risks that were present yet undiscoverable at the time the product was put into circulation.

To determine whether the product without the warning is defective, the Directive lays down a consumer expectations test. This leads here to answering the question whether the product that bears no warning failed to meet the safety level which a person is entitled to expect, taking all circumstances into account. Case law suggests that many courts in Europe have not stopped using risk-utility factors to establish whether the absence of the warning rendered the product defective, even though these factors originate from fault-based liability. These factors are especially of relevance in cases where the risk is of insignificant size and/or obvious, generally known and/or related to unreasonable misuse of the product. Many legal academics have submitted that the decision whether a producer failed his duty to warn under fault-liability is actually no different from deciding whether the absence of a warning rendered the product defective.[315] Hence, I argued here that it is allowed to conclude that the main part of assessing whether the warning's absence renders the product defective involves consideration of factors relating to the costs of adding a warning and to benefits/utility of adding the warning to the product. Are the costs of adding a warning to the product higher which leads to the conclusion that the product with a warning is not defective? Or does the warning have an expected added value as a result of which the product without the warning can be considered defective?

Some limits, like those with regard to risks arising from unreasonably expected use and undiscoverable risks are laid down by the Directive itself. These limits constitute a complete bar against liability. Albeit implicitly, it also follows from the Directive that producers are absolved from liability when it concerns risks of insignificant size, since the Directive does not demand of producers to design products that are absolutely safe to avoid liability. It must be borne in mind that in case of obvious and generally known risks, the finding of non-defectiveness on the basis of a no-warning claim does not have to automatically preclude a finding of defectiveness on other grounds. Because a warning can often be considered the least reliable measure to protect people from harm, other design measures that are more effective may as a result be required to render the product non-defective.

314 A warning for generally known risks can be required under the limited circumstances in which a reminder is needed.

315 A difference may lie in the approach to unknown and undiscoverable risks.

Three rationales were given to underpin the adoption of these limits. First of all, it was asserted that these limits to defectiveness do justice to the policy goal of the Directive, as European courts bring about a fair appropriate apportionment of the risks when giving decisions on the basis of these limits. As a result of not allowing defectiveness due to the absence of a warning of a risk that is of insignificant size, obvious, generally known, undiscoverable, or associated with unreasonably expected use, the burden of safety shifts from producer to consumer. Providing warnings for these risks would not lead to an additional substantial safety effect compared to the situation in which a warning is lacking.[316] As discussed in paragraph 4.3, it was argued here that European product liability law requires producers to provide warnings so that accidents can be prevented or reduced that would otherwise have occurred in the absence of a warning. Accordingly, on the basis of empirical research it can be presumed that the safety of products increases because having warnings present brings an added safety value to the situation. Hence, if it is expected that providing a warning does not substantially raise the safety of a product, a warning should not be required under law in those cases. In other words, if a warning's utility is minimal, because consumers are well able to avoid an accident without being given a printed warning, the warning can be considered unnecessary and should not be required under law. The costs attached to providing a warning can thus be considered of greater importance. Additionally, it could be argued that the limits to defectiveness also encourage consumers to act safely. Thirdly, by accepting these limits European product liability can also contribute to preventing or reducing the negative effects associated with 'overusing' warnings. As was shown in paragraph 3.8 of the previous chapter, a review of the warning research literature revealed that there are potential negative effects associated with providing warnings, especially in case of an 'overuse' of warnings. Overusing warnings refers to the circumstance that consumers are being overloaded with too many warnings that accompany the product and/or with too many products that bear a warning. Overusing warnings can adversely affect the cognitive processing of warnings in particular and in general in a number of ways. Concerns with respect to product liability providing incentives to overuse warnings have been noticed in the legal literature. It can be viewed as a common pitfall for courts to assess a no-warning claim in isolation. However, it is imperative that courts are cognisant of the potential negative effects on the effective processing of other warnings that can occur as a result of their decisions that impose defectiveness for not having warned against such risks as discussed above. Such decisions can trigger producers to overuse warnings, *i.e.* to

---

316 The limit pertaining to undiscoverable risks is special since this limit is the result of implementation of the development risk defence into national law. The main rationale for this limit is to shift the responsibility of these risks to consumers so as to encourage product innovation. If a Member State has chosen not to include the development risk defence into domestic law on the basis of Article 15(1)(b) EPLD, then a producer cannot escape liability for defectiveness as a result of the absence of an undiscoverable risk. In those cases, the responsibility for safety does not shift to consumers, but remains with producers.

provide warnings for every single risk out of fear for liability, whilst these warnings can be considered unnecessary. These negative effects should be viewed as social costs associated with adding an unnecessary warning to a product and these costs should play a role in the balancing process of courts in general, or even at a concrete level as a circumstance in a specific case. In sum, by providing clarity on the matter of what risks need no warning according to liability principles, European product liability can play a role in reducing the unintended effects of overwarning or overusing warnings.

Paragraphs 4.4.4-4.4.8 discussed the limits to defectiveness in more detail. Below, a summary is given of the five risk categories.

Risks of Insignificant Size

A most important bar against accepting liability for the absence of a warning concerns the limit with respect to risks of insignificant size. It was made clear that European product liability must not demand producers to warn of risks that have an insignificant size, because this would trigger producers to create a warning for every potential risk to minimise their liability. This trend would be detrimental to the effective processing of warnings. Especially with small risks, courts must be cautious not to be trapped into underestimating the costs of providing a warning for such risks and mistakenly rendering the product without the warning defective. The presumption that providing warnings is a cheap and easy measure to protect the interests of potential victims should not be generally held by judges. It was recommended that the social costs associated with overusing warnings need to have bearing too and tip the balance in favour of rejecting defectiveness when courts consider a risk small.

It is therefore of importance that the tool of a risk assessment, which was described in paragraph 4.4.2 and displayed in figure 7, plays a role within the decision making process of courts. A main lesson learned from this tool is that it is important to consider the risk, *i.e.* the combination of the probability and the severity of the injury, instead of putting too much focus on the single factor of the severity of the injury that was caused by the product hazard. A correct focus is of judicial importance, since in the latter situation courts are likely to be more eager to render the product without the warning defective than in the former situation. In addition, the results of a risk assessment can be used as input to select the risks that need to be approached by design, including warnings, and which not. The risk assessment method laid down in the RAPEX Guidelines of the European Commission may be useful to courts for determining when a risk of a non-food consumer product is of insignificant size, as it provides a table (see figure 8) that determines the level of product risk. It distinguishes four risk levels (serious, high, medium and low). It was asserted that courts can use the low risk level as a guiding principle for deciding that risks falling under this level generally require no warning.

Risks that Arise from Unreasonably Expected Use

The second limit to defectiveness that was described in paragraph 4.4.5 relates to the manner in which products that have hazardous properties are used. This limit that pertains to denying warnings for product risks that arise from unreasonably expected use can contribute to a reduction of superfluous warnings in the world. It follows from the wording of the Directive that risks resulting from use to which it could not reasonably be expected that the product would be put, do not need to be taken into account by producers, because this can be considered 'unreasonable misuse'. This means that product warnings are not required for risks that result from this kind of behavioural actions. Similarly, even without the printed warning of these risks, an acceptable level of safety is still achieved.

It is important that European courts take the interpretation of this factor seriously. If the term is consistently interpreted too widely and producers are being held liable because of the absence of a warning of a risk that was a result of behaviour of which it is considerably questionable that it is reasonably expected use, this trend can trigger producers to provide warnings that are actually unnecessary. Unnecessary warnings should be avoided as this may negatively affect the effective processing of other warnings that are indeed necessary in a number of ways.

To help draw the line between reasonably expected use and those interactions that can be deemed unreasonable misuse, some guidance was found in EU product safety legislation. EU product safety legislation has used the requirement of use that results from 'readily predictable' human behaviour to define the concept of 'reasonably foreseeable misuse', a concept that closely corresponds with the Directive's requirement of use that is reasonably expected. It was argued that the requirement of 'readily predictable' entails a useful restriction with regard to the range of behavioural actions that producers need to anticipate by design. Whether use can be regarded readily predictable can be assessed on the basis of the information sources that have been consulted. Using various information sources is of importance to become knowledgeable about the range of likely behavioural user actions and other factors that may lead to an accident. Generating accident scenarios behind your work desk with your imagination as the only source of information is in all likelihood not sufficient in the eyes of courts to get a good picture of what accidents can happen. Especially ergonomic methods such as qualitative observational research provide a clearer picture of what use actions people do with a product. This enables producers to better anticipate accidents and to improve the safety of the product by design, and in the aftermath of an accident it helps producers explain why this use should fall outside the scope of the Directive. Furthermore the type of product hazard also plays a role in establishing whether a certain use action can be viewed as easy to predict. Some hazards that are intrinsic to a product have more powerful health effects than others, as a result of which use actions even though not intended still fall under the scope of reasonably expected. This may consequently entail that producers are required to do more research into the occurrence of accidents and to collect more comprehensive data than just injury registration data. Naturally, courts should consider the value of the types of information sources used, since different sources provide different or more comprehensive results. It could even be argued that the use of ergonomics for the design of products is required under the Directive, since ergonomic methods in particular offer ways to gather information on how accidents occur and what product-use interactions are plausible. This would better equip producers to determine the boundaries between reasonably expected use and those interactions that can be deemed unreasonable and outside the scope of liability. Admittedly, the mere application of ergonomic principles does not constitute a barring effect to liability. Ergonomic methods also have their weaknesses. Nonetheless, it can be said that it generally offers producers a reasonably solid escape route.

It was also argued in this paragraph that producers should have an open mind when generating various accident scenarios. Expectations about correct behaviour of relevant product users should not be too high. Even though a producer is quite certain that the use action seems far-fetched, it remains important that such a use action is identified and taken into consideration during the identification of accident scenarios and the assessment of the risks, since the results of the risk assessment and evaluation can ultimately confirm whether it was justified to ignore the risk and to leave any risk-reduction measures aside. Risks arising from unreasonably expected use can often be considered of minimal size. Because certain use actions have little to do with the intended use, they are assessed as highly unlikely, which ultimately influences the size of the risk. However, if the severity of the injury can be estimated as very high then its combination with a low probability of the use action may still result into a risk of significant size that needs to be controlled. Showing courts documentation of their risk assessment that explains why they were not required to take such a risk arising from that particular use action into account will provide a good defence in court to contest the allegations made by the injured party.

Obvious Risks

Another limit to defectiveness that pertains to no-warning claims concerns obvious risks. European product liability should not encourage producers to provide printed warnings for obvious risks as it was argued that adding these warnings will have little safety utility compared to the situation in which no warning was given. Inspired by psychological notions, a legal test for determining whether a risk can be considered obvious was proposed in paragraph 4.4.6. On the basis of this test, Member State courts can rule in a consistent manner that a risk is obvious and adhere to this limit.

Two elements are of significance for determining whether there is an obvious risk: (1) the product hazard must be considered obvious, which means that the hazardous characteristic of the product is obvious; and (2) the interactions of users with the hazard to avoid or reduce the risk of damage must be considered obvious.

It was suggested here to consider a product hazard 'obvious' within European product liability law, if it is easy to sense it with one or more of our human abilities, and easy to understand. Important indications of an obvious hazard are the cue(s) given by the product's design to communicate to the user that there is a dangerous situation. An essential cue involves whether users can easily see the hazard and in consequence understand what behavioural actions need to be avoided. The reason for this is that sensing the hazard with your eyes triggers your awareness easily. Although sight is the most common and probably the best way of triggering awareness in relation to product hazards, it is not unimaginable that the danger reaches the user through another sensory modality. If this is the case, then this form of awareness should contribute to establishing that a risk is obvious. The other senses may not be as powerful as vision, but their role in triggering awareness should not be overlooked. Secondly, something is obvious if it is also easy to understand. I think it is important to make a distinction here between knowing/understanding and seeing, because seeing a hazard (or smelling, hearing, touching or tasting it) triggers awareness more easily than having knowledge about it. It seems that the label 'obvious' is often used in conjunction with the term 'known'. I believe that the obviousness of a risk should not be put in the same box as generally known risks. Of course, it is not denied here that these concepts are interrelated to each other. If something is obvious, it is usually well-known too. However, something that is known does not have to be obvious.

Generally Known Risks

The third risk category of which it was recommended that the absence of a warning does not render the product defective under the Directive concerns risks that are generally known. European product liability law should generally proceed from the viewpoint that the consumers' knowledge will normally lead to awareness as a result of which consumers are well able to act carefully. It is therefore expected that adding a warning will not make a substantial difference to the situation in which the product carries no warning.

Notwithstanding the rationales that support having such a limit in European product liability law, it was argued that a blanket rule consisting of denying warnings for generally known risks takes insufficient notice of the psychological distinction that can be made between awareness and knowledge. I therefore recommended that an exception to this limit should be made for *reminder warnings*. That consumers have existing knowledge in long-term memory of a certain hazard does not mean that they are always aware of it at the time they are at risk. The purpose of a reminder warning is thus to cue people to remember the needed information at the right time. It was argued in paragraph 4.4.7 that for reasons of product safety, warnings that act as reminders should be required in European product liability law, but only under those special circumstances in which it is expected that awareness is not triggered at the moment users need it to avoid the risk. In these cases, it was contended that a warning of a generally known risk has substantial utility.

This paragraph also provided a framework for deciding whether a risk is 'generally known'. The basis should be that the risk can be considered known to the 'public at large', which requirement is indicated by the text of the Directive. More specifically, this suggests that to consider it 'generally known' a large majority of the target audience of the product must have knowledge of the risk. This involves the assessment of who are the foreseeable persons that interact with the product and what is their knowledge level, and also the evaluation that a large majority of them possesses this knowledge. A factor that indicates whether a risk can be considered generally known is the circumstance that without the hazardous product property the product's functionality is severely impaired. Another indication that was mentioned is whether the information is also delivered through other media and/or transmitted by other sources.

Undiscoverable risks

The final limit, that was discussed in paragraph 4.4.8, entails that there is no liability for the absence of a warning with respect to risks that caused damage albeit undiscoverable at the time the product was put into circulation. This limit should not be interpreted to mean that if a risk were discoverable, a warning is required. The size of the risk can be determining for not requiring a warning for risks that were discoverable.

In accordance with the Directive's preamble, the defence forms part of creating a fair apportionment of risk between the injured person and the producer. The main rationale for adopting this defence in domestic law is that otherwise, out of fear of increased liability, producers would be discouraged to innovate.

The limit is embodied in the development risk defence of Article 7(e) of the Directive. It is an optional defence that protects producers against liability for damage caused by a risk which is present, but not yet discoverable on the basis of the most advanced state of scientific and technical knowledge at the time the product was put into circulation. Hence, producers can escape liability for defectiveness for not having warning of such a risk when they successfully invoke the defence. This means that they must prove that even with the objective state of scientific and technical knowledge that was accessible, it was impossible to detect the defect at the time the product was put into circulation. Because of the ECJ's narrow interpretation of this defence, it will be difficult to escape liability.

Legal Analysis of the Role of Need to Warn in Relation to Other Product Design Solutions

The legal analysis of the fourth important warning issue concerned the role of warnings in relation to other design solutions in a legal context.

This paragraph called for attention to treating warnings in European product liability law as a last-resort measure to reduce a product's risk and avoid accidents. This expression stems from the ergonomic design perspective of the basic hazard control hierarchy model that provides prevention methods in order of preference: design out, guard, warn. An appeal was made to European courts and injured litigants to recognise and adhere to this principle, which means that having provided a warning against a specific product risk does not necessarily absolve producers from a finding of defectiveness for that risk. Even though paragraph 4.5.3 showed that EU product safety policy and European product liability law already more or less recognise the principle of the last-resort measure, a contribution of this paragraph lies in providing the underlying explanations regarding why warnings should be viewed in such a manner.

There is firm empirical justification for this principle. Although the warning research findings generally show that the presence of warnings leads to more safe behaviour than in the absence of warnings, the degree to which warnings have an added value is often not huge. Research indicates that warnings have a moderate role as accident prevention mechanisms. Three interrelated explanations were offered in paragraph 4.5.2 with respect to why warnings are no safety panacea and they form the foundations of the last-resort-measure principle that governs the need for warnings. The first explanation can be offered by the information processing perspective. Before a warning can end in safe behaviour, several cognitive stages of information processing need to be successfully passed. Hence, behavioural compliance will be the least reliably observed effect of a warning. Furthermore, this corresponds with the viewpoint that warnings can be considered active protective measures that require users to perform or refrain from certain user actions each time the product is used and are not passive measures that require no behavioural input of people. Lastly, a consideration of the various factors (relating to the design, the person and the environment of use) that influence the warning process also elucidates why following warnings is generally a complex task.

The first recommendation with regard to this warning issue in a legal context was presented in paragraph 4.5.4. It was suggested that European product liability law must adopt the last-resort-measure principle of warnings. The last-resort-measure principle should not be interpreted as an automatic way of finding defectiveness in situations where a warning against the risk was provided whilst a safer design solution was available. In addition, the principle does not entail that a warning can never free a producer from a judgement of defectiveness. The main rationale underlying the adoption of the principle in law is that it provides an incentive to producers to improve the safety of a product by design. If there is no explicit acceptance of such a rule in product liability law, it is expected that producers will be less driven to increase the level of safety of their products through implementing design solutions that are more effective for protecting consumers against risks than warnings. A second important argument entails that it contributes to discouraging producers from providing unnecessary warnings that lead to an overuse of warnings and the occurrence of the associated negative consequences. A major argument likely to be used against the principle is that it places too little responsibility on consumers. Although this point may seem valid at first sight, the principle of warnings as a last-resort measure does not object to a responsibility of users for safety, it simply embraces human limitations and errors and consequently grounds the perspective of safe design on the notion that it is far more difficult to prevent accidents by influencing behavioural patterns of users than by a product's design that provides automatic protection. As human factor specialists Sanders and McCormick have advocated: "It is easier to bend metal than to twist arms".

I recommended in paragraph 4.5.5 that warnings should be viewed as part of the design process of products. To emphasise that warnings are design measures that have their own contribution in achieving safety by design, it should be recognised by European courts and litigants that the absence of a warning or the presence of an inadequate warning can be viewed as a defect in the design of the product, just as with other design measures, since all products of the line have this defect (*i.e.* the absence of a design feature or presence of an inadequate design feature). It is the result of the producers' conscious decisions during the design process of products. So, placing a warning on a product that tells you of the risk and how to avoid it should not permit the conclusion that a material or procedural guarding mechanism that blocks any contact between the hazardous product property and people is not required to render the product non-defective. Similarly, a producer that has used a guard to provide protection ought not to be automatically insulated from defectiveness because it may still be established for the absence of a design solution that designed the hazard itself out of the product. To put it in another way, even if the hazard has been designed away, that does not preclude the conclusion that the hazard should actually have been designed out of the product to render the product reasonably safe. Of course, for a finding of defectiveness other circumstances also play a role, *e.g.* that this latter change must not unduly restrict the functioning of the product. Hence, I agreed that these are distinct obligations that pertain to designing safe products. These are distinct because the design solutions based on these controlling methods offer different levels of protection. They are all independent which means that one design solution that reduces the risk does not preclude the necessity of another design solution. Defectiveness can always be imposed on another design defect ground, although some are more likely than others; it is more likely to establish defectiveness because only a warning is present and not a safety guard, than finding defectiveness because there is a guard and no warning. But these claims about design are also related because they all address how safety can be improved by the design of the product.

Special attention was called for the defect claim that centres on a producer's misuse of a warning, *i.e.* that even though a warning has been provided, the product can still be found defective because, in view of the circumstances of the case, a more effective design measure, which was adopted from the hazard control hierarchy model, was required to render the product non-defective. In consequence, lawyers of injured victims involved in a product liability case carry the important task of instituting legal actions under the Directive that are inspired by this preferred way of approaching product risks. Because warnings are related to design defects and can be viewed as part of the design of a product, it is often possible for claimants to assert that the product is defective because of its flawed design with regard to the risk in question and because of the presence of an inadequate warning or a lack of a warning against the same risk. If the circumstances of the case permit it to allege both deficiencies, it is of importance that claimants take account of the last-resort-measure principle of warnings and first claim that the alternative design solution that is more effective should have been implemented. Only after having pointed out to the court that there was a safer version of the product possible and feasible, claimants should focus on the warning defect and contend that this rendered the product defective. In contrast, if they allege defectiveness because of the inadequate design of the warning that was provided, whilst the heart of the problem concentrates on the absence of an alternative design solution, the outcome of the decision (inadequate warning) would not advance the incentive to design products that have an optimal safety level.

The final recommendation was made in paragraph 4.5.6 and concerned the approach of applying the defectiveness standard of the Directive to products that warned against the risk that allegedly caused the damage. It suggested to claimants who use the claim centred on a producer's misuse of the warning which factors should be presented to the court, and it recommended to European courts a consistent manner of applying these circumstances in order to reach a satisfying decision about a product that already contains a warning against the materialised risk.

The defectiveness test of the EPLD requires a determination of the level of safety persons generally are entitled to expect of a product. A serious flaw of this test is that it can be problematic to utilise in cases of products that have a warning against the risk that materialised. Because expectations about a product and its safety are often formed on the basis of a product's appearance, courts could be eager to conclude that the test permits rejecting defectiveness, because the warning enables consumers to form a good idea about the dangerousness of the product and to know what to expect. Although this flaw of the test is often cited in relation to its effect on obvious and generally known risks, this criticism is also of special concern for those cases in which it can be alleged that a producer has misused the warning in order to reduce his chances of being held liable, because obvious and generally known risks as well as warnings influence expectations.

Unlike manufacturing flaws that can be compared with the design standard intended by the producer and in consequence failed to meet consumer expectations, defects in the design of the product (including warnings) have no defined product standard against which the defective design of the product can be judged. The integrity of the design of the product itself is being challenged in court. Therefore, as regards the assessment of the defectiveness of a product with a warning, concentrating on factors that indicate the desirability of a safer design solution over a warning is hardly inescapable.

I believe that the product with the warning can be considered defective under the Directive, if the comparison shows that the balance of costs and benefits of the alternative design solution is better than the balance of the costs and benefits of the warning. In other words, a product that has a warning to reduce the risk does not provide the safety consumers are entitled to expect, because the warning can be considered less cost-effective than the absent design solution. Non-consideration of risk-utility (or cost-benefit) factors when determining what level of safety consumers are entitled to expect of products with risks that have been warned of may lead to wrong judgements, because the test would be too narrowly focused on the fact that the safety expectations are lowered because of the presence of the warning. Instead, the test should focus on assessing the level of safety consumers are 'entitled' to have. This directs the defect assessment towards analysing whether an available design solution that is more effective in reducing risk than a warning should have been present in the product's design to render it non-defective. For this reason, risk-utility factors should be allowed within the liability test of the Directive to determine the defectiveness of a product that warned against the materialised risk as they enable a comparison between the given warning and a proposed safer design solution.

More specifically, when asserting that the product with a warning is defective, claimants should include factors that indicate that providing a warning was an inappropriate way to control the risk. In other words, that there is a mismatch between the size of the risk (*i.e.* combination of severity of injury and probability) and the warning's safety benefits that were assessed by the producer. Factors indicating that the warning is an inappropriate precautionary measure are those that suggest that the likelihood that the warning will be followed is low and that therefore the safety benefits of the warning are insufficient to reduce a sufficient number of accidents. The warning research literature is instructive in this respect as it offers a list of factors that have empirically proven to influence the effective processing of warnings. Other relevant factors relate to the cost-benefit analysis of proposing another design solution to control the risk that is more effective, based on the hazard control hierarchy model. These include the availability of a design solution/feature, the effectiveness of it in controlling the risk (*i.e.* the safety benefits that arise from avoiding those type of accidents), the feasibility of implementing this design solution in terms of the cost of implementing this design solution. It was argued that cost should be interpreted widely; it entails the financial costs of adopting the design solution, but also the decreased usefulness of the product as a result of the adoption, and also the creation of new hazards.

Legal Analysis of the Way in which Warnings Should Be Designed

The final paragraph 4.6 addressed the major warning issue of 'How to warn?' in light of the findings from the warning research literature. It provided recommendations pertaining to the adjudication of the liability yardstick of the adequacy of a product warning according to European product liability law. More specifically, it discussed what the legal test is and should be for warning adequacy under the Directive, and in view of this proposed test what the requirements with respect to an adequate warning are and which relevant factors should be used to assess these requirements.

The paragraph started with a discussion of how courts commonly apply the test of the Directive to evaluate the content and form of a given product warning. The Directive's test requires that the product must provide the level of safety the general public is entitled to expect. The allegation that the product is defective, because the product warning provided by the producer should be considered inadequate is a popular allegation in product liability litigation. Although 'adequacy' appears to be the magic word in European product liability law for the assessment of a warning, the Directive provides little guidance on what governs the adequacy of a warning. Because the safety expectations that people have of a product are influenced by the warnings that accompany the product, it is important to have clarity about how to legally measure the adequacy of warnings.

It seems that there is no real, coherent test present in European product liability law to determine the adequacy of a product warning. Key terms that have been used by European courts and academic commentators to interpret the concept of warning adequacy are whether the warning is sufficiently visible, legible, or comprehensible. In addition, it was shown in chapter 2 that from a number of decisions of Member State courts, a (non-limitative) catalogue could be formed of relevant considerations with regard to a warning's adequacy. Hence, these appear to be of importance and include: the probability that a product danger emerges; the degree of harm arising from that; visibility; legibility; unambiguous language; complete information; nature of the hazard; location; language; and knowledge/expertise of the user and public law legislation or standards.

I argued that the ambiguity that governs the standard of warning adequacy could be explained by the uncertainty with regard to the warning issues of what is a product warning and what should it accomplish according to European product liability law. A link was made with the analyses of these warning issues in paragraphs 4.2 and 4.3 of this chapter. Paragraph 4.3 already noted that European product liability law should be explicit about the rationale and legal goal of product warnings, because this exercise is helpful for getting grip on the legal building blocks of an adequate warning.

It was asserted that it appears to be standard practice in European product liability law at this moment to consider warnings legally adequate insofar as they provide relevant information to product users in a noticeable, legible and understandable way. This test does, however, not entirely correspond with the viewpoint held in European product liability law that warnings are precautionary measures. As advocated in paragraph 4.3.4, the warning research literature has taught us that for a warning to lead to safe behaviour providing a comprehensible warning is not sufficient. In consequence, it looks like the current test for warning adequacy presumes that warning receivers will be persuaded by the warning and motivated by it to act safely. Hence, the effectiveness of these stages is borne fully by consumers.

In view of these observations, it was subsequently recommended in paragraph 4.6.2.2 to determine the adequacy of a product warning on the basis of whether the design of the warning has the potential to sufficiently influence behaviour in the way as prescribed by the warning. If this can be answered in the affirmative, then the responsibility for safe use fully shifts from producer to consumer and any losses should be borne by the latter party.

To further determine whether a warning meets the general test, the stages of information processing should act as requirements to interpret the legal concept of warning adequacy. It was concluded that the test for determining whether a product warning is legally adequate depends on the evaluation of the following warning adequacy requirements: (1) the product warning is sufficiently salient; (2) the product warning is sufficiently legible; (3) the product warning is sufficiently comprehensible and memorable; (4) the product warning is sufficiently persuasive. Because of the great overlap between the stages of beliefs and attitudes and motivation of the C-HIP model and their corresponding factors, the stages were combined to generate the legal requirement of a persuasive warning.

Several considerations were given why claims of defectiveness based on the inadequacy of a given product warning should be applied in this way. First, the test synchronises the law with the psychological model of the warning process; it reflects the insights on how individuals process warning information and make choices on the basis of that processing. Having legal requirements for product warnings that facilitate the occurrence of behavioural compliance by consumers is likely to reduce the number of accidents involving products with inadequate warnings. Another reason for advocating this test was that it elaborates further on the traditional view of warnings as precautionary measures in civil liability law. A warning is by its very nature a design measure that aims at protecting the safety interests of potential victims. It can be argued that consumers are entitled to expect that warnings do in fact have a design that invites consumer protection. Merely giving comprehensible hazard-related information to consumers to educate them on unknown risks or reminding them of familiar risks is not sufficient for safe behaviour. The adverse consequence of shifting the burden of complying with the warning from producer to consumer when the warning has reached the stage of comprehension is that important factors that can affect the subsequent stages including behaviour do not have to be taken into account by the producer when designing the warning. These factors that are largely dominant in the final stages can also influence the effectiveness of earlier stages, such as attention. As a result, the effectiveness of these stages is borne fully by consumers, and this should not be viewed as a proper allocation of the responsibility for safety.

It was submitted that the test probably imposes a more stringent outlook on warnings in European product liability law. Nevertheless, the use of persuasive techniques is very popular in the field of consumer products. Why can and should the employment of these insights from persuasion theory and literature not be applied to product warnings? In addition, another plausible argument raised against the test can be that it leaves little scope for the users' responsibility for safety. The producers' responsibility is to assist this decision making process by providing a warning that is designed in such a way that it sufficiently persuades people to decide to comply with the warning. There still rests a responsibility on users to use the product safely in accordance with the warning, as it is ultimately up to the consumer to produce the compliant behaviour. I prefer to view the warning process as a shared responsibility. There should be a remaining responsibility for producers to take the factors into account that affect the later stages in the warning process that succeed comprehension, because these are essential to the warning process. Consequently, the effectiveness of the later stages is also a shared responsibility between producer and consumer. Nevertheless, the responsibility in the later stages rests for a large part with the users, since it is the receiver characteristics (*i.e.* their expectations about the hazard, its severity and the behavioural actions of the warning) that are generally prevailing in these stages and secondly because they need to take the final step themselves of following the precautionary statements expressed in the warning.

Another argument that is likely to be raised against the application of this yardstick, concerns the interpretation of the test in such a way that the inadequacy of a warning in a concrete case is established by the mere fact that this warning did not have a positive effect on the behaviour of the injured claimant. This interpretation is incorrect and undesirable. The test of warning adequacy should not be interpreted in black and white; a non-effective warning is not tantamount to a finding of defectiveness as no single warning, how well-designed it may be, can guarantee 100% compliant behaviour. Such an application of the test would stand in the way of the incentive effect of European product liability law to create well-designed warnings that can protect the health and safety of consumers. The Directive does not demand all warnings to be 100% effective in protecting consumers.

Following the discussion on the relevant warning adequacy requirements, paragraph 4.6.2.4 addressed the warning adequacy factors that should be considered when evaluating whether a warning adequacy requirement has been met.

It followed from chapter 2 that the adequacy of a warning has often been determined on the basis of factors that have previously been identified in case law or factors that intuitively are considered to be of importance by legal parties or courts. In view of this arbitrary process of selecting factors, it is possible that the weight of a factor is misjudged or that an influential factor has inadvertently been excluded from the assessment. It was argued in this paragraph that guidance can be found in the warning research literature, as it provides influential factors that can be used as a basis. The use of this list should enhance the use of a consistent approach among European courts when applying the defectiveness test to warning adequacy. Moreover, because this list is supported by empirical evidence with respect to how people process warnings, it is argued here that the use of these factors in the assessment of warning adequacy is expected to improve the safety level of products. As a result, figure 10 was drawn up which provides a list of the factors identified by the warning research literature as potentially influential and organised on the basis of the warning adequacy requirements. The factors pertaining to actual behavioural compliance are included as well.

The paragraph also discussed how all the relevant research-based factors should be balanced. It was recommended that the starting point of the process of balancing the relevant circumstances of the case is to generally distinguish three important components of factors within the test. These factors should form the building blocks of claims that are based on the assertion that the product is defective because the concrete warning is inadequate. They entail (1) factors that relate to the level of risk involved; (2) factors that relate to the given warning; and (3) factors that relate to the availability of a better alternative design of the warning.

With regard to the first group of factors, the fault-based notion that the nature and size of the risk of injury determine the design of the warning should be maintained under the liability regime of the Directive. Various warning design features are available to producers to realise a warning that appropriately connotes the risky situation.

Naturally, the factors indicating the inadequacy of the given warning are of paramount importance. Analysing the warning adequacy requirements can pinpoint a probable cause for the inadequacy of the concrete warning. This helps to determine whether the research-based factors for this particular warning adequacy requirement have been adequately taken into account by the defendant. Figure 6.1 can be used as an investigative tool for claimants and courts in this respect. If a factor is absent or poorly taken into account, claimants can argue that the warning is defective since empirical evidence supports the influence of this factor on the adequacy of a warning. As regards the adequacy of a reminder warning, one can primarily zoom in on the factors relating to the warning adequacy requirements of salience and legibility. As far as a warning can be classified as a purchase warning, of a risk which occurrence is not dependent on how the user uses the product, the adequacy should mainly be determined by whether the warning is sufficiently salient, as the warning normally affects the purchase decision. Naturally, the comprehensibility of the information should also be considered vital as consumers need to decide on the basis of this information whether or not they want to expose themselves to this risk. In light of these functions, the requirement of persuasiveness should not be considered of relevance for both types of warnings.

Finally, the factors that relate to a better, alternative version of the warning also deserve attention. These factors should relate to the costs and safety benefits of a changed warning design. Evidently, the alternative design of the warning is proposed by claimants because of its expected safety benefits. By having pointed out the flaws of the given warning that consists of missing factors or poorly addressed factors, it normally follows what the improvements should be for the alternative warning design. However, because it can be asserted that in the aftermath of an accident it is a rather easy task for claimants to make suggestions for a warning that is likely to be better in avoiding injuries associated with the materialised risk, claimants and courts ought to be alert in considering the costs of changing the warning too. These costs may entail the design trade-offs that are involved when adding extra information to a warning or making the warning more impressive.

Having explained the basis for determining the defectiveness of a warning by specifying the warning adequacy requirements and the warning adequacy factors, the final recommendation in paragraph 4.6.4 offered a toolkit filled with a number of special design principles. Courts can use these to facilitate the evaluation of the adequacy of a warning. The tools are derived from the warning research literature and involve design principles that facilitate the effectiveness of warning designs. These include matching the warning's design to the risk, prioritisation of the available warnings, brevity of a warning, the warning as a warning system containing several warning components, and the judicial value of having designed warnings on the basis of testing results rather than only design guidelines.

List of references

Abroms e.a. 2003
L. Abroms e.a., 'Gender differences in young adults' beliefs about sunscreen use', *Health Education & Behavior* 2003-1, p. 29-43.

Adams & Edworthy 1995
A.S. Adams & J. Edworthy, 'Quantifying and predicting the effects of basic text display variables on the perceived urgency of warning labels: Tradeoffs involving font size, border weight and colour', *Ergonomics* 1995-11, p. 2221-2237.

Adler 1995
R.S. Adler, 'Redesigning people versus redesigning products: The Consumer Product Safety Commission addresses misuse', *Journal of Law & Politics* 1995-1, p. 79-128.

Ajzen 1991
I. Ajzen, 'The theory of planned behavior', *Organizational Behavior & Human Decision Processes* 1991-2, p. 179-212.

Ajzen & Fishbein 1980
I. Ajzen & M. Fishbein, *Understanding attitudes and predicting social behavior*, Upper Saddle River, N.J.: Prentice-Hall 1980.

Van Aken 1996
D. Van Aken, *Handboek ontwerpen van veilige producten*, Utrecht: LEMMA 1996.

Van Aken 1997
D. Van Aken, 'Consumer products: Hazard analysis, standardization and (re)design', *Safety Science* 1997-1-2, p. 87-94.

Amer & Maris 2007
T.S. Amer & J.B. Maris, 'Signal words and signal icons in application control and information technology exception messages. Hazard matching and habituation effects', *Journal of Information Systems* 2007-2, p. 1-26.

American Law Institute 1998
American Law Institute, *Restatement of the Law Third, Torts: Products Liability: As adopted and promulgated by the American Law Institute at Washington, D.C., May 20, 1997 §§ 1-end,* St. Paul, Minn.: American Law Institute Publishers 1998.

ANEC & BEUC 2008
ANEC & BEUC, *Booklet WARNING(s) you might be shocked*, Brussels, 2008.

ANSI 2002a
ANSI, *American national standard for product safety signs and labels: Z535.4-2002*, Rosslyn, VA: National Electrical Manufacturers Association 2002.

ANSI 2002b
ANSI, *American national standard criteria for safety symbols: ANSI Z535.3-2002*, Rosslyn, VA: National Electrical Manufacturers Association 2002.

Argo & Main 2004
J.J. Argo & K.J. Main, 'Meta-analyses of the effectiveness of warning labels', *Journal of Public Policy & Marketing* 2004-2, p. 193-208.

Arnett 2000
J.J. Arnett, 'Optimistic bias in adolescent and adult smokers and nonsmokers', *Addictive Behaviors* 2000-4, p. 625-632.

Van Assema e.a. 2001
P. Van Assema e.a., 'Framing of nutrition education messages in persuading consumers of the advantages of a healthy diet', *Journal of Human Nutrition and Dietetics* 2001-14, p. 435-442.

Ausness 1999
R.C. Ausness, 'When warnings alone won't do: A reply to Professor Phillips', *Northern Kentucky Law Review* 1999, p. 627-654.

Ayres e.a. 1989
T.J. Ayres e.a., 'What is a warning and when will it work', *Human Factors and Ergonomics Society Annual Meeting Proceedings* 1989-6, p. 426-430.

Bakken 2005
G.M. Bakken, 'Human-centric approach to forensic analysis for system liability', in: Y.I. Noy & W. Karwowski (eds.), *Handbook of human factors in litigation*, Boca Raton: CRC Press 2005, p. 28-1-28-21.

Bandura 1986
A. Bandura, *Social foundations of thought and action: A social cognitive theory*, Englewood Cliffs, N.J.: Prentice-Hall 1986.

Banks e.a. 1995
S.M. Banks e.a., 'The effects of message framing on mammography utilization', *Health Psychology* 1995-2, p. 178-184.

Barlow & Wogalter 1991
T. Barlow & M.S. Wogalter, 'Increasing the surface area on small product containers to facilitate communication of label information and warnings', *Proceedings of Interface* 1991, p. 88-93.

Barlow & Wogalter 1993
T. Barlow & M.S. Wogalter, 'Alcoholic Beverage Warnings in Magazine and Television Advertisements', *The Journal of Consumer Research* 1993-1, p. 147-156.

Becker 1974
M.H. Becker, *The health belief model and personal health behavior*, Thorofare, NJ: Slack 1974.

Better Regulation Executive & National Consumer Council July 2007
Better Regulation Executive & National Consumer Council, *Warning: Too much information can harm. An interim report on maximising the positive impact of regulated information for consumers and markets*, URN 07/1272, UK, July 2007.

Bettman, Payne & Staelin 1986
J.R. Bettman, J.W. Payne & R. Staelin, 'Cognitive considerations in designing effective labels for presenting risk information', *Journal of Public Policy & Marketing* 1986-1, p. 1-28.

Birren, Schaie & Abeles 2006
J.E. Birren, K.W. Schaie & R.P. Abeles, *Handbook of the psychology of aging*, Amsterdam: Elsevier Academic Press 2006.

Bohme & Egilman 2006
S.R. Bohme & D. Egilman, 'Consider the source: Warnings and anti-warnings in the tobacco, automobile, beryllium and pharmaceutical industries', in: M.S. Wogalter (ed.), *Handbook of Warnings*, Mahwah, New Jersey: Lawrence Erlbaum Associates 2006, p. 635-644.

Bolt, Spier & Haazen 1996
A.T. Bolt, J. Spier & O.A. Haazen, *De uitdijende reikwijdte van de aansprakelijkheid uit onrechtmatige daad*, Zwolle: W.E.J. Tjeenk Willink 1996.

Van Boom 2003
W.H. Van Boom, *Structurele fouten in het aansprakelijkheidsrecht (inaugural lecture Tilburg University)*, Den Haag: Boom Juridische uitgevers 2003.

Van Boom 2005
W.H. Van Boom, 'Fault under Dutch law', in: P. Widmer (ed.), *Unification of tort law: Fault*, The Hague: Kluwer Law International 2005, p. 167-178.

Van Boom 2006a
W.H. Van Boom, 'Compensating and preventing damage: Is there any future left for tort law?', in: H. Tiberg & B.W. Dufwa (eds.), *Festskrift till Bill W. Dufwa - Essays on tort, insurance law and society in honour of Bill W. Dufwa*, Stockholm: Jure Förlag AB 2006, p. 287-293.

Van Boom 2006b
W.H. Van Boom, *Efficacious enforcement in contract and tort (inaugural lecture Erasmus University Rotterdam)*, Den Haag: Boom Juridische uitgevers 2006.

Van Boom 2007
W.H. Van Boom, 'On the intersection between tort law and regulatory law – A comparative analysis', in: W.H. Van Boom, M. Lukas & C. Kissling (eds.), *Tort and regulatory law*, Wien: Springer 2007, p. 419-448.

Van Boom & Van Doorn 2006
W.H. Van Boom & C.J.M. Van Doorn, 'Productaansprakelijkheid en productveiligheid', in: E.H. Hondius & G.J. Rijken (eds.), *Handboek consumentenrecht: Een overzicht van de rechtspositie van de consument*, Zutphen: Paris Uitgevers 2006, p. 261-280.

Van Boom, Giesen &Verheij 2008
W.H. Van Boom, I. Giesen & A. Verheij, *Gedrag en privaatrecht: Over gedragspresumpties en gedragseffecten bij privaatrechtelijke leerstukken*, Den Haag: Boom Juridische uitgevers 2008.

Van Boom, Tuil & Van der Zalm 2010
W.H. Van Boom, M.L. Tuil & I. Van der Zalm, 'Feiten van algemene bekendheid en ervaringsregels - Virtuele werkelijheid?', *Nederlands Tijdschrift Voor Burgerlijk Recht* 2010, p. 36-43.

Bowbeer, Lumish & Cohen 2000-2001
H. Bowbeer, W.F. Lumish & J.A. Cohen, 'Warning – Failure to read this article may be hazardous to your failure to warn defense', *William Mitchell Law Review* 2000-2001, p. 439-466.

Braun e.a. 1994
C.C. Braun e.a., 'Signal word and color specifications for product warnings: An isoperformance application', *Human Factors and Ergonomics Society Annual Meeting Proceedings* 1994-17, p. 1104-1108.

Braun, Mine & Clayton Silver 1995
C.C. Braun, P.B. Mine & N. Clayton Silver, 'The influence of color on warning label perceptions', *International Journal of Industrial Ergonomics* 1995-3, p. 179-187.

Braun & Silver 1995
C.C. Braun & N.C. Silver, 'Interaction of signal word and colour on warning labels: Differences in perceived hazard and behavioural compliance', *Ergonomics* 1995-11, p. 2207-2210.

Braun & Shaver 1999
C.C. Braun & E.F. Shaver, 'Warning sign components and hazard perceptions', *Human Factors and Ergonomics Society Annual Meeting Proceedings* 1999-16, p. 878-882.

Breakwell e.a. 2006
G.M. Breakwell e.a., *Research methods in psychology*, London: Sage 2006.

Brug, Ruiter & Van Assema 2003
J. Brug, R. Ruiter & P. Van Assema, 'The (ir)relevance of framing nutrition education messages', *Nutrition and Health* 2003, p. 9-20.

Bzostek & Wogalter 1999
J.A. Bzostek & M.S. Wogalter, 'Measuring visual search time for a product warning label as a function of icon, color, column and vertical placement', *Human Factors and Ergonomics Society Annual Meeting Proceedings* 1999-16, p. 888-892.

Calabresi 1970
G. Calabresi, *The cost of accidents: A legal and economic analysis*, Yale University Press 1970.

Cameron & DeJoy 2006
K.A. Cameron & D.M. DeJoy, 'The persuasive functions of warnings: Theory and models', in: M.S. Wogalter (ed.), *Handbook of Warnings*, Mahwah, New Jersey: Lawrence Erlbaum Associates 2006, p. 301-312.

Campbell 2007
C. Campbell, *International product liability*, Austria: Yorkhill Law Publishing 2007.

Cane 2006
P.F. Cane, *Atiyah's accidents, compensation and the law*, Cambridge: Cambridge University Press 2006.

Cane & Kritzer 2010
P. Cane & H.M. Kritzer, *The Oxford handbook of empirical legal research*, Oxford: Oxford University Press 2010.

CEN/CENELEC 2006
CEN/CENELEC, *CEN/CENELEC guide 11 product information relevant to consumers – Guidelines for standard developers*, 2006.

Chapanis 1994
A. Chapanis, 'Hazards associated with three signal words and four colours on warning signs', *Ergonomics* 1994-2, p. 265-275.

Chen, Gilson & Mouloua 1997
J.Y.C. Chen, R.D. Gilson & M. Mouloua, 'Perceived risk dilution with multiple warnings', *Human Factors and Ergonomics Society Annual Meeting Proceedings* 1997-16, p. 831-835.

Chen, Gilson & Wang 1999
J.Y.C. Chen, R.D. Gilson & M.C. Wang, 'Gender differences in perceived hazardousness of warning messages', *Human Factors and Ergonomics Society Annual Meeting Proceedings* 1999-24, p. 1401-1401.

Chy-Dejoras 1992
E.A. Chy-Dejoras, 'Effects of an aversive vicarious experience and modelling on perceived risk and self-protective behavior', *Human Factors and Ergonomics Society Annual Meeting Proceedings* 1992-8, p. 603-607.

Clark 1989
A.M. Clark, *Product liability*, London: Sweet & Maxwell 1989.

Cohen e.a. 2006
H.H. Cohen e.a., 'Warning channel: Modality and media', in: M.S. Wogalter (ed.), *Handbook of Warnings*, Mahwah, New Jersey: Lawrence Erlbaum Associates 2006, p. 123-134.

Collins & Lerner 1982
B.L. Collins & N.D. Lerner, 'Assessment of fire-safety symbols', *Human Factors* 1982-1, p. 75-84.

Conzola & Wogalter 1999
V.C. Conzola & M.S. Wogalter, 'Using voice and print directives and warnings to supplement product manual instructions', *International Journal of Industrial Ergonomics* 1999-5-6, p. 549-556.

Cox III 1999
E.P. Cox III, 'Source', in: M.S. Wogalter, D.M. DeJoy & K.R. Laughery (eds.), *Warnings and risk communication*, London: Taylor & Francis 1999, p. 85-97.

Cox III 2006
E.P. Cox III, 'Marketing versus warning', in: M.S. Wogalter (ed.), *Handbook of Warnings*, Mahwah, New Jersey: Lawrence Erlbaum Associates 2006, p. 645-652.

Cox III e.a. 1997
E.P. Cox III e.a., 'Do product warnings increase safe behavior? A meta-analysis', *Journal of Public Policy & Marketing* 1997-2, p. 195-204.

Cox III & Wogalter 2006
E.P. Cox III & M.S. Wogalter, 'Warning source', in: M.S. Wogalter (ed.), *Handbook of Warnings*, Mahwah, New Jersey: Lawrence Erlbaum Associates 2006, p. 111-122.

Van Dam 2000
C.C. Van Dam, *Aansprakelijkheidsrecht: Een grensoverschrijdend handboek*, Den Haag: Boom Juridische Uitgevers 2000.

Van Dam 2005
C.C. Van Dam, 'Dutch case law on the EU Product Liability Directive', in: D. Fairgrieve (ed.), *Product liability in comparative perspective*, Cambridge: Cambridge University Press 2005, p. 126-137.

Van Dam 2006
C.C. Van Dam, *European tort law*, Oxford: Oxford University Press 2006.

Davies e.a. 1998
S. Davies e.a., 'Safety pictograms: Are they getting the message across?', *Applied Ergonomics* 1998-1, p. 15-23.

Van Duijne 2005
F.H. Van Duijne, *Risk perception in product use (dissertation TU Delft)*, S.I.: s.n 2005.

Van Duijne e.a. 2007
F. Van Duijne e.a., 'Design for safety: Involving users' perspectives: Redesign proposals for gas lamps using a pierceable cartridge', *Safety Science* 2007-1-2, p. 253-281.

Van Duijne e.a. 2008
F.H. Van Duijne e.a., 'Risk perception in the usage of electrically powered gardening tools', *Safety Science* 2008-1, p. 104-118.

Van Duijne, Van Aken & Schouten 2008
F.H. Van Duijne, D. Van Aken & *E.G.* Schouten, 'Considerations in developing complete and quantified methods for risk assessment', *Safety Science* 2008-2, p. 245-254.

Van Duijne, Green & Kanis 2001
F.H. Van Duijne, W.S. Green & H. Kanis, 'Risk perception: Let the user speak', *Contemporary Ergonomics* 2001, p. 297-302.

Deards & Twigg-Flesner 2001
E. Deards & C. Twigg-Flesner, 'Consumer protection act 1987: Proof at last that it is protecting consumers?', *Nottingham Law Journal* 2001, p. 1-19.

DeJoy 1989
D.M. DeJoy, 'Consumer product warnings: Review and analysis of effectiveness research', *Human Factors and Ergonomics Society Annual Meeting Proceedings* 1989-15, p. 936-940.

DeJoy 1991
D.M. DeJoy, 'A revised model of the warnings process derived from value-expectancy theory', *Human Factors and Ergonomics Society Annual Meeting Proceedings* 1991-15, p. 1043-1047.

DeJoy 1999a
D.M. DeJoy, 'Attitudes and beliefs', in: M.S. Wogalter, D.M. DeJoy & K.R. Laughery (eds.), *Warnings and risk communication*, London: Taylor & Francis 1999, p. 189-219.

DeJoy 1999b
D.M. DeJoy, 'Motivation', in: M.S. Wogalter, D.M. DeJoy & K.R. Laughery (eds.), *Warnings and risk communication*, London: Taylor & Francis 1999, p. 221-243.

DeJoy, Cameron & Della 2006
D.M. DeJoy, K.A. Cameron & L.J. Della, 'Postexposure evaluation of warning effectiveness: A review of field studies and population-based research', in: M.S. Wogalter (ed.), *Handbook of Warnings*, Mahwah, New Jersey: Lawrence Erlbaum Associates 2006, p. 35-48.

Deppa 2006
S. Deppa, 'U.S. and international standards for safety symbols', in: M.S. Wogalter (ed.), *Handbook of Warnings*, Mahwah, New Jersey: Lawrence Erlbaum Associates 2006, p. 477-486.

Desaulniers 1987
D.R. Desaulniers, 'Lay out, organization and the effectiveness of consumer product warnings', *Human Factors and Ergonomics Society Annual Meeting Proceedings* 1987-1, p. 56-60.

DeTurck 2002
M.A. DeTurck, 'Persuasive effects of product warning labels', in: J.P. Dillard & M. Pfau (eds.), *The persuasion handbook: Developments in theory and practice*, Thousand Oaks, California: Sage Publications 2002, p. 345-367.

DeTurck, Rachlin & Young 1994
M.A. DeTurck, R.A. Rachlin & M.L. Young, 'Effects of a role model and fear in warning label on perceptions of safety and safety behavior', *Advances in Consumer Research* 1994-1, p. 208-212.

DeTurck, Chih & Hsu 1999
M.A. DeTurck, I. Chih & Y. Hsu, 'Three studies testing the effects of role models on product users' safety behavior', *Human Factors* 1999-3, p. 397-412.

DeTurck & Goldhaber 1989
M.A. DeTurck & G.M. Goldhaber, 'Effectiveness of product warning labels: Effects of consumers' information processing objectives', *Journal of Consumer Affairs* 1989-1, p. 111-126.

Detweiler e.a. 1999
J.B. Detweiler e.a., 'Message framing and sunscreen use: Gain-framed messages motivate beach-goers', *Health Psychology* 1999-2, p. 189-196.

Dewees, Duff & Trebilcock 1996
D. Dewees, D. Duff & M. Trebilcock, *Exploring the domain of accident law: Taking the facts seriously*, New York: Oxford University Press 1996.

Dingus, Hathaway & Hunn 1991
T.A. Dingus, J.A. Hathaway & B.P. Hunn, 'A most critical warning variable: Two demonstrations of the powerful effects of cost', *Human Factors and Ergonomics Society Annual Meeting Proceedings* 1991, p. 1034-1038.

Dingus, Wreggit & Hathaway 1993
T.A. Dingus, S.S. Wreggit & J.A. Hathaway, 'Warning variables affecting personal protective equipment use', *Safety Science* 1993-5-6, p. 655-673.

Dirken 2004
H. Dirken, *Productergonomie. Ontwerpen voor gebruikers*, Delft, The Netherlands: VSSD 2004.

Dommering-Van Rongen 2000
L. Dommering-Van Rongen, *Productaansprakelijkheid: Een rechtsvergelijkend overzicht*, Deventer: Kluwer 2000.

Donner & Brelsford 1988
K.A. Donner & J.W. Brelsford, 'Cueing hazard information for consumer products', *Human Factors and Ergonomics Society Annual Meeting Proceedings* 1988-9, p. 532-535.

Drake, Conzola & Wogalter 1998
K.L. Drake, V.C. Conzola & M.S. Wogalter, 'Discrimination among sign and label warning signal words', *Human Factors and Ergonomics in Manufacturing* 1998-4, p. 289-301.

Duffy, Kalsher & Wogalter 1995
R.R. Duffy, M.J. Kalsher & M.S. Wogalter, 'Increased effectiveness of an interactive warning in a realistic incidental product-use situation', *International Journal of Industrial Ergonomics* 1995-3, p. 159-166.

Dunn 2009
D.S. Dunn, *Research methods for social psychology*, Malden, MA: Wiley-Blackwell 2009.

Dutcher 2006
J.S. Dutcher, 'Caution: This superman suit will not enable you to fly – Are consumer product warning labels out of control?', *Arizona State Law Journal* 2006, p. 633-660.

Easterby & Hakiel 1981
R.S. Easterby & S.R. Hakiel, 'Field testing of consumer safety signs: The comprehension of pictorially presented messages', *Applied Ergonomics* 1981-3, p. 143-152.

Edwards 1954
W. Edwards, 'The theory of decision making', *Psychological Bulletin* 1954-4, p. 380-417.

Edworthy 1994
J. Edworthy, 'The design and implementation of non-verbal auditory warnings', *Applied Ergonomics* 1994-4, p. 202-210.

Edworthy 1998
J. Edworthy, 'Warnings and hazards: An integrative approach to warnings research', *International Journal of Cognitive Ergonomics* 1998-1/2, p. 3.

Edworthy 2000
J. Edworthy, 'An integrative approach to warnings research', *Human Factors and Ergonomics Society Annual Meeting Proceedings* 2000-28, p. 770-773.

Edworthy e.a. 2004
J. Edworthy e.a., 'Linguistic and location efects in compliance with pesticide warning labels for amateur and professional users', *Human Factors* 2004-1, p. 11-31.

Edworthy & Adams 1996
J. Edworthy & A. Adams, *Warning design: A research prospective*, London: Taylor and Francis 1996.

Edworthy & Dale 2000
J. Edworthy & S. Dale, 'Extending knowledge of the effects of social influence in warning compliance', *Human Factors and Ergonomics Society Annual Meeting Proceedings* 2000-25, p. 107-110.

Edworthy, Loxley & Dennis 1991
J. Edworthy, S. Loxley & I. Dennis, 'Improving auditory warning design: Relationship between warning sound parameters and perceived urgency', *Human Factors: The Journal of the Human Factors and Ergonomics Society* 1991-2, p. 205-231.

European Commission 1999
European Commission, *Green paper. Liability for defective products*, COM (1999) 396 final, Brussels, 1999.

European Commission Enterprise and Industry 2010
European Commission Enterprise and Industry, *Guide to the application of the Machinery Directive 2006/42/EC*, Brussels, 2010.

EuroSafe, KfV & DG Sanco November 2009
EuroSafe, KfV & DG Sanco, *2009-Report injuries in the European Union. Statistics summary 2005-2007*, Vienna, November 2009.

Eysenck & Keane 2000
M.W. Eysenck & M.T. Keane, *Cognitive psychology: A student's handbook*, New York: Psychology Press 2000.

Eysenck & Keane 2005
M.W. Eysenck & M.T. Keane, *Cognitive psychology: A student's handbook*, New York: Psychology Press 2005.

Fairgrieve 2005
D. Fairgrieve, *Product liability in comparative perspective*, Cambridge: Cambridge University Press 2005.

Faure 2000
M.G. Faure, 'Product liability and product safety in Europe: Harmonization or differentiation', *Kyklos* 2000-4, p. 467-508.

Ferrari & Chan 1991
J.R. Ferrari & L.M. Chan, 'Interventions to reduce high-volume portable headsets: "Turn down the sound"!', *Journal of Applied Behavior Analysis* 1991-4, p. 695-704.

Fischhoff e.a. 1998
B. Fischhoff e.a., 'What information belongs in a warning?', *Psychology & Marketing* 1998-7, p. 663.

Fischhoff, Slovic & Lichtenstein 1977
B.B. Fischhoff, P. Slovic & S. Lichtenstein, 'Knowing with certainty: The appropriateness of extreme confidence', *Journal of Experimental Psychology: Human Perception and Performance* 1977-4, p. 552-564.

Fisk 2004
A.D. Fisk, *Designing for older adults: Principles and creative human factors approaches*, London: Taylor & Francis 2004.

Food Information Regulation 2008
Food Information Regulation, *Proposal of the Commission of 30 January 2008 for a regulation of the European Parliament and of the Council on the provision of food information to consumers*, COM (2008) 40 final, 2008.

Franken 2009
A.C.H. Franken, 'Actualiteiten productaansprakelijkheid 2002-2008', *Aansprakelijkheid, Verzekering & Schade* 2009, p. 189-201.

Frantz 1993
J.P. Frantz, 'Effect of location and presentation format on attention to and compliance with product warnings and instructions', *Journal of Safety Research* 1993-3, p. 131-154.

Frantz 1994
J.P. Frantz, 'Effect of location and procedural explicitness on user processing of and compliance with product warnings', *Human Factors* 1994-3, p. 532-546.

Frantz e.a. 1999
J.P. Frantz e.a., 'Potential problems associated with overusing warnings', *Human Factors and Ergonomics Society Annual Meeting Proceedings* 1999-16, p. 916-920.

Frantz & Miller 1993
J.P. Frantz & J.M. Miller, 'Communicating a safety-critical limitation of an infant-carrying product: The effect of product design and warning salience', *International Journal of Industrial Ergonomics* 1993-1, p. 1-12.

Frantz & Rhoades 1993
J.P. Frantz & T.P. Rhoades, 'A task-analytic approach to the temporal and spatial placement of product warnings', *Human Factors: The Journal of the Human Factors and Ergonomics Society* 1993-4, p. 719-730.

Frantz, Rhoades & Lehto 1999
J.P. Frantz, T.P. Rhoades & M.R. Lehto, 'Practical considerations regarding the design and evaluation of product warnings', in: M.S. Wogalter, D.M. DeJoy & K.R. Laughery (eds.), *Warnings and risk communication*, London: Taylor & Francis 1999, p. 291-311.

Frascara 2006
J. Frascara, 'Typography and the visual design of warnings', in: M.S. Wogalter (ed.), *Handbook of Warnings*, Mahwah, New Jersey: Lawrence Erlbaum Associates 2006, p. 385-405.

Freeman 2006
R. Freeman, 'The manufacturer's duty to design against potential misuse. Palmer v Palmer and others', *European Product Liability Review* 2006-24, p. 26-28.

Friedmann 1988
K. Friedmann, 'The effect of adding symbols to written warning labels on user behavior and recall', *Human Factors* 1988-4, p. 507-515.

Giesen 2004
I. Giesen, 'Van een openstaand kelderluik naar een gekantelde vrachtwagen', *Aansprakelijkheid, Verzekering & Schade* 2004, p. 35-39.

Giesen 2005
I. Giesen, *Handle with care!: De waarschuwingsplicht in het buitencontractuele aansprakelijkheidsrecht (inaugural lecture Utrecht University)*, Den Haag: Boom Juridische uitgevers 2005.

Giesen & Loos 2001
I. Giesen & M. Loos, 'Liability for defective products and services: The Netherlands', *EJCL* 2001, p. 75-122.

Gill, Barbera & Precht 1987
R.T. Gill, C. Barbera & T. Precht, 'A comparative evaluation of warning label designs', *Human Factors and Ergonomics Society Annual Meeting Proceedings* 1987-4, p. 476-478.

Godfrey e.a. 1983
S.S. Godfrey e.a., 'Warning messages: Will the consumer bother to look?', *Human Factors and Ergonomics Society Annual Meeting Proceedings* 1983, p. 950-954.

Godfrey & Laughery 1984
S.S. Godfrey & K.R. Laughery, 'The biasing effects of product familiarity on consumers awareness of hazard', *Human Factors and Ergonomics Society Annual Meeting Proceedings* 1984-5, p. 483-486.

Godfrey, Rothstein & Laughery 1985
S.S. Godfrey, P.R. Rothstein & K.R. Laughery, 'Warnings: Do they make a difference?', *Human Factors and Ergonomics Society Annual Meeting Proceedings* 1985-7, p. 669-673.

Goldhaber & DeTurck 1988a
G.M. Goldhaber & M.A. DeTurck, 'Effects of consumers' familiarity with a product on attention to and compliance with warnings', *Journal of Products Liability* 1988-3, p. 29-37.

Goldhaber & DeTurck 1988b
G.M. Goldhaber & M.A. DeTurck, 'Effectiveness of warning signs: Gender and familiarity effects', *Journal of Products Liability* 1988-3, p. 271-284.

Graves 1993
K.L. Graves, 'An evaluation of the alcohol warning label: A comparison of the United States and Ontario, Canada in 1990 and 1991', *Journal of Public Policy & Marketing* 1993-1, p. 19-29.

Green, Dierckins & Nyberg 1998
J. Green, T. Dierckins & T. Nyberg, *The warning label book*, New York: St. Martin's Griffin 1998.

Griffith & Leonard 1997
L.J. Griffith & S.D. Leonard, 'Association of colors with warning signal words', *International Journal of Industrial Ergonomics* 1997-4, p. 317-325.

Grubb & Howells 2007
A. Grubb & G. Howells, *The law of product liability*, UK: LexisNexis Butterworths 2007.

Haak 2006
K.F. Haak, 'Gemankeerde rechters?', *Nederlands Tijdschrift Voor Burgerlijk Recht* 2006, p. 39.

Haanappel & MacKaay 1990
P.P.C. Haanappel & E. MacKaay, *Nieuw Nederlands Burgerlijk Wetboek: Het vermogensrecht (zakenrecht, verbintenissenrecht en bijzondere overeenkomsten) = New Netherlands Civil Code: patrimonial law (property, obligations and special contracts) = Nouveau Code Civil Néerlandais: le droit patrimonial (les biens, les obligations et les contrats particuliers)*, Deventer: Kluwer 1990.

Hale e.a. 1990
A.R. Hale e.a., 'Safety standards, risk analysis and decision making on prevention measures: Implications of some recent European legislation and standards', *Journal of Occupational Accidents* 1990-3, p. 213-231.

Hale, Kirwan & Kjellén 2007
A. Hale, B. Kirwan & U. Kjellén, 'Safe by design: Where are we now?', *Safety Science* 2007-1-2, p. 305-327.

Hancock e.a. 2004
H.E. Hancock e.a., 'Safety symbol comprehension: Effects of symbol type, familiarity, and age', *Human Factors* 2004-2, p. 183-195.

Hancock e.a. 2006
H.E. Hancock e.a., 'Comprehension and retention of warning information', in: M.S. Wogalter (ed.), *Handbook of Warnings*, Mahwah, New Jersey: Lawrence Erlbaum Associates 2006, p. 267-277.

Hancock, Fisk & Rogers 2005
H.E. Hancock, A.D. Fisk & W.A. Rogers, 'Comprehending product warning information: Age-related effects and the roles of memory, inferencing, and knowledge', *Human Factors: The Journal of the Human Factors and Ergonomics Society* 2005-2, p. 219-234.

Hancock, Rogers & Fisk 2001
H.E. Hancock, W.A. Rogers & A.D. Fisk, 'An evaluation of warning habits and beliefs across the adult life span', *Human Factors: The Journal of the Human Factors and Ergonomics Society* 2001-3, p. 343-354.

Hannes 2003
M. Hannes, 'Stressing consumers' responsibility, Higher Regional Court of Düsseldorf dismisses appeal claim for damages against Masterfood', *European Product Liability Review* 2003-10, p. 31-33.

Hannes 2004
M. Hannes, 'Form and content of warning signs', *European Product Liability Review* 2004-16, p. 36-37.

Harrell 2003
W.A. Harrell, 'Effect of two warning signs on adult supervision and risky activities by children in grocery shopping carts', *Psychological Reports* 2003-3 Pt 1, p. 889-898.

Harris & Wiklund 1989
J.E. Harris & M.E. Wiklund, 'Consumer acceptance of threatening warnings in the residential environment', *Human Factors and Ergonomics Society Annual Meeting Proceedings* 1989, p. 989-993.

Hartlief 2004
T. Hartlief, 'Kelderluik revisited. De kracht van een waarschuwing', *Ars Aequi* 2004, p. 866-873.

Heaps & Henley 1999
C.M. Heaps & T.B. Henley, 'Language matters: Wording considerations in hazard perception and warning comprehension', *Journal of Psychology* 1999-3, p. 341-351.

Helander 2005
M. Helander, 'The discipline of human factors engineering and ergonomics', in: Y.I. Noy & W. Karwowski (eds.), *Handbook of human factors in litigation*, Boca Raton: CRC Press 2005, p. 1-3-1-24.

Helander 2006
M. Helander, *A guide to human factors and ergonomics*, Boca Raton, Fla.: CRC Press 2006.

Hellier e.a. 2000
E. Hellier e.a., 'On the stability of the arousal strength of warning signal words', *Applied Cognitive Psychology* 2000-6, p. 577-592.

Hellier & Edworthy 2006
E. Hellier & J. Edworthy, 'Signal words', in: M.S. Wogalter (ed.), *Handbook of Warnings*, Mahwah, New Jersey: Lawrence Erlbaum Associates 2006, p. 407-417.

Hellier, Edworthy & Dennis 1993
E. Hellier, J. Edworthy & I. Dennis, 'Improving auditory warning design: Quantifying and predicting the effects of different warning parameters on perceived urgency', *Human Factors: The Journal of the Human Factors and Ergonomics Society* 1993-4, p. 693-706.

Henderson & Twerski 1990
J.A.J. Henderson & A.D. Twerski, 'Doctrinal collapse in products liability: The empty shell of failure to warn', *New York University Law Review* 1990, p. 265-328.

Henderson & Twerski 1999
J.A.J. Henderson & A.D. Twerski, 'What Europe, Japan, and other countries can learn from the new American Restatement of Products Liability', *Texas International Law Journal* 1999, p. 1-20.

Henderson & Twerski 2000-2001
J.A.J. Henderson & A.D. Twerski, 'Products Liability Restatement in the Courts: An Initial Assessment', *William Mitchell Law Review* 2000-2001, p. 7-32.

Hodges 1993a
C.J.S. Hodges, *Product liability: European laws and practice*, London: Sweet & Maxwell 1993.

Hodges 1993b
C.J.S. Hodges, 'Product liability in practice', in: C.J.S. Hodges (ed.), *Product liability: European laws and practice*, London: Sweet & Maxwell 1993, p. 93-117.

Hodges 1998
C. Hodges, 'Development risks: Unanswered questions', *Modern Law Review* 1998-4, p. 560-570.

Hodges 2005
C.J.S. Hodges, *European regulation of consumer product safety*, Oxford: Oxford University Press 2005.

Hovland, Janis & Kelley 1953
C.I. Hovland, I.L. Janis & H.H. Kelley, *Communication and persuasion: Psychological studies of opinion change*, New Haven: Yale University Press 1953.

Howells 1993
G. Howells, *Comparative product liability*, Aldershot: Dartmouth 1993.

Howells 2005a
G. Howells, 'Defect in English law – Lessons for the harmonisation of European product liability', in: D. Fairgrieve (ed.), *Product liability in comparative perspective*, Cambridge: Cambridge University Press 2005, p. 138-152.

Howells 2005b
G. Howells, 'Information and product liability – A game of Russian roulette?', in: G. Howells, A. Janssen & R. Schulze (eds.), *Information rights and obligations: A challenge for party autonomy and transactional fairness*, Aldershot: Ashgate 2005, p. 155-169.

Howells 2005c
G. Howells, 'Product liability – A history of harmonisation', in: D. Fairgrieve (ed.), *Product liability in comparative perspective*, Cambridge: Cambridge University Press 2005, p. 202-217.

Howells & Borghetti 2010
G. Howells & J. Borghetti, 'Product liability', in: H.W. Micklitz, J. Stuyck & E. Terryn (eds.), *Cases, materials and text on consumer law*, Oxford: Hart Publishing 2010, p. 439-498.

Howells, Janssen &Schulze 2005
G. Howells, A. Janssen & R. Schulze, *Information rights and obligations: A challenge for party autonomy and transactional fairness*, Aldershot: Ashgate 2005.

Howells & Mildred 2002
G. Howells & M. Mildred, 'Infected blood defect and discoverability. A first exposition of the EC product liability directive', *The Modern Law Review* 2002-1, p. 95-105.

Hunn & Dingus 1992
B.P. Hunn & T.A. Dingus, 'Interactivity, information, and compliance cost in a consumer product warning scenario', *Accident Analysis & Prevention* 1992-5, p. 497-505.

ISO 2004
ISO, *ISO 3864-2 Graphical symbols safety colours and safety signs Part 2: Design principles for product safety labels*, IHS 2004.

ISO/IEC 1999
ISO/IEC, *ISO-IEC Guide 51 safety aspects – Guidelines for their inclusion in standards*, 1999.

Jankowski 1995
T.S. Jankowski, 'Focusing on quality and risk: The central role of reasonable alternatives in evaluating design and warning decisions special issue: Review of the system of products liability law', *South Texas Law Review* 1995, p. 283-352.

Jaynes & Boles 1990
L.S. Jaynes & D.B. Boles, 'The effect of symbols on warning compliance', *Human Factors and Ergonomics Society Annual Meeting Proceedings* 1990-14, p. 984-987.

Johnson 2006
D.A. Johnson, 'Pratical aspects of graphics related to safety instructions and warnings', in: M.S. Wogalter (ed.), *Handbook of Warnings*, Mahwah, New Jersey: Lawrence Erlbaum Associates 2006, p. 463-476.

Kahneman 1973
D. Kahneman, *Attention and effort*, Englewood Cliffs; Hemel Hempstead: Prentice-Hall 1973.

Kahneman, Slovic & Tversky 1982
D. Kahneman, P. Slovic & A. Tversky, *Judgment under uncertainty: Heuristics and biases*, Cambridge: Cambridge University Press 1982.

Kahneman & Tversky 1979
D. Kahneman & A. Tversky, 'Prospect theory: An analysis of decision under risk', *Econometrica* 1979-2, p. 263-291.

Kalsher e.a. 1994
M.J. Kalsher e.a., 'Enhancing the perceived readability of pharmaceutical container labels and warnings: The use of alternative designs and pictorials', *Proceedings of the Human Factors and Ergonomics Society* 1994 p. 384-388.

Kalsher & Williams 2006
M.J. Kalsher & K.J. Williams, 'Behavioural compliance: Theory, methodology, and results', in: M.S. Wogalter (ed.), *Handbook of Warnings*, Mahwah, New Jersey: Lawrence Erlbaum Associates 2006, p. 313-331.

Kalsher, Wogalter & Racicot 1996
M.J. Kalsher, M.S. Wogalter & B.M. Racicot, 'Pharmaceutical container labels: Enhancing preference perceptions with alternative designs and pictorials', *International Journal of Industrial Ergonomics* 1996-1, p. 83-90.

Kanis 1998
H. Kanis, 'Usage centred research for everyday product design', *Applied Ergonomics* 1998-1, p. 75-82.

Kanis 1999
H. Kanis, 'Design centred research into use activities', in: W.S. Green & P.W. Jordan (eds.), *Human factors in product design: Current practice and future trends*, London: Taylor & Francis 1999, p. 36-46.

Karnes Edward, Leonard & Rachwal 1986
W. Karnes Edward, S.D. Leonard & G. Rachwal, 'Effects of benign experiences on the perception of risk', *Human Factors and Ergonomics Society Annual Meeting Proceedings* 1986-2, p. 121-125.

Karwowski 2006
W. Karwowski, 'The discipline of ergonomics and human factors', in: G. Salvendy (ed.), *Handbook of human factors and ergonomics*, Hoboken, New Jersey: John Wiley & Sons 2006, p. 3-31.

Kaskutas & Greenfield 1992
L. Kaskutas & T.K. Greenfield, 'First effects of warning labels on alcoholic beverage containers', *Drug and Alcohol Dependence* 1992-1, p. 1-14.

Kellogg 2007
R.T. Kellogg, *Fundamentals of cognitive psychology*, Los Angeles, CA: Sage 2007.

Kiene e.a. 2005
S.M. Kiene e.a., 'Why are you bringing up condoms now? The effect of message content on framing effects of condom use messages', *Health Psychology* 2005-3, p. 321-326.

Kim, Cowley & Wogalter 2007
S. Kim, J. Cowley & M.S. Wogalter, 'Emphasis terms for warning directives on compliance intent', *Human Factors and Ergonomics Society Annual Meeting Proceedings* 2007-9, p. 569-573.

Kim & Hunter 1993a
M.S. Kim & J.E. Hunter, 'Attitude-behavior relations: A meta-analysis of attitudinal relevance and topic', *The Journal of Communication* 1993-1, p. 101-142.

Kim & Hunter 1993b
M.S. Kim & J.E. Hunter, 'Relationships among attitudes, behavioral intentions, and behavior: A meta-analysis of past research, part 2', *Communication Research* 1993-3, p. 331-364.

Kline e.a. 1990
T.J.B. Kline e.a., 'Visibility distance of highway signs among young, middle-aged, and older observers: Icons are better than text', *Human Factors: The Journal of the Human Factors and Ergonomics Society* 1990-5, p. 609-619.

Kline e.a. 1993
P.B. Kline e.a., 'The impact of color on warnings research', *Human Factors and Ergonomics Society Annual Meeting Proceedings* 1993-14, p. 940-944.

Kline & Fuchs 1993
D.W. Kline & P. Fuchs, 'The visibility of symbolic highway signs can be increased among drivers of all ages', *Human Factors: The Journal of the Human Factors and Ergonomics Society* 1993-1, p. 25-34.

LaRue & Cohen 1987
C. LaRue & H.H. Cohen, 'Factors affecting consumers perceptions of product warnings: An examination of the differences between male and female consumers', *Human Factors and Ergonomics Society Annual Meeting Proceedings* 1987-5, p. 610-614.

Latin 1994
H. Latin, 'Good warnings, bad products, and cognitive limitations', *UCLA Law Review* 1994, p. 1193-1296.

Laughery 1993
K.R. Laughery, 'Everybody knows – or do they?', *Ergonomics in Design* 1993, p. 8-13.

Laughery 2006
K.R. Laughery, 'Safety communications: warnings', *Applied Ergonomics* 2006, p. 467-478.

Laughery e.a. 1991
K.R. Laughery e.a., 'Effects of explicitness in conveying severity information in product warnings', *Human Factors and Ergonomics Society Annual Meeting Proceedings* 1991-6, p. 481-485.

Laughery e.a. 1993a
K.R. Laughery e.a., 'Explicitness of consequence information in warnings', *Safety Science* 1993-5-6, p. 597-613.

Laughery e.a. 1993b
K.R. Laughery e.a., 'The noticeability of warnings on alcoholic beverage containers', *Journal of Public Policy & Marketing* 1993-1, p. 38-56.

Laughery & Brelsford 1991
K.R. Laughery & J.W. Brelsford, 'Receiver characteristics in safety communications', *Human Factors and Ergonomics Society Annual Meeting Proceedings* 1991-15, p. 1068-1072.

Laughery & Hammond 1999
K.R. Laughery & A. Hammond, 'Overview', in: M.S. Wogalter, D.M. DeJoy & K.R. Laughery (eds.), *Warnings and risk communication*, London: Taylor & Francis 1999, p. 3-13.

Laughery & Smith 2006
K.R. Laughery & D.P. Smith, 'Explicit information in warnings', in: M.S. Wogalter (ed.), *Handbook of Warnings*, Mahwah, New Jersey: Lawrence Erlbaum Associates 2006, p. 419-428.

Laughery & Stanush 1989
K.R. Laughery & J.A. Stanush, 'Effects of warning explicitness on product perceptions', *Human Factors and Ergonomics Society Annual Meeting Proceedings* 1989-6, p. 431-435.

Laughery & Wogalter 2005
K.R. Laughery & M.S. Wogalter, 'The warning expert', in: Y.I. Noy & W. Karwowski (eds.), *Handbook of human factors in litigation*, Boca Raton: CRC Press 2005, p. 30-1-30-14.

Laughery & Wogalter 2008
K.R. Laughery & M.S. Wogalter, 'On the symbiotic relationship between warnings research and forensics', *Human Factors* 2008-3, p. 529-533.

Laughery & Young 1991a
K.R. Laughery & S.L. Young, 'An eye scan analysis of accessing product warning information', *Human Factors and Ergonomics Society Annual Meeting Proceedings* 1991-9, p. 585-589.

Laughery & Young 1991b
K.R. Laughery & S.L. Young, 'Consumer product warnings: Design factors that influence noticeability', *Proceedings of the Congress of the International Ergonomics Association* 1991-11, p. 1104-1106.

Laux & Brelsford 1989
L. Laux & J.W. Brelsford, 'Locus of control, risk perception, and precautionary behavior', *Proceedings of Interface* 1989, p. 121-124.

Lehto 1991
M.R. Lehto, 'A proposed conceptual model of human behavior and its implications for the design of product warnings', *Perceptual and Motor Skills* 1991, p. 595-611.

Lehto 1992
M.R. Lehto, 'Designing warning signs and warning labels: Part II – Scientific basis for initial guidelines', *International Journal of Industrial Ergonomics* 1992-1-2, p. 115-138.

Lehto 2006
M. Lehto, 'Human factors models', in: M.S. Wogalter (ed.), *Handbook of Warnings*, Mahwah, New Jersey: Lawrence Erlbaum Associates 2006, p. 63-87.

Lehto & Foley 1991
M.R. Lehto & J.P. Foley, 'Risk-taking, warning labels, training, and regulation: Are they associated with the use of helmets by all-terrain vehicle riders?', *Journal of Safety Research* 1991-4, p. 191-200.

Lehto & Miller 1986
M.R. Lehto & J.M. Miller, *Warnings, Volume I: Fundamentals, design, and evaluation methodologies*, Ann Arbor, MI: Fuller Technical Publications 1986.

Lehto & Miller 1988
M.R. Lehto & J.M. Miller, 'The effectiveness of warning labels', *Journal of Products Liability* 1988-3, p. 225-270.

Lehto & Papastavrou 1993
M.R. Lehto & J.D. Papastavrou, 'Models of the warning process: Important implications towards effectiveness', *Safety Science* 1993-5-6, p. 569-595.

Lehto & Salvendy 1995
M. Lehto & G. Salvendy, 'Warnings: A supplement not a substitute for other approaches to safety', 1995.

Lenze 2003a
S. Lenze, 'A poor design is a poor design – or is it?', *European Product Liability Review* 2003-13, p. 44-46.

Lenze 2003b
S. Lenze, 'Failure to warn and the concept of defect', *European Product Liability Review* 2003-12, p. 44-45.

Lenze 2003c
S. Lenze, 'Take it from a duck – Overlap between negligence and defective products', *European Product Liability Review* 2003-13, p. 40-41.

Lenze 2005
S. Lenze, 'German product liability law: Between European directives, American Restatements and common sense', in: D. Fairgrieve (ed.), *Product liability in comparative perspective*, Cambridge: Cambridge University Press 2005, p. 100-125.

Lenze 2006
S. Lenze, 'Product safety regulations and defect', *European Product Liability Review* 2006-13, p. 20-21.

Leonard 1999
S.D. Leonard, 'Does color of warnings affect risk perception?', *International Journal of Industrial Ergonomics* 1999-5-6, p. 499-504.

Leonard e.a. 1989
D.C. Leonard e.a., 'Pest-control products: Reading warnings and purchasing intentions', *Human Factors and Ergonomics Society Annual Meeting Proceedings* 1989-6, p. 436-440.

Leonard, Creel & Karnes 1991
S.D. Leonard, E. Creel & E.W. Karnes, 'Adequacy of responses to warning terms', *Human Factors and Ergonomics Society Annual Meeting Proceedings* 1991-6, p. 1024-1028.

Leonard, Otani &Wogalter 1999
S.D. Leonard, H. Otani & M.S. Wogalter, 'Comprehension and memory', in: M.S. Wogalter, D.M. DeJoy & K.R. Laughery (eds.), *Warnings and risk communication*, London: Taylor & Francis 1999, p. 149-187.

Leonard & Wogalter 2000
S.D. Leonard & M.S. Wogalter, 'What you don't know can hurt you: Household products and events', *Accident Analysis & Prevention* 2000-3, p. 383-388.

Lesch 2003
M.F. Lesch, 'Comprehension and memory for warning symbols: Age-related differences and impact of training', *Journal of Safety Research* 2003-5, p. 495-505.

Lesch 2006
M.F. Lesch, 'Consumer product warnings', in: M.S. Wogalter (ed.), *Handbook of Warnings*, Mahwah, New Jersey: Lawrence Erlbaum Associates 2006, p. 137-146.

Lesch 2008a
M.F. Lesch, 'A comparison of two training methods for improving warning symbol comprehension', *Applied Ergonomics* 2008-2, p. 135-143.

Lesch 2008b
M.F. Lesch, 'Warning symbols as reminders of hazards: Impact of training', *Accident Analysis & Prevention* 2008-3, p. 1005-1012.

Lesch e.a. 2009
M.F. Lesch e.a., 'A cross-cultural comparison of perceived hazard in response to warning components and configurations: US vs. China', *Applied Ergonomics* 2009-5, p. 953-961.

Lichtenstein e.a. 1978
S. Lichtenstein e.a., 'Judged frequency of lethal events', *Journal of Experimental Psychology: Human Learning and Memory* 1978-6, p. 551-578.

Likert 1932
R. Likert, 'A technique for the measurement of attitudes', *Archives of Psychology* 1932-140, p. 1-55.

Lin & Salvendy 2000
H.X. Lin & G. Salvendy, 'Warning effect on human error reduction', *International Journal of Cognitive Ergonomics* 2000-2, p. 145-161.

Lindenbergh 2007
S.D. Lindenbergh, *Alles is betrekkelijk: Over de relatie tussen normschending en sanctie in het aansprakelijkheidsrecht (inaugural lecture Erasmus University Rotterdam)*, Den Haag: Boom Juridische uitgevers 2007.

Lipstein & McGuire 1978
B. Lipstein & W.J. McGuire, *Evaluating advertising: A bibliography of the communications process*, New York: Advertising Research Foundation 1978.

Lirtzman & Shuv-Ami 1986
S.I. Lirtzman & A. Shuv-Ami, 'Credibility of sources of communication on products' safety hazards', *Psychological Reports* 1986-3, p. 707-718.

Lord Griffiths, De Val & Dormer 1987-1988
Lord Griffiths, P. De Val & R.J. Dormer, 'Developments in English product liability law: A comparison with the American system', *Tulane Law Review* 1987-1988, p. 353-404.

Lovells 2003
Lovells, *Product liability in the European Union: A report for the European Commission*, MARKT/2001/11D, London, 2003.

Lueder & Rice 2008
R. Lueder & V.J.B. Rice, *Ergonomics for children: Designing products and places for toddlers to teens*, New York: Taylor & Francis 2008.

Van Maanen 2008
G.E. Van Maanen, 'Europese klassiekers – De Nederlandse kelderluikarresten. Al meer dan honderd jaar rechtseconomisch (!) – Op de goede weg in Europa! ', *Nederlands Tijdschrift Voor Burgerlijk Recht* 2008, p. 42-49.

MacKinnon & Lapin 1998
D.P. MacKinnon & A. Lapin, 'Effects of alcohol warnings and advertisements: A test of the boomerang hypothesis', *Psychology & Marketing* 1998-7, p. 707-726.

Magat, Viscusi & Huber 1988
W.A. Magat, W.K. Viscusi & J. Huber, 'Consumer processing of hazard warning information', *Journal of Risk and Uncertainty* 1988-2, p. 201-232.

Main, Frantz & Rhoades 1993
B.W. Main, J.P. Frantz & T.P. Rhoades, 'Do consumers understand the difference between "flammable" and "combustible"?', *Ergonomics in Design: The Quarterly of Human Factors Applications* 1993-3, p. 14-32.

Malouff & Schutte 1992
J. Malouff & N. Schutte, 'Preventing smoking: Evaluating the potential effectiveness of cigarette warnings', *Journal of Psychology* 1992-4, p. 371.

Martin 2000
B.J. Martin, 'The value of explicit hazard and consequence warnings for products with hidden hazards', *Human Factors and Ergonomics Society Annual Meeting Proceedings* 2000-27, p. 302-305.

Martin, Smith-Jackson & Artis 2003
L.F. Martin, T.L. Smith-Jackson & S. Artis, 'Cultural differences in risk perception: Comparison of USA and Ghanaian workers', *Human Factors and Ergonomics Society Annual Meeting Proceedings* 2003-14, p. 1762-1766.

Mayhorn & Podany 2006
C.B. Mayhorn & K.I. Podany, 'Warnings and aging: Describing the receiver characteristics of older adults', in: M.S. Wogalter (ed.), *Handbook of Warnings*, Mahwah, New Jersey: Lawrence Erlbaum Associates 2006, p. 355-361.

Mazis, Morris & Swasy 1991
M.B. Mazis, L.A. Morris & J.L. Swasy, 'An evaluation of the alcohol warning label: Initial survey results', *Journal of Public Policy & Marketing* 1991-1, p. 229-241.

Mazis & Morris 1999
M.B. Mazis & L.A. Morris, 'Channel', in: M.S. Wogalter, D.M. DeJoy & K.R. Laughery (eds.), *Warnings and risk communication*, London: Taylor & Francis 1999, p. 99-121.

McCarthy e.a. 1982
R.L. McCarthy e.a., 'Warnings on consumer products: Objective criteria for their use', *Human Factors and Ergonomics Society Annual Meeting Proceedings* 1982, p. 98-102.

McCarthy e.a. 1984
R.L. McCarthy e.a., 'Product information presentation, user behavior, and safety', *Human Factors and Ergonomics Society Annual Meeting Proceedings* 1984-1, p. 81-85.

McCarthy e.a. 1995
R.L. McCarthy e.a., 'Risk and effectiveness criteria for using on-product warnings', *Ergonomics* 1995-11, p. 2164-2175.

McCormick & Sanders 1992
E.J. McCormick & M.S. Sanders, *Human factors in engineering and design*, New Delhi: McGraw-Hill Companies 1992.

McGinnies & Ward 1974
E.E. McGinnies & C.D. Ward, 'Persuasibility as a function of source credibility and locus of control: Five cross cultural experiments', *Journal of Personality* 1974-3, p. 360-371.

McKenna 1993
F.P. McKenna, 'It won't happen to me: Unrealistic optimism or illusion of control?', *British Journal of Psychology* 1993-1, p. 39.

Meijer 2004
J. Meijer, 'Asbestos producer liable for consumer's cancer', *European Product Liability Review* 2004-16, p. 38-39.

Meyerowitz & Chaiken 1987
B.E. Meyerowitz & S. Chaiken, 'The effect of message framing on breast self-examination attitudes, intentions, and behavior', *Journal of Personality & Social Psychology* 1987-3, p. 500-510.

Mildred 2005
M. Mildred, 'The development risks defence', in: D. Fairgrieve (ed.), *Product liability in comparative perspective*, Cambridge: Cambridge University Press 2005, p. 167-191.

Miller 1956
G.A. Miller, 'The magic number seven, plus or minus two: Some limits on our capacity for processing information', *Psychological Review* 1956, p. 81-93.

Miller 1980
G.R. Miller, 'On being persuaded: Some basic distintions', in: M.E. Roloff & G.R. Miller (eds.), *Persuasion: New directions in theory and research*, Beverly Hills: Sage publications 1980, p. 11-28.

Miller & Goldberg 2004
C.J. Miller & R.S. Goldberg, *Product liability*, Oxford: Oxford University Press 2004.

Morrow e.a. 1998
D.G. Morrow e.a., 'The Influence of List Format and Category Headers on Age Differences in Understanding Medication Instructions', *Experimental Aging Research* 1998-3, p. 231-256.

De Mot, Canta & Gangapersadsing 2004
J.P.B. De Mot, A. Canta & V. Gangapersadsing, 'The Learned Hand formula: The case of the Netherlands', *Global Jurist Advances* 2004-2, p. 1-14.

Myers 2004
D.G. Myers, *Social psychology*, Boston, MA: McGraw-Hill 2004.

Nilsson & Kaiserman 2005
T. Nilsson & M. Kaiserman, 'Legibility of warnings in color', in: Y.I. Noy & W. Karwowski (eds.), *Handbook of human factors in litigation*, Boca Raton: CRC Press 2005, p. 32-1-32-17.

Noah 1994
L. Noah, 'Imperative to warn: disentangling the right to know from the need to know about consumer product hazards', *Yale Journal on Regulation* 1994, p. 293-400.

Norman 1981
D.A. Norman, 'Categorization of action slips', *Psychological Review* 1981-1, p. 1-15.

Norman 2002
D.A. Norman, *The design of everyday things*, New York: Basic Books 2002.

Norris & Wilson 1999
B. Norris & J.R. Wilson, 'Ergonomics and safety in consumer product design', in: W.S. Green & P.W. Jordan (eds.), *Human factors in product design: Current practice and future trends*, London: Taylor & Francis 1999, p. 73-84.

Ortiz, Resnick & Kengskool 2000
J. Ortiz, M.L. Resnick & K. Kengskool, 'The effects of familiarity and risk perception in workplace warning compliance', *Human Factors and Ergonomics Society Annual Meeting Proceedings* 2000, p. 826-829.

Otsubo 1988
S.M. Otsubo, 'A behavioral study of warning labels for consumer products: Perceived danger and use of pictographs', *Human Factors and Ergonomics Society Annual Meeting Proceedings* 1988-9, p. 536-540.

Owen 1996-1997
D.G. Owen, 'Risk-utility balancing in design defect cases', *University of Michigan Journal of Law Reform* 1996-1997, p. 239-260.

Owen 2008
D.G. Owen, *Products liability law*, St. Paul, MN: Thomson/West 2008.

Page 1998
M. Page, 'Consumer products – More by accident than design?', in: N. Stanton (ed.), *Human factors in consumer products*, London: Taylor & Francis 1998, p. 127-146.

Paivio 1971
A. Paivio, *Imagery and verbal processes*, New York: Holt, Rinehart and Winston 1971.

Pape 2006
S.B. Pape, 'De betekenis van het Jetblast-arrest voor de waarschuwing in het productaansprakelijkheidsrecht', *Nederlands Tijdschrift Voor Burgerlijk Recht* 2006, p. 374-382.

Pape 2009
S.B. Pape, '"May Contain" Labelling: Adequate consumer warning or unnecessarily defensive manufacturer behaviour?', *Journal of Consumer Policy* 2009, p. 165-188.

Payne & Wenger 1998
D.G. Payne & M.J. Wenger, *Cognitive psychology*, Boston: Houghton Mifflin 1998.

Purswell, Schlegel & Kejriwal 1986
J.L. Purswell, R.E. Schlegel & S.K. Kejriwal, 'A prediction model for consumer behavior regarding product safety', *Human Factors and Ergonomics Society Annual Meeting Proceedings* 1986-12, p. 1202-1205.

Pyrczak & Roth 1976
F. Pyrczak & D.H. Roth, 'The readability of directions on non-prescription drugs', *Journal of the American Pharmaceutical Association* 1976-5, p. 242-43, 267.

Racicot & Wogalter 1992
B.M. Racicot & M.S. Wogalter, 'Warning compliance: Effects of a video warning sign and modeling on behavior', *Human Factors and Ergonomics Society Annual Meeting Proceedings* 1992-8, p. 608-610.

Racicot & Wogalter 1995
B.M. Racicot & M.S. Wogalter, 'Effects of a video warning sign and social modeling on behavioral compliance', *Accident Analysis & Prevention* 1995-1, p. 57-64.

Rajneri 2005
E. Rajneri, 'Interaction between the European Directive on Product Liability and the former liability regime in Italy', in: D. Fairgrieve (ed.), *Product liability in comparative perspective*, Cambridge: Cambridge University Press 2005, p. 67-83.

Ramsey 1985
J.D. Ramsey, 'Ergonomic factors in task analysis for consumer product safety', *Journal of Occupational Accidents* 1985-2, p. 113-123.

Ramsey 1989
J.D. Ramsey, 'Assessment of warnings based on an ergonomic accident sequence model', *International Journal of Industrial Ergonomics* 1989-3, p. 195-199.

Rashid & Wogalter 1997
R. Rashid & M.S. Wogalter, 'Effects of warning border color, width, and design on perceived effectiveness', in: B. Das & W. Karwowski (eds.), *Advances in occupational ergonomics and safety II*, Louisville, KY: IOS Press, and Ohmsha 1997, p. 455-458.

Rasmussen 1986
J. Rasmussen, *Information processing and human-machine interaction*, New York: North-Holland 1986.

Reason 1990
J. Reason, *Human error*, Cambridge: Cambridge University Press 1990.

Reich 1986
N. Reich, 'Product safety and product liability – An analysis of the EEC Council Directive of 25 July 1985 on the approximation of the laws, regulations, and administrative provisions of the Member States concerning liability for defective products', *Journal of Consumer Policy* 1986-2, p. 133-154.

Rider e.a. 2009
G. Rider e.a., 'Framework model of product risk assessment', *International Journal of Injury Control and Safety Promotion* 2009-2, p. 73.

Riley 2006
D.M. Riley, 'Beliefs, attitudes, and motivation', in: M.S. Wogalter (ed.), *Handbook of Warnings*, Mahwah, New Jersey: Lawrence Erlbaum Associates 2006, p. 289-300.

Rogers 1975
R. Rogers, 'A protection motivation theory of fear appeals and attitude change', *The Journal of Psychology* 1975-1, p. 93.

Rogers, Lamson & Rousseau 2000
W.A. Rogers, N. Lamson & G.K. Rousseau, 'Warning research: An integrative perspective', *Human Factors: The Journal of the Human Factors and Ergonomics Society* 2000-1, p. 102-139.

Rother 2008
H. Rother, 'South African farm workers' interpretation of risk assessment data expressed as pictograms on pesticide labels', *Environmental Research* 2008-3, p. 419-427.

Rothman & Salovey 1997
A.J. Rothman & P. Salovey, 'Shaping perceptions to motivate healthy behavior: The role of message framing', *Psychological Bulletin* 1997-1, p. 3-19.

Rothstein 1985
P.R. Rothstein, 'Designing warnings to be read and remembered', *Human Factors and Ergonomics Society Annual Meeting Proceedings* 1985-7, p. 684-688.

Rotter 1975
J.B. Rotter, 'Some problems and misconceptions related to the construct of internal versus external control of reinforcement', *J.CONSULT.CLIN.PSYCHOL.* 1975-1, p. 56-67.

Rousseau, Lamson & Rogers 1998
G.K. Rousseau, N. Lamson & W.A. Rogers, 'Designing warnings to compensate for age-related changes in perceptual and cognitive abilities', *Psychology & Marketing* 1998-7, p. 643-662.

Rousseau & Wogalter 2006
G.K. Rousseau & M.S. Wogalter, 'Research on warning signs', in: M.S. Wogalter (ed.), *Handbook of Warnings*, Mahwah, New Jersey: Lawrence Erlbaum Associates 2006, p. 147-158.

Rudin-Brown e.a. 2004
C.M. Rudin-Brown e.a., 'The design of child restraint system (CRS) labels and warnings affects overall CRS usability', *Traffic Injury Prevention* 2004-1, p. 8.

SCENIHR 2008
SCENIHR, *Potential health risks of exposure to noise from personal players and mobile phones including a music playing function*, Preliminary report, 2008.

Schank & Abelson 1977
R.C. Schank & R.P. Abelson, *Scripts, plans, goals and understanding: an inquiry into human knowledge structures*, Hillsdale, N.J.: Lawrence Erlbaum Associates 1977.

Schneider 1977
K.C. Schneider, 'Prevention of accidental poisoning through package and label design', *Journal of Consumer Research* 1977-2, p. 67-74.

Schommer, Doucette & Worley 2001
J.C. Schommer, W.R. Doucette & M.M. Worley, 'Processing prescription drug information under different conditions of presentation', *Patient Education and Counseling* 2001-1, p. 49-59.

Schwartz 1999-2000
V.E. Schwartz, 'See no evil, hear no evil: When clear and adequate warnings do not prevent the imposition of product liability', *University of Cincinnati Law Review* 1999-2000, p. 47-64.

Schwartz & Driver 1983
V.E. Schwartz & R.W. Driver, 'Warnings in the workplace: The need for a synthesis of law and communication theory', *University of Cincinnati Law Review* 1983, p. 38-83.

Shaver & Braun 2000
E.F. Shaver & C.C. Braun, 'Effects of warning symbol explicitness and warning color on behavioral compliance', *Human Factors and Ergonomics Society Annual Meeting Proceedings* 2000-27, p. 290-293.

Silver & Braun 1993
N.C. Silver & C.C. Braun, 'Perceived readability of warning labels with varied font sizes and styles', *Safety Science* 1993-5-6, p. 615-625.

Silver & Braun 1999
N.C. Silver & C.C. Braun, 'Behavior', in: M.S. Wogalter, D.M. DeJoy & K.R. Laughery (eds.), *Warnings and risk communication*, London: Taylor & Francis 1999, p. 245-262.

Silver & Wogalter 1989
N.C. Silver & M.S. Wogalter, 'Broadening the range of signal words', *Human Factors and Ergonomics Society Annual Meeting Proceedings* 1989-9, p. 555-559.

Silver & Wogalter 1991
N.C. Silver & M.S. Wogalter, 'Strength and understanding of signal words by elementary and middle school students', *Human Factors and Ergonomics Society Annual Meeting Proceedings* 1991-9, p. 590-594.

Slovic, Fischhoff & Lichtenstein 1977
P. Slovic, B. Fischhoff & S. Lichtenstein, 'Behavioral decision theory', *Annual Review of Psychology* 1977-1, p. 1-39.

Slovic, Fischhoff & Lichtenstein 1979
P. Slovic, B. Fischhoff & S. Lichtenstein, 'Rating the risks', *Environment* 1979, p. 14-39.

Slovic, Fischhoff & Lichtenstein 1980
P. Slovic, B. Fischhoff & S. Lichtenstein, 'Facts versus fears: Understanding perceived risk', in: R.C. Schwing & W.A. Albers Jr (eds.), *Societal risk assessment: How safe is safe enough?*, New York: Plenum Press 1980, p. 181-214.

Smith-Jackson 2006a
T.L. Smith-Jackson, 'Receiver characteristics', in: M.S. Wogalter (ed.), *Handbook of Warnings*, Mahwah, New Jersey: Lawrence Erlbaum Associates 2006, p. 335-344.

Smith-Jackson 2006b
T.L. Smith-Jackson, 'Culture and warnings', in: M.S. Wogalter (ed.), *Handbook of Warnings*, Mahwah, New Jersey: Lawrence Erlbaum Associates 2006, p. 363-371.

Smith-Jackson & Essuman-Johnson 2002
T.L. Smith-Jackson & A. Essuman-Johnson, 'Cultural ergonomics in Ghana, West Africa: A descriptive survey of industry and trade workers' interpretations of safety symbols', *International Journal of Occupational Safety and Ergonomics: JOSE* 2002-1, p. 37-50.

Smith-Jackson & Wogalter 2006
T.L. Smith-Jackson & M.S. Wogalter, 'Methods and procedures in warning research', in: M.S. Wogalter (ed.), *Handbook of Warnings*, Mahwah, New Jersey: Lawrence Erlbaum Associates 2006, p. 23-33.

Snijders 1984
G.M.F. Snijders, 'Productaansprakelijkheid: Gebrekkigheid en (on)zorgvuldigheid', *NJB* 1984, p. 477-482.

Snyder & Blood 1992
L.B. Snyder & D.J. Blood, 'Caution: Alcohol advertising and the surgeon general's alcohol warnings may have adverse effect on young adults', *Journal of Applied Communication Research* 1992-1, p. 37.

Solso, Maclin & Maclin 2008
R.L. Solso, O.H. Maclin & M.K. Maclin, *Cognitive psychology*, Boston, MA: Pearson 2008.

Spier 1998
J. Spier, 'The Netherlands. Wrongfulness in the Dutch context', in: H. Koziol & F.D. Busnelli (eds.), *Unification of tort law: Wrongfulness*, The Hague: Kluwer Law International 1998, p. 87-100.

Spier e.a. 2009
J. Spier e.a., *Verbintenissen uit de wet en schadevergoeding*, Deventer: Kluwer 2009.

Stanton 1998a
N. Stanton, 'Product design with people in mind', in: N. Stanton (ed.), *Human factors in consumer products*, London: Taylor & Francis 1998, p. 1-17.

Stanton 1998b
N.A. Stanton, *Human factors in consumer products*, London: Taylor & Francis 1998.

Stapleton 1994
J. Stapleton, *Product liability*, London: Butterworths 1994.

Stapleton 1999
J. Stapleton, 'Products liability in the United Kingdom: The myths of reform', *Texas International Law Journal* 1999, p. 45-70.

Statler 2005
S.M. Statler, 'Preventing "accidental" injury: Accountability for safer products by anticipating product risks and user behaviors', in: Y.I. Noy & W. Karwowski (eds.), *Handbook of human factors in litigation*, Boca Raton: CRC Press 2005, p. 25-1-25-14.

Stewart, Folkes & Martin 2001
D.W. Stewart, V.S. Folkes & E.G. Martin, 'Consumer response to warnings and other types of product hazard information', in: P.N. Bloom & G.T. Gundlach (eds.), *Handbook of marketing and society*, Thousand Oaks: Sage 2001, p. 335-371.

Stewart & Martin 1994
D.W. Stewart & I.M. Martin, 'Intended and unintended consequences of warning messages: A review and synthesis of empirical research', *Journal of Public Policy & Marketing* 1994-1, p. 1-19.

Stolker & Westerdijk 1984
C.J.J.M. Stolker & R.J.J. Westerdijk, 'Is productenaansprakelijkheid nog een risicoaansprakelijkheid? Nieuwe ontwikkelingen in het Amerikaanse recht', *A&V* 1984, p. 59-65.

Stoppa 1992
A. Stoppa, 'Concept of defectiveness in the consumer protection act 1987: A critical analysis', *Legal Studies* 1992, p. 210-226.

Strawbridge 1986
J.A. Strawbridge, 'The influence of position, highlighting, and imbedding on warning effectiveness', *Human Factors and Ergonomics Society Annual Meeting Proceedings* 1986-7, p. 716-720.

Taschner 2005
H.C. Taschner, 'Product liability: Basic problems in a comparative law perspective', in: D. Fairgrieve (ed.), *Product liability in comparative perspective*, Cambridge: Cambridge University Press 2005, p. 155-166.

Taylor 2007
C. Taylor, 'Warning: Too much information can harm', *European Product Liability Review* 2007-29, p. 4-5.

Thorley, Hellier &Edworthy 2001
P. Thorley, E.J. Hellier & J. Edworthy, 'Habituation effects in visual warnings', in: M.A. Hanson (ed.), *Contemporary ergonomics 2001*, London: Taylor and Francis 2001, p. 223-231.

Trommelen 1997
M. Trommelen, 'Effectiveness of explicit warnings', *Safety Science* 1997-1-3, p. 79-88.

Trommelen & Zwaga 1998
M. Trommelen & H.J. Zwaga, 'Development of warning symbols', in: N. Stanton (ed.), *Human factors in consumer products*, London: Taylor & Francis 1998, p. 223-238.

Twerski e.a. 1975-1976
A.D. Twerski e.a., 'Use and abuse of warnings in products Liability – Design defect litigation comes of age', *Cornell Law Review* 1975-1976, p. 495-540.

Tversky & Kahneman 1973
A. Tversky & D. Kahneman, 'Availability: A heuristic for judging frequency and probability', *Cognitive Psychology* 1973-2, p. 207-232.

Tversky & Kahneman 1981
A. Tversky & D. Kahneman, 'The framing of decision and the psychology of choice', *Science* 1981, p. 453-458.

Ursic 1984
M. Ursic, 'The impact of safety warnings on perception and memory', *Human Factors: The Journal of the Human Factors and Ergonomics Society* 1984-6, p. 677-682.

Vanderplas & Vanderplas 1980
J.M. Vanderplas & J.H. Vanderplas, 'Some factors affecting legibility of printed materials for older adults', *Perceptual and Motor Skills* 1980-3, p. 923-932.

Vanilla Research 2007
Vanilla Research, *Consumer information and regulation*, UK, 2007.

Vardavas e.a. 2009
C.I. Vardavas e.a., 'Adolescents perceived effectiveness of the proposed European graphic tobacco warning labels', *The European Journal of Public Health* 2009-2, p. 212-217.

Vaubel 1990
K.P. Vaubel, 'Effects of warning explicitness on consumer product purchase intentions', *Human Factors and Ergonomics Society Annual Meeting Proceedings* 1990-5, p. 513-517.

Vaubel & Brelsford 1991
K.P. Vaubel & J.W. Brelsford, 'Product evaluations and injury assessments as related to preferences for explicitness in warnings', *Human Factors and Ergonomics Society Annual Meeting Proceedings* 1991-15, p. 1048-1052.

Vigilante & Wogalter 1997
W.J. Vigilante & M.S. Wogalter, 'On the prioritization of safety warnings in product manuals', *International Journal of Industrial Ergonomics* 1997-4, p. 277-285.

Viscusi 1995-1996
W.K. Viscusi, 'Individual rationality, hazard warnings, and the foundations of tort law', *Rutgers Law Review* 1995-1996, p. 625-672.

Visscher 2005
L.T. Visscher, *Een rechtseconomische analyse van het Nederlandse onrechtmatigedaadsrecht (dissertatie EUR)*, Rotterdam: Erasmus Universiteit 2005.

Visschers e.a. 2004
V. Visschers e.a., 'The effects of warnings and an educational brochure on computer working posture: A test of the C-HIP model in the context of RSI-relevant behaviour', *Ergonomics* 2004-14, p. 1484-1498.

Vredenburgh e.a. 2005
A.G. Vredenburgh e.a., 'Evaluating latex glove container warnings in a realistic setting', *International Journal of Industrial Ergonomics* 2005-6, p. 559-568.

Vredenburgh & Cohen 1993
A.G. Vredenburgh & H.H. Cohen, 'Compliance with warnings in high risk recreational activities: Skiing and scuba', *Human Factors and Ergonomics Society Annual Meeting Proceedings* 1993-14, p. 945-949.

Vredenburgh & Helmick-Rich 2006
A.G. Vredenburgh & J. Helmick-Rich, 'Extrinsic nonwarning factors', in: M.S. Wogalter (ed.), *Handbook of Warnings*, Mahwah, New Jersey: Lawrence Erlbaum Associates 2006, p. 373-382.

Vredenburgh & Zackowitz 2006
A.G. Vredenburgh & I.B. Zackowitz, 'Expectations', in: M.S. Wogalter (ed.), *Handbook of Warnings*, Mahwah, New Jersey: Lawrence Erlbaum Associates 2006, p. 345-354.

VWA 2009
VWA, *Verkenning van de veiligheid van wensballonnen*, ZW 09231d, Den Haag, 2009.

VWA november 2009
VWA, *Advies van de directeur bureau risicobeoordeling*, VWA/BuR/2009/29730, Den Haag, november 2009.

VWA december 2009
VWA, *Afspraken gericht op het veiliger maken van wensballonnen*, VWA/C-2009-42035, Den Haag, december 2009.

Wagenaar 1992
W.A. Wagenaar, 'Risk taking and accident causation', in: J.F. Yates (ed.), *Risk-taking behavior*, Chichester: Wiley 1992, p. 257-281.

Weaver e.a. 2003
J.L. Weaver e.a., 'Individual differences in behavioral compliance to warnings representing varying degrees of threat', *International Journal of Occupational Safety and Ergonomics: JOSE* 2003-2, p. 149-160.

Weaver, Helmick & Burke 2003
J.L. Weaver, J.A. Helmick & K.A. Burke, 'An investigation of individual differences in relation to warning compliance and user affect: Preliminary findings', *Human Factors and Ergonomics Society Annual Meeting Proceedings* 2003-9, p. 1102-1106.

Weegels 1996
M.F. Weegels, *Accidents involving consumer products (dissertation TU Delft)*, 1996.

Weegels 1998
M.F. Weegels, 'Retrospective research into accidents', *International Journal of Consumer & Product Safety* 1998-4, p. 173.

Weegels & Kanis 2000
M.F. Weegels & H. Kanis, 'Risk perception in consumer product use', *Accident Analysis & Prevention* 2000-3, p. 365-370.

Weinstein e.a. 1978
A.S. Weinstein e.a., *Products liability and the reasonably safe product: A guide for management, design and marketing*, New York: Wiley 1978.

Weinstein 1989
N.D. Weinstein, 'Optimistic biases about personal risks', *Science (New York , N.Y.)* 1989-4935, p. 1232-1233.

Weinstein 1993
N.D. Weinstein, 'Testing four competing theories of health-protective behavior', *Health Psychology* 1993-4, p. 324-333.

Weinstein 1998
N. Weinstein, 'Accuracy of smokers' risk perceptions', *Annals of Behavioral Medicine* 1998-2, p. 135-140.

Van der Wiel 2007
B.T.M. Van der Wiel, 'Kroniek algemeen deel aansprakelijkheidsrecht', *Aansprakelijkheid, Verzekering & Schade* 2007, p. 175-191.

Whittaker 2005
S. Whittaker, *Liability for products: English law, French law, and European harmonisation*, Oxford: Oxford University Press 2005.

Williams & Noyes 2007
D.J. Williams & J.M. Noyes, 'How does our perception of risk influence decision-making? Implications for the design of risk information', *Theoretical Issues in Ergonomics Science* 2007-1, p. 1-35.

Witte 1992
K. Witte, 'Putting the fear back into fear appeals: The extended parallel process model', *Communication Monographs* 1992-4, p. 329-349.

Wogalter 2006
M.S. Wogalter, *Handbook of Warnings*, Mahwah: Lawrence Erlbaum Associates 2006.

Wogalter 2006a
M.S. Wogalter, 'Purposes and scope of warnings', in: M.S. Wogalter (ed.), *Handbook of Warnings*, Mahwah, New Jersey: Lawrence Erlbaum Associates 2006, p. 3-9.

Wogalter 2006b
M.S. Wogalter, 'Communication-human information processing (C-HIP) model', in: M.S. Wogalter (ed.), *Handbook of Warnings*, Mahwah, New Jersey: Lawrence Erlbaum Associates 2006, p. 51-61.

Wogalter e.a. 1987
M.S. Wogalter e.a., 'Effectiveness of warnings', *Human Factors* 1987-5, p. 599-612.

Wogalter e.a. 1991a
M.S. Wogalter e.a., 'Consumer product warnings: The role of hazard perception', *Journal of Safety Research* 1991-2, p. 71-82.

Wogalter e.a. 1991b
M.S. Wogalter e.a., 'Evaluating the behavioral effectiveness of a multi-modal voice warning sign in a visually cluttered environment', *Human Factors and Ergonomics Society Annual Meeting Proceedings* 1991-10, p. 718-722.

Wogalter e.a. 1994
M.S. Wogalter e.a., 'Personalization of warning signs: The role of perceived relevance on behavioral compliance', *International Journal of Industrial Ergonomics* 1994-3, p. 233-242.

Wogalter e.a. 1995
M.S. Wogalter e.a., 'Hazard associations of warning header components', *Human Factors and Ergonomics Society Annual Meeting Proceedings* 1995-15, p. 979-983.

Wogalter e.a. 1998
M.S. Wogalter e.a., 'The influence of time stress and location on behavioral warning compliance', *Safety Science* 1998-2, p. 143-158.

Wogalter e.a. 1999a
M.S. Wogalter e.a., 'Enhancing information acquisition for over-the-counter medications by making better use of container surface space', *Experimental Aging Research* 1999-1, p. 27-48.

Wogalter e.a. 1999b
M.S. Wogalter e.a., 'The relative contributions of injury severity and likelihood information on hazard-risk judgments and warning compliance', *Journal of Safety Research* 1999-3, p. 151-162.

Wogalter e.a. 2002
M.S. Wogalter e.a., 'Comprehension of different types of prohibitive safety symbols with glance exposure', *Human Factors and Ergonomics Society Annual Meeting Proceedings* 2002-19, p. 1753-1757.

Wogalter e.a. 2006
M.S. Wogalter e.a., 'Warning symbols', in: M.S. Wogalter (ed.), *Handbook of Warnings*, Mahwah, New Jersey: Lawrence Erlbaum Associates 2006, p. 159-176.

Wogalter, Allison & McKenna 1989
M.S. Wogalter, S.T. Allison & N.A. McKenna, 'Effects of cost and social influence on warning compliance', *Human Factors* 1989-2, p. 133-140.

Wogalter & Barlow 1990
M.S. Wogalter & T. Barlow, 'Injury severity and likelihood in warnings', *Human Factors and Ergonomics Society Annual Meeting Proceedings* 1990-8, p. 580-583.

Wogalter, Barlow & Murphy 1995
M.S. Wogalter, T. Barlow & S. Murphy, 'Compliance to owner's manual warnings: Influence of familiarity and the placement of a supplemental directive', *Ergonomics* 1995-6, p. 1081-1092.

Wogalter, Brems & Martin 1993
M.S. Wogalter, D.J. Brems & *E.G.* Martin, 'Risk perception of common consumer products: Judgments of accident frequency and precautionary intent', *Journal of Safety Research* 1993-2, p. 97-106.

Wogalter, Conzola & Smith-Jackson 2002
M.S. Wogalter, V.C. Conzola & T.L. Smith-Jackson, 'Research-based guidelines for warning design and evaluation', *Applied Ergonomics* 2002-3, p. 219-230.

Wogalter, Conzola & Vigilante 2006
M.S. Wogalter, V.C. Conzola & W.J. Vigilante, 'Applying usability engineering principles to the design and testing of warning text', in: M.S. Wogalter (ed.), *Handbook of Warnings*, Mahwah, New Jersey: Lawrence Erlbaum Associates 2006, p. 487-498.

Wogalter, DeJoy & Laughery 1999a
M.S. Wogalter, D.M. DeJoy & K.R. Laughery, 'Organizing theoretical framework: A consolidated communication-human information processing (C-HIP) model', in: M.S. Wogalter, D.M. DeJoy & K.R. Laughery (eds.), *Warnings and risk communication*, London: Taylor & Francis 1999, p. 15-23.

Wogalter, DeJoy & Laughery 1999b
M.S. Wogalter, D.M. DeJoy & K.R. Laughery, *Warnings and risk communication*, London: Taylor & Francis 1999.

Wogalter, Desaulniers & Brelsford 1986
M.S. Wogalter, D.R. Desaulniers & J.W. Brelsford, 'Perceptions of consumer products: Hazardousness and warning expectations', *Human Factors and Ergonomics Society Annual Meeting Proceedings* 1986-12, p. 1197-1201.

Wogalter & Dingus 1999
M.S. Wogalter & T.A. Dingus, 'Methodological techniques for evaluating behavioral intentions and compliance', in: M.S. Wogalter, D.M. DeJoy & K.R. Laughery (eds.), *Warnings and risk communication*, London: Taylor & Francis 1999, p. 53-81.

Wogalter, Fontenelle & Laughery 1985
M.S. Wogalter, G.A. Fontenelle & K.R. Laughery, 'Behavioral effectiveness of warnings', *Human Factors and Ergonomics Society Annual Meeting Proceedings* 1985-7, p. 679-686.

Wogalter, Jarrard & Simpson 1992
M.S. Wogalter, S.W. Jarrard & S.N. Simpson, 'Effects of warning signal words on consumer-product hazard perceptions', *Human Factors and Ergonomics Society Annual Meeting Proceedings* 1992-2, p. 935-939.

Wogalter, Jarrard & Simpson 1994
M.S. Wogalter, S.W. Jarrard & S.N. Simpson, 'Influence of warning label signal words on perceived hazard level', *Human Factors: The Journal of the Human Factors and Ergonomics Society* 1994-3, p. 547-556.

Wogalter, Kalsher & Racicot 1992
M.S. Wogalter, M.J. Kalsher & B.M. Racicot, 'The influence of location and pictorials on behavioral compliance to warnings', *Human Factors and Ergonomics Society Annual Meeting Proceedings* 1992-13, p. 1029-1033.

Wogalter, Kalsher & Racicot 1993
M.S. Wogalter, M.J. Kalsher & B.M. Racicot, 'Behavioral compliance with warnings: Effects of voice, context, and location', *Safety Science* 1993-5-6, p. 637-654.

Wogalter, Kalsher & Rashid 1999
M.S. Wogalter, M.J. Kalsher & R. Rashid, 'Effect of signal word and source attribution on judgments of warning credibility and compliance likelihood', *International Journal of Industrial Ergonomics* 1999-2, p. 185-192.

Wogalter & Laughery 2006
M.S. Wogalter & K.R. Laughery, 'Warnings and hazard communications', in: G. Salvendy (ed.), *Handbook of human factors and ergonomics*, Hoboken, New Jersey: John Wiley & Sons 2006, p. 889-911.

Wogalter & Leonard 1999
M.S. Wogalter & S.D. Leonard, 'Attention capture and maintenance', in: M.S. Wogalter, D.M. DeJoy & K.R. Laughery (eds.), *Warnings and risk communication*, London: Taylor & Francis 1999, p. 123-148.

Wogalter, McKenna & Allison 1988
M.S. Wogalter, N.A. McKenna & S.T. Allison, 'Warning compliance: Behavioral effects of cost and consensus', *Human Factors and Ergonomics Society Annual Meeting Proceedings* 1988-15, p. 901-904.

Wogalter & Rashid 1998
M.S. Wogalter & R. Rashid, 'A Border Surrounding a Warning Sign Affects Looking Behavior: A Field Observational Study', *Human Factors and Ergonomics Society Annual Meeting Proceedings* 1998-23, p. 1628-1628.

Wogalter & Shaver 2001
M.S. Wogalter & E.F. Shaver, 'Evaluation of list vs. paragraph text format on search time for warning symptoms in a product manual', in: A.C. Bittner (ed.), *Advances in occupational ergonomics and safety*, The Netherlands: IOS Press 2001, p. 434-438.

Wogalter, Shaver & Chan 2002
M.S. Wogalter, E.F. Shaver & L.S. Chan, 'List vs. paragraph formats on time to compare nutrition labels', in: P.T. McCabe (ed.), *Contemporary ergonomics 2002*, UK: CRC Press 2002, p. 458-462.

Wogalter & Silver 1990
M.S. Wogalter & N.C. Silver, 'Arousal strength of signal words', *Forensic Reports* 1990, p. 407-420.

Wogalter & Silver 1995
M.S. Wogalter & N.C. Silver, 'Warning signal words: Connoted strength and understandability by children, elders, and non-native English speakers', *Ergonomics* 1995-11, p. 2188-2206.

Wogalter, Sojourner & Brelsford 1997
M.S. Wogalter, R.J. Sojourner & J.W. Brelsford, 'Comprehension and retention of safety pictorials', *Ergonomics* 1997-5, p. 531-542.

Wogalter & Usher 1999
M.S. Wogalter & M.O. Usher, 'Effects of concurrent cognitive task loading on warning compliance behavior', *Human Factors and Ergonomics Society Annual Meeting Proceedings* 1999-6, p. 525-529.

Wogalter & Vigilante 2003
M.S. Wogalter & J. Vigilante, 'Effects of label format on knowledge acquisition and perceived readability by younger and older adults', *Ergonomics* 2003-4, p. 327-344.

Wogalter & Vigilante 2006
M.S. Wogalter & J. Vigilante, 'Attention switch and maintenance', in: M.S. Wogalter (ed.), *Handbook of Warnings*, Mahwah, New Jersey: Lawrence Erlbaum Associates 2006, p. 245-265.

Wogalter & Young 1991
M.S. Wogalter & S.L. Young, 'Behavioural compliance to voice and print warnings', *Ergonomics* 1991-1, p. 79-89.

Wogalter & Young 1994
M.S. Wogalter & S.L. Young, 'The effect of alternative product-label design on warning compliance', *Applied Ergonomics* 1994-1, p. 53-57.

Wolff & Wogalter 1998
J.S. Wolff & M.S. Wogalter, 'Comprehension of pictorial symbols: Effects of context and test method', *Human Factors* 1998-2, p. 173-186.

Young 1998
S.L. Young, 'Connotation of hazard for signal words and their associated panels', *Applied Ergonomics* 1998-2, p. 101-110.

Young e.a. 1999
S.L. Young e.a., 'Receiver characteristics in safety communications', in: W. Karwowski & W.S. Marras (eds.), *The occupational ergonomics handbook*, Boca Raton: CRC Press 1999, p. 693-706.

Young e.a. 2006a
S.L. Young e.a., 'Development and objectives of the ANSI Z535 series of standards for safety signs and colors: A historical perspective', in: M.S. Wogalter (ed.), *Handbook of Warnings*, Mahwah, New Jersey: Lawrence Erlbaum Associates 2006, p. 445-454.

Young e.a. 2006b
S.L. Young e.a., 'Hazard analysis as part of the safety information development process', in: M.S. Wogalter (ed.), *Handbook of Warnings*, Mahwah, New Jersey: Lawrence Erlbaum Associates 2006, p. 431-436.

Young, Brelsford & Wogalter 1990
S.L. Young, J.W. Brelsford & M.S. Wogalter, 'Judgments of hazard, risk, and danger: Do they differ?', *Human Factors and Ergonomics Society Annual Meeting Proceedings* 1990-5, p. 503-507.

Young & Lovvoll 1999
S.L. Young & D.R. Lovvoll, 'Intermediate processing stages: Methodological considerations for research on warnings', in: M.S. Wogalter, D.M. DeJoy & K.R. Laughery (eds.), *Warnings and risk communication*, London: Taylor & Francis 1999, p. 27-52.

Young, Martin & Wogalter 1989
S.L. Young, E.G. Martin & M.S. Wogalter, 'Gender differences in consumer product hazard perceptions', *Proceedings of Interface* 1989, p. 73-78.

Young & Wogalter 1988
S.L. Young & M.S. Wogalter, 'Memory of instruction manual warnings: Effects of pictorial icons and conspicuous print', *Proceedings of the Human Factors and Ergonomics Society* 1988-15, p. 905-909.

Young & Wogalter 1990
S.L. Young & M.S. Wogalter, 'Comprehension and memory of instruction manual warnings: Conspicuous print and pictorial icons', *Human Factors* 1990-6, p. 637-649.

Young, Wogalter & Brelsford 1992
S.L. Young, M.S. Wogalter & J.W. Brelsford, 'Relative contribution of likelihood and severity of injury to risk perceptions', *Human Factors and Ergonomics Society Annual Meeting Proceedings* 1992-36, p. 1014-1018.

Zeitlin 1994
L.R. Zeitlin, 'Failure to follow safety instructions: Faulty communication or risky decisions?', *Human Factors: The Journal of the Human Factors and Ergonomics Society* 1994-1, p. 172-181.

Zuckerman 1979
M. Zuckerman, *Sensation seeking: Beyond the optimal level of arousal*, Hillsdale, NJ: Lawrence Erlbaum Associates 1979.

Table of cases

Dutch case law

Hoge Raad (Supreme Court)

HR 25 november 2005, NJ 2009, 103 (*Eternit*)
HR 11 november 2005, NJ 2008, 460 (*Multivac-machine*)
HR 7 april 2006, *NJ* 2006, 244 (*Uienrot*)
HR 17 december 2004, *NJ* 2006, 147 (*Hertel/Van der Lugt*)
HR 28 mei 2005, *NJ* 2004, 105 (*Jetblast*)
HR 26 september 2003, *NJ* 2003, 660 (*Gekantelde vrachtwagen*)
HR 29 november 2002, *NJ* 2003, 50 (*Onkruidverdelger Thyram*)
HR 22 oktober 1999, *NJ* 2000, 159 (*Koolhaas/Rockwool*)
HR 22 september 2000, *NJ* 2000, 644 (*Vladeko/VSCI*)
HR 2 oktober 1998, *NJ* 1999, 683 (*De Schelde/Erven Cijsouw*)
HR 6 december 1996, *NJ* 1997, 219 (*Du Pont/Hermans*)
HR 9 december 1994, *NJ* 1996, 403 (*Zwiepende tak*)
HR 20 maart 1992, *NJ* 1993, 547 (*Bussluis*)
HR 22 maart 1991, *NJ* 1991, 420 (*Roeffen/Thijssen*)
HR 30 juni 1989, *NJ* 1990, 652 (*Halcion*)
HR 6 april 1990, *NJ* 1990, 573 (*Janssen/Nefabas*)
HR 14 april 1978, *NJ* 1979, 245 (*Messaoudi/Hoescht*)
HR 22 november 1974, *NJ* 1975, 149 (*Struikelende broodbezorger*)
HR 2 februari 1973, *NJ* 1973, 315 (*Warmwaterkruik*)
HR 5 november 1965, *NJ* 1966, 136 (*Kelderluik*)

Hof (Court of Appeal)

Gemeenschappelijk Hof Nederlandse Antillen en Aruba 18 maart 2005, *NJ* 2005, 302
Hof Arnhem 14 oktober 2003, *NJF* 2004, 46 (*Datafan*)
Hof Arnhem 13 april 2004, *NJ* 2004, 612
Hof 's-Hertogenbosch 15 mei 2007, *LJN* 2007, BA6838 (*Nagelstyling*)

Rechtbank (District Court)

Rb. Amsterdam 17 december 2008, *NJ* 2009, 311 (*Rokersclaim*)
Rb. Amsterdam 3 februari 1999, *NJ* 1999, 621 (*HIV*)
Rb. Maastricht 21 maart 2002, *LJN* 2002, AE0776 (*Versgeperste jus d'orange*)
Rb. Middelburg 13 juli 2005, *JA* 2005, 104 (*Betonmortel*)

Rb. Zwolle 24 april 2002, *Praktijkgids* 2002, 5921 (*Mini-tampon*)

EUROPEAN COURT OF JUSTICE CASE LAW

ECJ 9 February 2006 (Case C-127/04), *ECR* 2006, p. I-1313 (*Declan O'Byrne v. Sanofi Pasteur MSD Ltd and Sanofi Pasteur SA*)
ECJ 25 April 2002 (Case C-52/00), *ECR* 2002, p. I-3827 (*Commission v. France*)
ECJ 25 April 2002 (Case C-154/00), *ECR* 2002, p. I-3879 (*Commission v. Greece*)
ECJ 25 April 2002 (Case C-183/00), *ECR* 2002, p. I-3901 (*María Victoria González Sánchez v. Medicina Asturiana SA*)
ECJ 29 May 1997 (Case C-300/95), *ECR* 1997, p. I-2649 (*Commission v. United Kingdom*)

EUROPEAN CASE LAW

England

[2001] 3 *All ER* 289 (*A v. National Blood Authority*)
[2002] *EWHC* 490 (*Bogle v. Mc Donald's Restaurants*)
[2006] *EWHC* 1284 (QB) *All ER* (D) 86 (Jun) (*Klunk Klip device*)
[2000] *PIQR* 95 (*Worsely v. Tambrands Ltd*)

Germany

BGH 9 May 1995, *NJW* 1995, 2162 (*Exploded mineral bottle*)
OLG Düsseldorf 20 December 2002, U 99/02, *VersR* 2003, 912 (*Chocolate bar*)
OLG Celle 29 January 2003, 9 U 176/02, *VersR* 2004, 964 (*German mixed concrete*)
Landgericht Düsseldorf 30 November 2005, 10 O 144/04, *NJW-RR* 2006, 1033 ff. (*Floor panel stripper*)

Austria

OGH 13 November 2002, 7 Ob 245/02h (*Hammock*)
OGH 5 December 2002, 2 Ob 249/02k (*Stepladder*)

Index

Civilology | Civilologie

In de reeks Civilology | Civilologie staat de bestudering van het civiele recht centraal, niet vanuit zuiver juridisch perspectief maar met gebruikmaking van inzichten uit sociaalwetenschappelijke disciplines zoals economie, psychologie en sociologie. Die disciplines in onderlinge samenhang kunnen onze kennis vergroten van de gedragsassumpties waar het privaatrecht zich van bedient, de effecten die het privaatrecht heeft op het gedrag van individuen en organisaties en van de betekenis die deze assumpties en effecten hebben voor privaatrechtelijke beleidsvorming, regelstelling en -toepassing. De delen in de reeks Civilology | Civilologie beogen aldus een beter inzicht te geven in hoe privaatrecht écht werkt. De reeks staat onder redactie van Willem van Boom (EUR) en Ivo Giesen (UU).

In the Civilology | Civilologie series, the focus is on how private law works in real life. Rather than analysing private law from a strictly legal doctrinal perspective, this series encompasses the combined use of insights from social sciences – ranging from economics to psychology and sociology – in order to analyze the behavioural assumptions underlying private law, to understand the effects it has on individuals, organisations and businesses and to appreciate the impact of both assumptions and effects on policymakers, legislatures and courts. The Civilology | Civilologie series is edited by Willem van Boom (Erasmus Law School) and Ivo Giesen (Utrecht School of Law).